SLUG & SNAIL PESTS

BCPC Symposium

SLUG & SNAIL PESTS IN AGRICULTURE

BCPC Symposium Proceedings No. 66

Proceedings of a Symposium organised by
The British Crop Protection Council in conjunction
with The Association of Applied Biologists and
The Malacological Society of London

Held at the University of Kent
Canterbury, UK on 24-26 September 1996

Chaired by I F Henderson

BRITISH
CROP
PROTECTION
COUNCIL

BCPC Registered Office:
49 Downing Street, Farnham
Surrey GU9 7PH, UK.

© 1996 The British Crop Protection Council
49 Downing Street
Farnham, Surrey GU9 7PH

British Library Cataloguing in Publication Data.
A catalogue record for this book is available from the British Library.

British Crop Protection Council
Slug and Snail Pests in Agriculture
(Proceedings/Monograph Series, ISSN 0306-3941; No. 66)

ISBN 0 948404 96 5

Cover design by Major Design & Production Ltd, Nottingham
Printed in Great Britain by Major Design & Production Ltd, Nottingham

Contents

SESSION 4:
BEHAVIOUR AND ECOLOGY

SESSION 5:
CHEMICAL CONTROL

POSTER SESSION

Preface

At first sight, slugs and snails are unlikely pests of agriculture. Their large size, poor mobility and vulnerable water relations would seem insuperable barriers to the achievement of pest status in such a dynamic environment. But they do achieve it, over an impressively wide range of geographically, climatically and agronomically diverse situations, as the papers presented at this meeting well illustrate. Their versatility even extends to the manner in which they inflict damage. Problems are described in which slugs destroy the plant at the embryonic stage in the seed, in which snails compete with farmed fish for food and in which they reduce the value of harvested grain merely by their presence in it.

As a group they have proved difficult to control: this has given impetus to the process of refining existing methods of chemical control and has encouraged fundamental research aimed at finding new and more specific control measures. The development of bait additives to optimise efficacy and to reduce the amount of molluscicide required is described as are studies on the effects on non-target organisms, reflecting continuing concern that control methods should be environmentally responsible. The emergence of an new biocontrol agent and the identification of enzyme systems as potential targets for future control methods illustrate a innovative and forward-thinking spirit in current research.

The constraints imposed by the real world, the need to meet increasingly stringent regulatory criteria and the impact of the economics of the marketplace on the development and marketing of control products, are not ignored. Excellent papers on this aspect have been presented.

That the pace of research in applied malacology matches that of the animals under study is a canard sometimes advanced by mischievous colleagues - in this Symposium I believe we have finally nailed the lie!

Ian Henderson
IACR-Rothamsted Experimental Station
Harpenden

September 1996

Symposium Organising Committee

Programme Chairman

Dr I F Henderson IACR-Rothamsted Experimental Station, Harpenden, Herts AL5 2JQ

Vice Chairman

Dr A Cook University of Ulster, Coleraine, Co. Londonderry, Northern Ireland, BT52 1SA

Session Organisers

Dr S E R Bailey University of Manchester, Stopford Building, Oxford Road, Manchester M13 9PT

Prof I D Bowen University of Wales College of Cardiff, Cardiff CF1 3TL

Dr K A Evans Scottish Agricultural College, West Mains Road, Edinburgh EH9 3JG

Dr D M Glen IACR-Long Ashton Research Station, Long Ashton, Bristol BS18 9AF

Mr T J Martin 'Cydia', Gislingham Road, Finningham, Stowmarket, Suffolk IP14 4HZ

Dr L R Noble University of Aberdeen, Tillydrone Avenue, Aberdeen AB9 2TN

Mr J Oakley ADAS Bridgets, Martyr Worthy, Winchester, Hants SO21 1AP

Dr G R Port The University, Newcastle-upon-Tyne NE1 7RU

Dr N W Runham University of Wales Bangor, Gwynedd LL57 2UW

Administration Manager

Mr C J Siddall British Crop Protection Enterprises, 49 Downing Street, Farnham, Surrey GU9 7PH

Exhibition Manager

Mr G Beaumont British Crop Protection Enterprises, Linden House, Old Stowmarket Road, Woolpit, Bury St Edmunds, Suffolk IP30 9QS

Editor-in-Chief & Press Manager

Miss F M McKim British Crop Protection Enterprises, Foxhill, Stanford on Soar, Loughborough, Leics LE12 5PZ

Guest Speakers

Prof B C Clarke FRS University of Nottingham, Department of Genetics, Queens Medical Centre, Nottingham NG7 2UH

Prof T Lewis British Crop Protection Enterprises, IACR-Rothamsted Experimental Station, Harpenden Herts AL5 2JQ

Acknowledgement

This Symposium was conceived by the British Crop Protection Council and was greatly facilitated by the support of the Association of Applied Biologists and the Malacological Society of London.

It is a pleasure to acknowledge the efforts of all the members of the Programme Committee and in particular the enthusiasm and tenacity of the Session Organisers.

I am indebted to the staff of British Crop Protection Enterprises and Conference Associates & Services International Ltd for their patience, courtesy and efficiency in making the myriad arrangements required. The successful production of the Proceedings owes much to the dedication and sang froid of the Editor.

The contribution of the guest speakers, not only from the platform, but also over the duration of the meeting, is warmly appreciated.

The Reception and Symposium Dinner were generously underwritten by Bayer AG, Leverkusen, and the attendance of overseas contributors was assisted by Lonza Ltd., Basel.

Exhibitors

Bayer plc, Crop Protection Business Group, Eastern Way, Bury St Edmunds, Suffolk IP32 7AH

BCPC Publications Sales, Bear Farm, Binfield, Bracknell, Berks RG42 5QE

Chiltern Farm Chemicals Ltd, 11 High Street, Thornborough, Buckingham MK18 2DF

De Sangosse UK, PO Box 135, Market Weighton, York YO4 3YY

MicroBio Ltd, NPPL Building, IACR-Rothamsted, Harpenden, Herts AL5 2JQ

Rhône-Poulenc Agriculture Ltd, Fyfield Road, Ongar, Essex CM5 0HW

Abbreviations

Where abbreviations are necessary the following are permitted without definition

acceptable daily intake	ADI
acid equivalent	a.e.
active ingredient	a.i.
approximately	c.
body weight	b.w.
boiling point	b.p.
British Standards Institution	BSI
by the author last mentioned	*idem.*
centimetre(s)	cm
Chemical Abstracts Services Registry Number	CAS RN
compare	cf.
concentration x time product	ct
concentration required to kill 50% of test organisms	LC$_{50}$
correlation coefficient	*r*
cultivar	cv.
cultivars	cvs.
day(s)	d
days after treatment	DAT
degrees Celsius (centigrade)	°C
dose required to kill 50% of test organisums	LD$_{50}$
dry matter	d.m.
Edition	Edn
editor	ed.
editors	eds
emulsifiable concentrate	EC
freezing point	f.p.
for example	e.g.
gas chromatography-mass spectrometry	gc-ms
gas-liquid chromatography	glc
gram(s)	g
growth stage	GS
hectare(s)	ha
high performance (or pressure) liquid chromatography	hplc
hour	h
infrared	i.r.
integrated crop management	ICM
integrated pest management	IPM
International Standardisation Organisation	ISO
in the journal last mentioned	*ibid.*
Joules	J
Kelvin	K
kilogram(s)	kg
least significant difference	LSD
litre(s)	litre(s)
litres per hectare	litres/ha
mass	*m*
mass per mass	*m/m*
mass per volume	*m/V*
mass spectrometry	ms
maximum	max.
melting point	m.p.
metre(s)	m
milligram(s)	mg
milligrams per litre	mg/litre
milligrams per kg	mg/kg
millilitre(s)	ml
millimetre(s)	mm
minimum	min.

Ministry of Agriculture Fisheries and Food (England & Wales)	MAFF
minute (time unit)	min
molar concentration	M
no observed adverse effect level	NOAEL
no observed effect concentration	NOEC
no observed effect level	NOEL
nuclear magnetic resonance	nmr
number average diameter	n.a.d.
number median diameter	n.m.d.
organic matter	o.m.
page	p.
pages	pp.
parts per million	ppm
pascal	Pa
percentage	%
post-emergence	post-em.
power take off	p.t.o.
pre-emergence	pre-em.
pre-plant incorporated	ppi
probability (statistical)	*P*
relative humidity	r.h.
revolutions per minute	rev/min
second (time unit)	s
standard error	SE
standard error of means	SEM
soluble powder	SP
species (singular)	sp.
species (plural)	spp.
square metre	m²
subspecies	ssp.
surface mean diameter	s.m.d.
suspension concentrate	SC
systemic acquired resistance	SAR
technical grade	tech.
temperature	temp.
that is	*i.e.*
thin-layer chromatography	tlc
time for 50% loss; half life	DT$_{50}$
tonne(s)	t
ultraviolet	u.v.
United Kingdom	UK
United States Department of Agriculture	USDA
vapour pressure	v.p.
variety (wild plant use)	var.
volume	*V*
weight	*wt*
weight by volume (mass by volume is more correct)	*wt/v* (*m/V*)
weight by weight (mass by mass is more correct)	*wt/wt* (*m/m*)
wettable powder	WP

less than	<
more than	>
not less than	≮
not more than	≯

Multiplying symbols-		Prefixes
mega	(x 10⁶)	M
kilo	(x 10³)	k
milli	(x 10⁻³)	m
micro	(x 10⁻⁶)	μ
nano	(x 10⁻⁹)	n
pico	(x 10⁻¹²)	p

Session 1
Identification and Taxonomy

Session Organiser
and Chairman   Dr L R Noble

NEW APPROACHES TO THE IDENTIFICATION OF MEDICALLY IMPORTANT FRESHWATER SNAILS ASSOCIATED WITH IRRIGATED AGRICULTURE IN AFRICA

D ROLLINSON
The Natural History Museum, Cromwell Road, London SW7 5BD, UK

C S JONES, L R NOBLE
Zoology Department, Aberdeen University AB9 2TN, UK

ABSTRACT

Water development projects in Africa are essential for a wide range of human activities, notably agriculture and energy production, but all too often are associated with an increase in waterborne diseases such as schistosomiasis. Impounded lakes and especially irrigation canals provide favourable habitats for aquatic snails some of which may act as intermediate hosts for schistosomes, the trematode parasites responsible for the disease schistosomiasis (bilharzia). Snails of the genus *Bulinus* and *Biomphalaria* are responsible for the transmission of schistosomes in Africa. However the snail parasite relationship is an intimate one and only certain snail species are susceptible to infection. In order to improve methods of snail identification and our understanding of the epidemiology of schistosomiasis we have developed a number of molecular approaches. Consideration is given here to the study of ribosomal RNA genes, randomly amplified polymorphic DNA and mitochondrial DNA for the characterization of *Bulinus*. The value of an integrated molecular approach is illustrated by work on snails from West and East Africa.

INTRODUCTION

Irrigated agriculture contributes significantly to the world's food supply. Rice is the main food of 60% of the world's population and is grown on 35% of the irrigated land (Coosemans & Mouchet, 1990). Developing countries faced with an increasing population are required to expand agricultural production and in many areas this can only be achieved by intensification of water impoundment and irrigation. For example, in Africa, where the population may double in the next 25 years, only 30% of the land is suitable for food crops dependent on rainfall. There is therefore considerable pressure to increase the global area under surface irrigation in order to provide opportunities for greater food production and increased prosperity. However, in many developing countries water resource development projects may pose a threat to human health by the introduction or enhancement of waterborne diseases which may have serious implications for the welfare of rural communities. Increases in the prevalence of tropical parasitic diseases such as filariasis, malaria, and schistosomiasis often go hand in hand with the ecological and social changes that are associated with irrigation projects. The most significant reported increases in disease prevalence attributable to water development schemes concern schistosomiasis, a parasitic disease, which depends on freshwater snails for transmission. Snails are an essential component of the parasite life cycle, many species flourishing in the sites polluted and frequented by man. With over 200 million people in 74 countries throughout the developing world estimated to be infected, schistosomiasis is a serious public health problem.

Schistosomiasis is primarily a disease of children, the number of infected people declining after adolescence. The parasites, or schistosomes, are small flatworms that live inside the blood vessels of the gut, liver or bladder. The female worm can lay hundreds of eggs per day and it is the eggs that get trapped within tissues of the body that are responsible for the pathology associated with the disease. To continue the life cycle the egg must be voided from the body within urine or faeces and make contact with freshwater. The egg hatches in water and a free swimming larva emerges which must seek out and penetrate a particular kind of freshwater snail to continue its development. Water is the common factor that brings man and snails together and which allows the parasite to pass from the mammalian host to the mollusc and back. Within the snail asexual reproduction leads to a large increase in parasite numbers, so that some 30 days after invasion many thousands of free swimming larvae leave the snail in search of the mammalian host. Even short exposure to contaminated water may lead to infection as many unfortunate tourists to endemic areas have learnt to their cost. The cercariae penetrate directly through the skin and, in most cases, probably pass unnoticed into the body. In Africa two genera of snails, *Biomphalaria* and *Bulinus* (family Planorbidae) are associated with schistosomiasis transmission. Of the important parasites infecting man, *Biomphalaria* species act as intermediate hosts for *Schistosoma mansoni*, which causes intestinal schistosomiasis, whereas *Bulinus* species transmit *S. haematobium*, responsible for urinary schistosomiasis, and also *S. intercalatum* another intestinal form.

The relationship between the parasite and its intermediate snail host is an extremely intricate one with strains and species of the parasite depending on particular species of snail, which in turn may vary in their susceptibility to the parasites. It follows that the distribution of a suitable snail host is a limiting factor in the distribution of the parasite. In order to understand the epidemiology of the disease new methods are needed to identify snails in endemic areas of disease and to determine those species responsible for parasite transmission. In this paper we consider some of the recent research aimed at providing molecular methods for snail characterization and draw on examples from Senegal and Cameroon to illustrate the importance of "knowing your enemy".

SNAILS, SCHISTOSOMES AND WATER DEVELOPMENT

There are many examples of increased transmission of schistosomiasis as a result of irrigation, the most dramatic being found along the Nile Valley in Egypt and Sudan and more recently along the Senegal River Basin. Between 1974 and 1984 the rate of irrigation expansion in Madagascar, Mali and Nigeria exceeded 100% and vast irrigated areas have been developed in Brazil, China, Indonesia, Malaysia, the Philippines and Thailand (Hunter *et al.*, 1993). All too often the adverse health effects of much needed irrigation projects for agriculture can be underestimated and ironically the increase in standing or slow flowing water such as pools, dam, lakes and irrigation channels may create new breeding sites for intermediate snail hosts of schistosomes, as well as for various insect vectors of diseases such as malaria.

A major outbreak of schistosomiasis has recently been reported in the Senegal river basin (SRB), where 2 dams have been built as part of a large irrigation project. The Senegal river is the second largest river in West Africa and forms the border between Senegal and Mauritania. At Diama in Senegal, approximately 40 km from the mouth of the river, the first dam was constructed and became operational in 1985 to prevent salt water intrusion from the sea. The second dam, designed to provide hydroelectricity, was constructed in 1989 at Manantali in Mali. Prior to the water resource development, donor agencies assessed health risks but no means were provided for schistosomiasis control although some studies warned of a possible extension of schistosomiasis in the SRB. Both dams have had a major influence on the flow and availability of freshwater for irrigation of rice and sugar and at the same time have produced abundant new habitats for the intermediate snail hosts of schistosomes. A five-fold increase in rice cultivation has taken place from 12,000 hec at the end of 1983 to 67,788 hec in 1994 and further expansion is planned.

The increase in agriculture has caused population influx with the population of the main town, Richard Toll, 130 km upstream from the dam at Diama, presently estimated at >50,000. In 1990 an outbreak of intestinal schistosomiasis was reported at Richard Toll and provided the first indication that ecological changes were having an impact on schistosomiasis in the region (Talla *et al.*, 1990). Prior to 1988 *S. mansoni* had never been reported. A similar introduction of urinary schistosomiasis took place at the village of Mbodiene which lies between Richard Toll and the sea (Verlé *et al*, 1994). The recent nature of this infection is reflected in the high prevalence of schistosomiasis observed in all age groups, as without past experience of infection the population was immunologically naive. Intensity of infection, measured by the number of eggs per gram of faeces, was particularly high. In a cohort of 422 individuals in Richard Toll, positive egg counts were found in 91% of the subjects, with a mean egg count of 646 eggs/g. Forty one percent of the community excreted over 1000 eggs/g and individual egg counts were as high as 24,000 eggs/g (Gryseels *et al.*, 1994). To put this in context, a heavy *S. mansoni* infection is considered to be >400 eggs/g.

More recent survey work has shown that schistosomiasis is now becoming a serious health problem in other regions of the SRB and extraordinarily high levels of infection and prevalence are being recorded in the human population (Picquet *et al.*, 1996). Changes in water chemistry have enabled snails to colonise new habitats at the same time as the population increased due to the opportunities provided by agricultural expansion. In order to understand disease transmission precise identifications of the snails are required. The snail parasite interactions in this area are complex ((Picquet *et al.*, 1996), with five species of *Bulinus* recorded; *B. globosus* (Morelet, 1866), *B. umbilicatus* Mandahl-Barth, 1973, *B. senegalensis* Müller, 1781, *B. forskalii* (Ehrenberg, 1831) and *B. truncatus* (Audouin, 1827) but not all of these snails play a role in the transmission of *S. haematobium* and some act as hosts for closely related schistosomes from cattle, sheep and goats. *Biomphalaria pfeifferi* (Krauss, 1848) acts as the intermediate host of *S. mansoni*.

MOLECULAR CHARACTERIZATION OF BULINUS

The most recent taxonomic overview of *Bulinus* lists some 37 species which are commonly assigned to four groups: the *B.truncatus/tropicus* complex, the *B. forskalii* group, the *B. africanus* group and the *B. reticulatus* group (Brown, 1994), each containing species which are involved in the transmission of schistosomes. The genus *Bulinus* is a taxonomically difficult assemblage and, although some species can be clearly defined by morphological characters, problems remain in the characterization and identification of many taxa. The actual boundaries between snail species, which are potentially self-fertilizing hermaphrodites, are often hard to delineate on morphological grounds alone. To characterize snails a variety of characters may be used including shell morphometry, chromosome number, soft part anatomy, enzyme electrophoresis and more recently techniques of DNA analysis. The emphasis has been to provide reliable methods for snail differentiation and identification in order to determine those species and strains playing a major role in schistosomiasis transmission.

The potential of various molecular approaches for discriminating between species groups, species and individuals of *Bulinus* has been explored. The three approaches which have provided useful markers include: the examination of variation in ribosomal RNA (rRNA) genes, as determined by conventional restriction fragment length polymorphism (RFLP) analysis and by PCR-RFLP of the internal transcribed spacer (ITS), secondly the use of randomly amplified polymorphic DNA (RAPDs) and thirdly the analysis of mitochondrial DNA.

Ribosomal RNA genes

The genomic ribosomal RNA (rRNA) gene complex is common to all eukaryotes and is well suited for taxonomic studies as it contains regions which evolve at different rates, thus permitting analysis of relationships over a wide taxonomic level. DNA probes have been shown to be of value for distinguishing schistosome species and particular use has been made of the rRNA gene probes pSM889, pSM890 and pSM389 derived from *S. mansoni* (Simpson *et al.*, 1984). The probe pSM889 has been used to study restriction enzyme digests of various species of *Bulinus*. This probe is a 4.4kb fragment that encompasses part of the small rRNA, internal transcribed spacer (ITS) and large rRNA gene. As it contains highly conserved regions it shows homology with a wide variety of organisms and will readily bind to restricted fragments of snail DNA.

Clear differences in the sizes of restriction fragments between species of *Bulinus* representing the four species groups were observed when DNA was digested with either *Bam*HI or *Bgl*II and hybridised to pSM889. No differences were observed between samples of *B. tropicus* (Krauss, 1848) and *B. truncatus* but intraspecific variation was observed between samples of *B. forskalii* from São Tomé and Angola (Rollinson & Kane, 1991). This work suggested that sequence variation in the spacer regions may be useful for discrimination. The entire ITS region including the 5.8S rRNA gene has recently been amplified from taxa representing the four species groups following the methods of Kane & Rollinson (1994). The sequences for the primers used for PCR amplification were based on conserved regions of the 3' end of the 18S rRNA gene (ETTS1) and the 5' end of the 28S rRNA gene (ETTS2). The amplified product was digested with one of a number of 4-base or 6-base cutting restriction enzymes (*Rsa*I, *Alu*I, *Cfo*I, *Sma*I and *Sac*I). Characteristic RFLP patterns were obtained for the four species groups and, in the majority of cases, species within a group could be differentiated (Stothard *et al.*, 1996). The complete ITS1 spacer region has been sequenced for *B. globosus*, *B. cernicus* (Morelet, 1867) and *B. truncatus* and substantial nucleotide variation occurs, indicating considerable divergence between the species groups (Stothard *et al.*, 1996).

Randomly amplified polymorphic DNA (RAPD)

Unlike conventional PCR-based analyses, RAPD approaches use single oligonucleotide primers (5-20 bases) of arbitrary sequence to initiate DNA strand synthesis under conditions of low stringency at a number of complementary binding sites scattered throughout the genome (Welsh & McClelland, 1990; Williams *et al.*, 1990). Discrete amplification products form where primer sites are orientated in an inverted repeat and are within an amplifiable distance of each other. Several amplification fragments may be generated within a single RAPD reaction and, when separated by either agarose or polyacrylamide electrophoresis, give rise to a RAPD profile. Twenty-eight primers (Operon Technologies Ltd and British Biotechnology) were screened against genomic DNA extracted from *Biomphalaria sudanica* (Martens, 1870) to identify primers which gave reproducible and informative RAPD profiles. Of these primers, eight (4 x 10mers, 4 x 15mers) were subsequently selected to study genetic variation within and between nine species of *Bulinus* (Stothard & Rollinson, 1996). RAPD profiles were visualised using both silver stained polyacrylamide and ethidium stained agarose gel electrophoresis. This approach allowed a direct comparison of the two methods and therefore permitted an evaluation of the stability of the derived phylogenetic relationships.

RAPD profiles were highly divergent between the species of *Bulinus* and intra-specific variation was also observed. Hence, RAPD data have little value for comparisons between species groups and diagnostic RAPD profiles for a species throughout its range might be difficult to find. However, RAPDs do allow the identification of differences between species, populations and individuals on a regional basis. Using this technique Langand *et al.* (1993) were able to differentiate the two closely related species *B. globosus* and *B. umbilicatus*, and populations of *B. forskalii*. They also suggested that RAPDs may also be used to identify whether offspring are the result of cross- or self-fertilisation.

Mitochondrial DNA

The mitochondrion exists as an organelle within eukaryote cells and is responsible for aerobic respiration. In addition to the genetic material found within each cell nucleus, the mitochondrion also contains its own DNA which codes for several proteins and transfer RNA molecules as well as two ribosomal RNAs. The mitochondrial genome of molluscs is circular and of approximately 14 kb in length in the gastropods *Albinaria turrita* and *Cepaea nemoralis* (Linné 1758) (Lecanidou *et al.*, 1994; Terret *et al.*, 1994). As this molecule does not recombine and is usually maternally inherited, sequence variation within mitochondrial DNA provides excellent characters for phylogenetic studies. One such gene coding for the Cytochrome Oxidase subunit I (COI) has had increasing attention as a phylogenetic marker, as sequence variation within this gene is accumulating at a rate which allows the determination of relationships both within and between species (Bowles *et al.*, 1992; Juan *et al.*, 1995; Brown *et al.*, 1994). In addition, certain regions within COI are highly conserved enabling portions of this gene to be amplified by PCR using universal primers.

Using such primer sequences, a 450bp product has been amplified from three species within the *B. africanus* group and these partial COI sequences have been characterized by both restriction digestion and DNA sequencing. Double digestion, using *AluI* and *RsaI*, generates informative profiles which appear to discriminate *B. africanus* (Krauss, 1848), *B. globosus* and *B. nasutus* (Martens, 1979). In addition, a species-specific restriction site has been identified within *B. globosus* sequences from East Africa with the enzyme *SspI*. As these restriction assays are relatively simple and quick, they may be of value for routine identification purposes (Stothard & Rollinson, in press). DNA sequencing of PCR products of COI obtained from *B. globosus* and *B. nasutus* has identified further variation; a total, 33 variable sites within a 330bp region. Cladistic analysis of these COI sequences has shown that these taxa are evolutionary separate thereby firmly rejecting the possibility of conspecific taxa (Stothard & Rollinson, in press). It is hoped that further studies on this mitochondrial gene will shed light on the relationships between East and West African *B. globosus*. As the mtDNA sequence database expands, it will be interesting to see how the species relationships as well as those between the species groups unfold.

DEVELOPMENT OF MOLECULAR METHODS TO DIFFERENTIATE B.FORSKALII GROUP SNAILS: A CASE STUDY

Species of the *B. forskalii* group are not defined entirely satisfactorily. The present system is founded on characters of the shell and anatomy, particularly the shape of the mesocone on the first lateral radular tooth, and the presence or absence of a carina on the shell surface. However, the appearance of these features is variable and the absence of a carina is not always indicative. These morphological characters are not easily used by the non-specialist and many workers experience difficulty in differentiating taxa within the group. Currently there are eleven species recognised within the *B. forskalii* group, three of which are represented in Cameroon; *B. forskalii, B. senegalensis* and *B. camerunensis* Mandahl-Barth, 1957. Mimpfoundi & Greer (1989) compared allozyme patterns among members of the *B. forskalii* group from Cameroon. Clear differences were found between *B. forskalii* and *B. senegalensis* but allozyme patterns for *B. camerunensis* were identical to some populations of *B. forskalii*. A more detailed study of 32 *B. forskalii* populations revealed intrapopulation variation in 8 populations, but heterozygotes were present in only 2 of these and neither population was in Hardy-Weinberg equilibrium (Mimpfoundi & Greer, 1990). The results suggested low genetic diversity and indicated that *B. forskalii* reproduces principally by self-fertilisation.

In Cameroon, *B. forskalii* acts as a host for *S. intercalatum* whereas *B. senegalensis* and *B. camerunensis* have been implicated in the transmission of *S. haematobium*. Mixed populations of *B. forskalii* and *B. senegalensis* have been found in Northern Cameroon. Although *B. senegalensis* and *B. forskalii* have distinct enzyme loci detectable by electrophoresis, *B. senegalensis* does not survive well upon leaving the waterbody and snails of this species are extremely difficult to maintain in the laboratory. Freezing facilities, which are required to maintain the snails in a suitable condition for protein electrophoresis, are not always practical or easily available. The advent of simple molecular techniques allows ethanol preservation of field material and provides access to museum collections.

The RAPD assay has been used to investigate natural populations of *B. forskalii* group snails mainly from Cameroon and other West African localities (Jones *et al.*, in prep a). Forty primers were screened and amplification products visualised on ethidium stained agarose gels. Most primers gave reliable, reproducible profiles and were used routinely to produce diagnostic profiles which are of value for discriminating between *B. forskalii* group snails. Twenty primers have now been used to characterize, in detail, a minimum of 10 individuals from 21 populations of *B. forskalii* taxa, including populations of *B. forskalii* from throughout Cameroon, several from Senegal and one from São Tomé; *B. senegalensis* from the type locality in Senegal and other localities, north Cameroon and Mali; *B. cernicus* from Mauritius and *B. crystallinus* (Morelet, 1868) from Angola; *B. camerunensis* from S.W. Cameroon; a laboratory population of *B. wrighti* Mandahl-Barth, 1965 originally from Oman; and one population of *B. truncatus* from Cameroon used as an outgroup.

Few fragments were shared between *B. forskalii* and *B. senegalensis*, hence, unknown specimens conchologically assigned to these taxa may be identified to the specific level by RAPD profiling. No intraspecific polymorphisms were detected in the type locality material of *B. senegalensis* from Senegal. However, limited within and between population variation was found in *B. senegalensis* from north Cameroon. Similarly, limited within population variation was demonstrated in *B. forskalii*, but greater differences between populations were observed, especially between geographically separated localities.

Comparison of RAPD profiles is useful, although standardizing PCR parameters between laboratories and interpretation of the resulting banding patterns may be problematic. This difficulty may be resolved by identifying and excising 'unique' fragments which are then used as probes on full Southern blots of the RAPD gels, making species identification less ambiguous. Several such species-specific markers have been identified for *B. forskalii* and *B. senegalensis*. However, when hybridized to dot blots of total genomic DNA species-specificity was lost. In order to develop a simpler procedure, RAPD products which appeared to be diagnostic for either *B . senegalensis* or *B. forskalii* from West Africa have been cloned and sequenced and a panel of primers which allow the amplification of species-specific products have now been designed to differentiate these taxa (Jones *et al.*, submitted). Widespread sampling confirmed primer specificity throughout the sampling area of each species. Thus a simple, robust, unambiguous PCR-based method of identification is now available for use by the non-specialist.

RAPD fragments 'shared' between taxa were labelled as probes to check homology and were then used as characters in cluster analysis. Phylogenetic analyses, using the Fitch-Margoliash criterion, suggest that *B. forskalii* and *B. senegalensis* cluster on distinct branches, demonstrating that these morphologically similar species are clearly differentiated by molecular methods. Although identified as a separate species on morphology *B. camerunensis* clearly clusters with specimens identified as *B . forskalii*. In particular its affinity with *B. forskalii* from nearby Kumba, South West Cameroon (the closest *B. forskalii* population to the *B. camerunensis* site) suggests that *B. camerunensis* has probably evolved *in situ* from local *B. forskalii* which became isolated in the crater lake. Other *B. forskalii* group species, such as *B. cernicus* and *B. crystallinus*, cluster with *B. truncatus* to form a third distinct branch. Conversely, the molecular data suggest samples of what are tentatively described as an extreme geographical variant of *B. forskalii* from São Tomé on shell morphology, do in fact cluster with *B. crystallinus* from Angola. Given the historical links between these two places this seems plausible, but more samples are required to confirm this hypothesis. Additionally, preliminary work suggests an affinity between West and East African (Kenya) *B. forskalii*, which remain distinct from other East African *B . forskalii* taxa (such as *B. scalaris* [Dunker, 1845]).

Phylogenetic relationships within the group were investigated in more detail by sequencing a 350bp region of the mitochondrial COI gene for six *B. forskalii* group taxa, and the outgroup *B. truncatus*, representing 42 samples from West and East Africa (Jones *et al.*, in prep b). A total of 86 variable sites was detected within this region, with *B. senegalensis* and *B. forskalii* differing at 56 sites whereas *B. camerunensis* and its closest population of *B. forskalii* differed at a single site. Phenetic and cladistic analyses suggest relationships concordant with those based on RAPD data: the same clear separation of both West and East African *B. forskalii* from *B. senegalensis* is evident, as is the clustering of *B. cernicus* (Mauritius) with *B . scalaris* (Kenya), supporting the contention that this island species may be derived from mainland populations of *B. scalaris*. The close similarity of *B. forskalii* and *B . camerunensis* sequences again suggests they may be considered conspecifics.
Clearly, the value of an integrated molecular approach to problematic gastropod taxa lies in the unambiguous resolution of their relationships and the production of useful systematic tools with which to differentiate closely related self-fertile taxa.

CONCLUDING REMARKS

Progress is being made in the application of molecular methods for the characterization of *Bulinus* and it is clear that this avenue of research has considerable potential. By examining DNA directly and in detail more will be revealed about genetic diversity at the level of the genus and the population. The three approaches detailed for *Bulinus* show some promise for characterization studies: RFLP PCR of the internal transcribed spacer region of the rRNA gene allows differentiation of the species groups and shows that sequence variation exists between species; RAPD profiles indicate high divergence between *Bulinus* species groups and allow in some cases the recognition of taxa, as well as providing a means for examining intrapopulation variation. This technique can also lead to the isolation of DNA fragments that show different degrees of specificity, cloning and sequencing of these fragments can lead to the production of PCR based diagnostic tests. The preliminary results with the cytochrome oxidase gene indicate that mitochondrial DNA is worthy of further study and may help to elucidate the relationships of morphologically different forms in the *B. africanus* group and the anatomically similar *B. forskalii* taxa which continue to perplex investigators. One common theme to emerge is that there is extensive variation at the DNA level within the genus and high levels of variability have been observed within some species.

Molecular studies have advantages in that snail material can be stored in ethanol prior to DNA extraction, in some cases this will facilitate analysis of species which are difficult to collect and maintain. Snail collections held in ethanol in Museums and Institutes may be stored in a suitable way for future analysis. Unfortunately this is not true for enzymes which must be recovered from either fresh or frozen material. Formaldehyde preservation is also unsuitable for subsequent DNA extraction.

Molecular studies do require good laboratory facilities and access to the necessary reagents, the cost of which may seem prohibitive. However, protocols are readily available and techniques can be easily learnt by those with laboratory experience. Whereas development work might be confined to specialist laboratories it is to be hoped that most workers will be able to gain access to facilities that could be used for molecular studies, which are becoming an important part of research in tropical medicine. Moreover, it is important for molecular scientists to work closely with malacologists in the field in order to appreciate the problems encountered in snail identification. Emphasis should be given to the development of identification tests and procedures which are both cost-effective and suitable for use in basic laboratories.

It is hoped that the future application of molecular approaches will lead to a better understanding of the snails responsible for the transmission of schistosomiasis and in the longer term contribute to the control of schistosomiasis in the developing world.

ACKNOWLEDGEMENTS

Dr C.S. Jones holds a Taxonomy and Systematics Fellowship from the Wellcome Trust.

REFERENCES

Bowles, J; Blair, D; McManus, D (1992) Genetic variants within the genus *Echinococcus* identified by mitochondrial DNA sequencing. *Molecular and Biochemical Parasitology*. **54**, 165-174.

Brown, D S (1994). *Freshwater snails of Africa and their Medical Importance*. 2nd edition. Taylor and Francis. 609pp.

Brown, J M; Pellmyr, O; Thompson, J N; Harrison, R G (1994) Phylogeny of *Greya* (Lepidoptera: Prodoxidae), based on nucleotide sequence variation in mitochondrial cytochrome oxidase I and II: congruence with morphological data. *Molecular Biology and Evolution*. **11**, 128-141.

Coosemans, M; Mouchet, J (1990) Consequences of rural development on vectors and their control. *Annales de la Société Belge de Médicine Tropicale*. **70**, 5-23.

Gryseels, B; Stelma, F F; Talla, I; van Dam, G J; Polman, K; Sow, S; Diaw, M; Sturrock, R. F; Doehring-Schwerdtfeger, E; Kardoff, R; Decam, C; Niang, M; Deelder, A M (1994) Epidemiology, immunology and chemotherapy of *Schistosoma mansoni* infections in a recently exposed community in Senegal. *Tropical and Geographical Medicine*. **46**, 209-219.

Hunter, J M; Rey, L, Chu K Y; Adekolu-John, E O; Mott K E (1993) *Parasitic diseases in water resources development: the need for intersectoral negotiation*. World Health Organization, Geneva, 152 pp.

Juan, C; Oromi, P; Hewitt, M G (1995) Mitochondrial DNA phylogeny and sequential colonisation of Canary Islands by darkling beetles of the genus *Pimelia* (Tenebrionidae). *Proceedings of the Royal Society of London B*. **261**, 173-180.

Kane R A; Rollinson D (1994) Repetitive sequences in the ribosomal DNA internal transcribed spacer of *Schistosoma haematobium, Schistosoma intercalatum* and *Schistosoma mattheei*. *Molecular and Biochemical Parasitology*. **63**, 153-156.

Langand, J; Barral, V; Delay, B; Jourdane, J (1993) Detection of genetic diversity within snail intermediate hosts of the genus *Bulinus* by using Random Amplified Polymorphic DNA markers (RAPDs). *Acta Tropica*. **55**, 205-215.

Lecanidou, R; Douris, V; Rodakis, G C (1994) Novel features of metazoan mtDNA revealed from sequence analysis of three mitochondrial DNA segments of the land snail *Albinaria turrita* (Gastropoda: Clausiliidae). *Journal of Molecular Evolution*. **38**, 369-382.

Mimpfoundi, R; Greer G J (1989) Allozyme comparisons among species of the *Bulinus forskalii* group (Gastropoda: Planorbidae) in Cameroon. *The Journal of Molluscan Studies*. **55**, 405-410.

Mimpfoundi, R; Greer G J (1990) Allozyme variation among populations of *Bulinus forskalii* (Ehrenberg, 1831) (Gastropoda: Planorbidae) in Cameroon. *The Journal of Molluscan Studies*. **56**, 363-372.

Picquet, M; Ernould J C; Vercruysse, J; Southgate, V R; Mbaye, A; Sambou, B; Niang, M; Rollinson, D (in press) The epidemiology of human schistosomiasis in the Senegal river basin.*Transactions of the Royal Society of Tropical Medicine and Hygiene*. **90**.

Rollinson, D; Kane R A (1991) Restriction enzyme analysis of DNA from species of *Bulinus* (Basommatophora: Planorbidae) using a cloned ribosomal RNA gene probe. *Journal of Molluscan Studies*. **57**, 93-98.

Simpson, A J G; Dame, J B; Lewis F A; McCutchan T F (1984) The arrangement of ribosomal RNA genes in *Schistosoma mansoni*. Identification of polymorphic structural variants. *European Journal of Biochemistry*. **139**, 41-45.

Stothard, J R; Rollinson, D (1996) An evaluation of Random Amplified Polymorphic DNA (RAPD) for the identification and phylogeny of freshwater snails of the genus *Bulinus* (Gastropoda: Planorbidae). *Journal of Molluscan Studies.* **62**, 165-176.

Stothard, J R; Hughes, S; Rollinson, D (1996) Variation within the ribosomal DNA internal transcribed spacer (ITS) of intermediate snail hosts within the genus *Bulinus* (Gastropoda:Planorbidae).*Acta Tropica.* **61**, 19-29.

Stothard, J R; Rollinson, D (in press) Partial sequences from the mitochondrial cytochrome oxidase subunit I can differentiate the intermediate snail hosts *Bulinus globosus* and *B. nasustus* (Gastropoda: Planorbidae). *Journal of Natural History.*

Talla, I; Kongs, A; Verlé, P; Belot, J; Sarr, S; Coll, A. M. (1990) Outbreak of intestinal schistosomiasis in the Senegal River Basin. *Annales de la Sociéte Belge de Médecine Tropicale.* **70**, 173-180.

Terret, J; Miles, S; Thomas, R.H. (1994) The mitochondrial genome of *Cepaea nemoralis* (Gastropoda: Stylommatophora): gene order, base composition and heteroplasmy. *Nautilus supplement.* **2**, 79-84.

Verlé, P; Stelma, F; Desreumaux, P; Dieng, A; Diaw, O; Kongs, A; Niang, M; Sow,S; Talla, I; Sturrock, R. F; Gryseels, B; Capron, A (1994) Preliminary study of urinary schistosomiasis in a village in the delta of the Senegal river basin, Senegal. *Transactions of the Royal Society of Tropical Medicine and Hygiene.* **88**, 401-405.

Welsh, J; McClelland, M. (1990) Fingerprinting genomes using PCR with arbitrary primers. *Nucleic Acids Research.* **18**, 7213-7218.

Williams, J G K; Kubelik, A R; Livak, K J; Rafolski, J A; Tingey, S V (1990) DNA polymorphisms amplified by arbitrary primers are useful as genetic markers. *Nucleic Acids Research*, **18**, 6531-6535.

POSSIBLE OUTCROSSING IN NATURAL *CARINARION* POPULATIONS (MOLLUSCA, PULMONATA)

K JORDAENS, H DE WOLF, R VERHAGEN

University of Antwerp (RUCA), Evolutionary Biology Group, Groenenborgerlaan 171, B-2020 Antwerp, Belgium; E-mail: Jordaens@ruca.ua.ac.be

T BACKELJAU

Royal Belgian Institute of Natural Sciences, Vautierstraat 29, B-1000 Brussels, Belgium

ABSTRACT

Currently, the land slug subgenus *Carinarion* consists of three species, viz. *Arion (Carinarion) fasciatus*, *A. (C.) circumscriptus* and *A. (C.) silvaticus*. Allozyme electrophoresis shows that the three species reproduce uniparentally, most probably by self-fertilization, suggesting that each species is a collection of fixed homozygous lines (multilocus genotypes or strains). Using vertical polyacrylamide gel electrophoresis we surveyed the population genetic structure of the species in Europe and found 24 strains among 630 individuals. Yet, seven individuals could not be assigned to a specific strain because they were heterozygous, either for phosphoglucomutase (five *A. silvaticus* and one *A. fasciatus*) or for lactate dehydrogenase (one *A. silvaticus*).

Hitherto, at least 18 individuals were reported to contain a spermatophore. In addition, we found three spermatophores in *A. silvaticus*. The occurrence of spermatophores and heterozygotes suggests that *Carinarion* spp. are capable of outcrossing, although the three species predominantly reproduce uniparentally. It remains to be investigated whether this uniparental mode of reproduction implies self-fertilization or some kind of parthenogenesis. Finally, the systematic implications of these findings are discussed and it is tentatively argued that the colour differences between *Carinarion* sp. perhaps reflect a fixation of different alleles caused by selfing, rather than a specific difference.

INTRODUCTION

The land slug genus *Arion* is usually divided into four or five subgenera (see Backeljau & De Bruyn, 1990). One of these is the subgenus *Carinarion* which comprises three species, viz. *Arion (Carinarion) fasciatus* (Nilsson, 1823), *A. (C.) circumscriptus* Johnston, 1828 and *A. (C.) silvaticus* Lohmander, 1937. The three species are believed to reproduce uniparentally, most probably through self-fertilization [= autogamy *sensu* Mogie (1986)] (e.g. McCracken & Selander, 1980, Foltz *et al.*, 1982), although parthenogenesis (automixis) can not yet be excluded (e.g. Runham, 1993). Using starch gel electrophoresis, McCracken & Selander (1980) concluded that in North America each *Carinarion* species is represented by a single homozygous line (multilocus genotype or strain). In Ireland, Foltz *et al.* (1982) observed one strain of *A. circumscriptus* and two strains of *A. silvaticus*. However, using vertical polyacrylamide gel electrophoresis (PAGE), Backeljau *et al.* (1996) and Jordaens *et al.* (unpublished data) showed that, at least in NW Europe, each species is a collection of several strains, which regularly co-occur.

Despite the evidence of selfing, *Carinarion* spp. sometimes produce spermatophores (e.g. Backeljau & De Bruyn, 1989). This suggests that copulations may occur in natural populations, so that there is a theoretical possibility of outcrossing too. In order to investigate this possibility, we used allozyme electrophoresis to conduct a genetic survey of natural *Carinarion* populations in Europe.

MATERIAL AND METHODS

In total we surveyed 630 specimens (Table 1). Species identifications were based on Lohmander (1937) and Waldén (1955). In case of doubt allozyme patterns at the loci phosphoglucomutase, fumarase and alanine aminotransferase were considered decisive (Backeljau *et al.*, 1996).

Table 1. *Carinarion* material examined. Figures refer to the numbers of specimens and populations (in parentheses) studied.

Country	A. silvaticus	A. fasciatus	A. circumscriptus
Belgium/The Netherlands	384 (75)		134 (32)
Germany	17 (3)	9 (3)	
Poland	17 (2)	46 (4)	1 (1)
Sweden		22 (1)	
Czechia	12 (2)	3 (1)	

Individual digestive gland homogenates were screened for seven polymorphic loci (Backeljau *et al.*, 1996). A continuous Tris/Citric acid buffer was used to resolve alanine aminotransferase (*Alat*, E.C. 2.6.1.2), esterase Q (*EsQ*, E.C. 3.1.1.1 [see Backeljau *et al.*, 1987]) and fumarate hydratase (*Fumh*, E.C. 4.2.1.2). A discontinuous buffer (gel: Tris-HCl [pH 9.0]; tray: Tris/Glycine [pH 9.0]) was used to resolve lactate dehydrogenase (*Ldh*, E.C. 1.1.1.27), phosphoglucomutase (*Pgm*, E.C. 2.7.5.1) and leucylalanine aminopeptidase (*Pep-2* and *Pep-3*, E.C. 3.4.11 [see Backeljau *et al.*, 1996]). For sample preparation, methods of PAGE and staining procedures we refer to Backeljau (1987) and Backeljau *et al.* (1996). Individual multilocus genotypes were assigned to specific strains as defined by Backeljau *et al.* (1996). The remaining parts of the animals were stored in 70% ethanol for later anatomical study and to check for the presence of spermatophores.

RESULTS

A total of 24 strains were detected: two for *A. fasciatus*, 20 for *A. silvaticus* and two for *A. circumscriptus*. Comparing our results with those of Backeljau *et al.* (1996) and Jordaens *et al.* (unpublished data), one new strain was recorded for *A. silvaticus*, while two strains reported by Backeljau *et al.*, 1996) were not found, viz. one strain of *A. silvaticus* (strain G) and one of *A. fasciatus* (strain C). However, seven individuals could not be assigned to a specific strain as they were heterozygous. In The Netherlands one specimen of *A. silvaticus* was heterozygous for *Ldh*. In Poland, one specimen of *A. fasciatus* was heterozygous for *Pgm*, one of the alleles being typical for *A. fasciatus*, the other for *A. silvaticus* (Backeljau *et al.*, 1996). Finally, in Germany, five *A. silvaticus* specimens were heterozygous for *Pgm* and appeared to be of the *A. fasciatus/A. silvaticus* type (Fig. 1). All Swedish animals belonged to a single strain and no heterozygotes were observed in animals from the Czech Republic, although in each population several strains co-occurred.

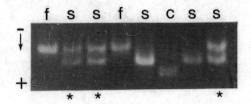

Figure 1.
Genetic variation at phosphoglucomutase in *A. fasciatus* (f) (Görlitz, Germany), *A. silvaticus* (s) (Görlitz, Germany) and *A. circumscriptus* (c) (Jawar, Poland) as resolved with vertical polyacrylamide gel electrophoresis. Note the hybrid appearance of the *A. silvaticus* heterozygotes (marked by *). The arrow indicates the direction of migration.

Other polymorphic loci (*EsQ*, *Fumh*, *Alat*, *Pep-1* and *Pep-2*) did not yield heterozygotes. However, esterase patterns sometimes revealed remarkable results, especially among the anodal bands which may represent a single locus with (at least) three alleles. However, we are still undecided about the genetic interpretation of these bands. Therefore, we hesitate to regard the two-band esterase profiles as heterozygotes (in contrast to the single band profiles which would represent homozygotes). We did not observe heterozygous *A. circumscriptus.*

Only three *A. silvaticus* out of the 630 specimens surveyed contained a spermatophore (Fig. 2). The spermatophores were more or less complete and were found in animals collected in April (Table 2). Two spermatophores were observed in animals from a population where four strains co-occurred. The third spermatophore was found in an animal from a population where three strains co-occurred.

Figure 2.
Scanning electron micrograph of the spermatophore of *A. silvaticus* (Kontich, Belgium). Scale bar = 1 mm.

DISCUSSION

It is commonly believed that the three *Carinarion* species self-fertilize (McCracken & Selander, 1980; Foltz *et al.*, 1982, 1984; Selander & Ochman, 1983; Backeljau *et al.*, 1996). This is perhaps not unexpected as *Carinarion* species lack a ligula, a stimulator used during copulation (Runham & Hunter, 1970). In addition, the oviduct, which in *Arion* generally acts as copulatory organ, is only weakly developed in *Carinarion* (e.g. Lohmander, 1937). A similar situation is found in *Arion intermedius* Normand, 1852, which also reproduces uniparentally (e.g. Backeljau & De Bruyn, 1990). In spite of the weak development of the copulatory organs in *Carinarion*, some specimens do occassionally contain a spermatophore, produced during copulation to transfer sperm from one individual to another (e.g. Runham & Hunter, 1970; Lind, 1973). Hence, the presence of a spermatophore suggests the occurrence of mating behaviour and the possibility of outcrossing.

Ökland (1922) was the first to describe a spermatophore in *Carinarion*. Since then, spermatophores have been recorded regularly (Lohmander, 1937; Quick, 1960; Lupu, 1974; Backeljau & De Bruyn, 1989) (Table 2). Also the mating behaviour of *Carinarion* has been described, although it was observed only twice (Gerhardt, 1935) in Germany. Based on this superficial and vague description, it seems that copulation is very quick (approximately 10 minutes) and that there is no visible eversion of the copulatory organs as in e.g. *Arion ater* (Linnaeus, 1758) or *Arion hortensis* de Férussac, 1819. Although the two copulations Gerhardt (1935) observed took place in January, all spermatophores for which collecting data are available were recorded in the period from April-July (Table 2). During our fieldwork we did not observe copulation in *Carinarion*, although these slugs are said to copulate frequently in Sweden (Von Proschwitz, pers.comm.).

Table 2. Records of spermatophores found in *Carinarion*.

Country	Species	Locality (Date)
Belgium	A. silvaticus	Bossut-Gottechain (03/IV/1978)[*]
	A. circumscriptus	Sougné-Remouchamps (22/V/1979)[*]
	A. silvaticus	Vielsalm (04/V/1985)[*]
	A. circumscriptus	Vielsalm (04/V/1985)[*]
	A. silvaticus	Kontich (05/IV/1994)[$]
	A. silvaticus	Kontich (05/IV/1994)[$]
	A. silvaticus	Flaawers (18/IV/1994)[$]
Czech Republic	A. fasciatus	Teplice (22/V/1984)[*]
Denmark	A. silvaticus (?)	Holsteinborg (28/IV/1945)[*]
	A. silvaticus	Tystrup (04/VII/1942)[*]
Sweden	A. silvaticus	Tierps (23/V/1955)[*]
	A. silvaticus	Overgrans (09/V/1955)[*]
	A. silvaticus	Overgrans (09/V/1955)[*]
	A. silvaticus	Overgrans (09/V/1955)[*]
	A. silvaticus	Arnas (12/VI/1958)[*]
	A. silvaticus	Arnas (12/VI/1958)[*]
	A. fasciatus	Dalby Hage (?/V/1937)[*]
Scandinavia	A. fasciatus (several)	? (end May - end June)[**]
	A. silvaticus	? (end May - end June)[**]

references: ([$]: this study; [*]: Backeljau & De Bruyn, 1989; [**]Lohmander, 1937)

Neither the ability to produce sperm (Luchtel, 1972), nor the occurrence of mating behaviour or the transfer of spermatophores, implies that allosperm will be used for fertilization. It is, indeed, theoretically possible that allosperm only serve to initiate egg development, but do not contribute any genetic material to the offspring (e.g. pseudogamy or gynogenesis). Yet, Backeljau *et al.* (1996) recorded two malate dehydrogenase (*Mdh*) heterozygotes in *A. circumscriptus* among 229 *Carinarion* specimens examined. Here, we report seven more heterozygotes in *A. fasciatus* and *A. silvaticus*. The occurrence of possible heterozygotes in the esterases requires confirmation as esterase patterns can be age and/or food dependent (Oxford, 1978).

Whether the breeding system of *Carinarion* involves outcrossing and autogamy or some form of parthenogenesis remains to be investigated by breeding experiments with known genetic markers (e.g. Jarne & Städler, 1995). Nevertheless, it is argued that parthenogenesis is rare among pulmonates (Nicklas & Hoffmann, 1981; Hoffmann, 1983; Runham, 1993), whereas self-fertilization (facultative or obligatory) seems to be more common (e.g. Ikeda, 1937; Selander & Ochman, 1983).

Finally, the possibility of outcrossing is supported by additional information from Austrian *Carinarion* populations, where heterozygotes seem to be more frequent and *A. silvaticus* and *A. fasciatus* may hybridize (Fig. 1) (Jordaens *et al.* unpublished data). If hybridization is more widespread, then one should have serious doubts as to the species status of at least *A. fasciatus* and *A. silvaticus*. Moreover, the two *Mdh* heterozygotes of *A. circumscriptus* reported by Backeljau *et al.* (1996) were also of a putative "hybrid" nature since they contained an *A. silvaticus* allele. Backeljau *et al.* (1996) further remarked that (1) many other terrestrial pulmonates are (facultatively) selfing multistrain complexes that often consist of "discrete" morphotypes based on colour, size and shell form (e.g. *Rumina decollata*, *Chondrina clienta*, *Cochlicopa* spp.), that (2) Nei's genetic identity between the *Carinarion* strains is comparable to that of conspecific strains in *Chondrina clienta* or *Arion intermedius*, and (3) phylogenetic trees based on allozyme data do not consistently separate *A. silvaticus* from *A. fasciatus* when SE European populations are included. In view of this we suggest that *Carinarion* might very well represent a single "species", which because of its mainly uniparental breeding system is divided into a series of morphotypes and genetic strains. After all, inasmuch as arionid colour characteristics are genetically determined, the genes involved should be subject to the same reproductive mode as the allozyme loci. Thus in case of selfing strains, so-called species-specific colour patterns (such as those reported for *Carinarion*), may just as well reflect a fixation of different alleles, rather than a taxonomic difference. We will expand further on this issue in a forthcoming paper.

ACKNOWLEDGEMENTS

The authors wish to thank Dr. Ted von Proschwitz (Natural History Museum, Göteborg, Sweden), Heike Reise (Staatliches Museum für Naturkunde, Görlitz, Germany) and Beata Pokryszko (University of Wroclaw, Wroclaw, Poland) for providing us with part of the material. Hans De Wolf is funded by an IWT scholarship. This work was supported by F.J.B.R. grants 2.0004.91 and 2.0128.94.

REFERENCES

Backeljau, T (1987) Electrophoretic distinction between *Arion hortensis*, *A. distinctus* and *A. owenii*. *Zoologischer Anzeiger*. **219**, 33-39.

Backeljau, T; Ahmadyar S Z; Selens M; Van Rompaey J; Verheyen W (1987) Comparative electrophoretic analyses of three European *Carinarion* species (Mollusca, Pulmonata, Arionidae). *Zoologica Scripta*. **16**, 209-222.

Backeljau, T; De Bruyn L (1989) Notes on *Arion hortensis s.l.* and *A. fasciatus s.l.* in Denmark. *Apex*. **4**, 41-48.

Backeljau, T; De Bruyn L (1990) On the infrageneric systematics of the genus *Arion* Férussac, 1819 (Mollusca, Pulmonata). *Bulletin de l'Institut Royal des Sciences Naturelles de Belgique, Biologie*. **60**, 35-68.

Backeljau, T; De Bruyn L; De Wolf H; Jordaens K; Van Dongen S; Winnepenninckx B (1996) Allozyme diversity in slugs of the *Carinarion* complex (Mollusca, Pulmonata). *Heredity*. in press.

Foltz, D W; Ochman H; Jones J S; Evangelisti S M; Selander R K (1982) Genetic population structure and breeding systems in arionid slugs (Mollusca: Pulmonata). *Biological Journal of the Linnean Society*. **17**, 225-241.

Foltz, D W; Ochman H; Selander R K (1984) Genetic diversity and breeding systems in terrestrial slugs of the families Limacidae and Arionidae. *Malacologia*. **25**, 593-605.

Gerhardt, U (1935) Weitere Untersuchungen zur Kopulation der Nacktschnecken. *Zeitschrift für Morphologie und Ökologie der Tiere*. **30**, 297-232.

Hoffmann, R J (1983) The mating system of the terrestrial slug *Deroceras laeve*. *Evolution*. **37**, 423-425.

Ikeda, K (1937) Cytogenetic studies on the self-fertilization of *Phylomycus bilineatus* Benson. *Journal of Science of the Hirosima Universtity, Serie B*. **15**, 67-123.

Jarne, P; Städler T (1995) Population genetic structure and mating system evolution in freshwater pulmonates. *Experientia*. **51**, 482-497.

Lind, H (1973) The functional significance of the spermatophore and the fate of spermatozoa in the genital tract of *Helix pomatia* (Gastropoda: Stylommatophora). *Journal of Zoology (London)*. **169**, 39-64.

Lohmander, H (1937) Über die nordischen Formen von *Arion circumscriptus* Johnston. *Acta Societatis Pro Fauna et Flora Fennica*. **60**, 90-112.

Luchtel, D (1972) Gonadal development and sex determination in pulmonate molluscs. I. *Arion circumscriptus*. *Zeitschrift für Zellforschung und mikroskopische Anatomie*. **130**, 279-301.

Lupu, D (1974) La révision des représentants de la famille des Arionidae (Gastropoda - Pulmonata) de Roumanie. *Travaux du Muséum d'Histoire Naturelle "Grigore Antipa"*. **15**, 31-44.

McCracken, G F; Selander R K (1980) Self-fertilization and monogenic strains in natural populations of terrestrial slugs. *Proceedings of the National Academy of Sciences of the USA*. **77**, 684-688.

Mogie, M (1986) Automixis: its distribution and status. *Biological Journal of the Linnean Society*. **28**, 321-329.

Nicklas, N L; Hoffmann R J (1981) Apomictic parthenogenesis in a hermaphroditic terrestrial slug, *Deroceras laeve* (Müller). *Biological Bulletin*. **160**, 123-135.

Ökland, F (1922) Arionidae of Norway. *Skrifter udgivne af Videnskabsselskabet i Christiana.* **5**, 1-61.

Oxford, G S (1978) The nature and distribution of food-induced esterases in Helicid snails. *Malacologia.* **17**, 331-339.

Quick, H E (1960) British slugs (Pulmonata; Testacellidae, Arionidae, Limacidae). *Bulletin of The British Museum (Natural History).* **6**, 103-226.

Runham, N W; Hunter P J (1970) *Terrestrial slugs.* London: Hutchinson Library, pp. 1-184.

Runham, N W (1993) Mollusca. In: *Reproductive Biology of Invertebrates*, K G Adiyodi& R G Adiyodi (eds), Chichester: John Wiley & Sons Ltd., pp. 311-383.

Selander, R K; Ochman H (1983) The genetic structure of populations as illustrated by molluscs. In: *Isozymes: Current Topics in Biological and Medical Research*, M C Rattazzi, C Scandalios & G S Whitt (eds), New York: Alan R. Liss, Inc., pp. 93-123.

Waldén, H (1955) The land Gastropoda of the vicinity of Stockholm. *Arkiv för Zoologi.* **7**, 391-448.

PROTEIN ELECTROPHORESIS IN ARIONID TAXONOMY

T BACKELJAU, B WINNEPENNINCKX
Royal Belgian Institute of Natural Sciences, Vautierstraat 29, B-1000 Brussels, Belgium;
E-mail: TBackeljau@kbinirsnb.be

K JORDAENS, H DE WOLF, K BREUGELMANS
University of Antwerp (RUCA), Evolutionary Biology Group, Groenenborgerlaan 171,
B-2020 Antwerp, Belgium

C PAREJO
Universidad Complutense, Departamento de Biología Animal I, E-28040 Madrid, Spain

T RODRÍGUEZ
Universidad de Santiago de Compostela, Departamento de Biología Animal, E-15706
Santiago de Compostela, Spain

ABSTRACT

The aim of this paper is to illustrate how protein electrophoresis can contribute to a better understanding of arionid systematics and biology. To this end we provide a brief literature review (excluding *Arion ater* and *A. lusitanicus*), after which new comparative electrophoretic data on four poorly known species (*A. franciscoloi*, *A. paularensis*, *A. wiktori* and *Geomalacus malagensis*) are presented. The utility of protein electrophoresis in arionid species identification is emphasized throughout the paper. Finally, arionid systematics are challenged by the discovery that several species consist of homozygous strains.

INTRODUCTION

The terrestrial slug family Arionidae comprises four genera in Europe and North Africa (*Arion* Férussac, 1819, *Ariunculus* Lessona, 1881, *Geomalacus* Allman, 1843 and *Letourneuxia* Bourguignat, 1866) with currently about 30-50 species. Many of these species can only be distinguished by subtle morphological and anatomical differences, mainly involving body colour and genital structures. Arionid taxonomy is therefore still problematic. Several authors have tried to improve species descriptions by studying spermatophores and mating behaviour. Yet, these data have hitherto not provoked major progress, except perhaps in the taxonomy of the *Arion hortensis* complex (e.g. Davies, 1977, 1979). Moreover, spermatophores and mating behaviour are features with limited value for species identification, for they can only be observed at particular moments in the life cycle. To a lesser extent this drawback also applies to genital and colour features, for genital structures are normally only fully developed in sexually mature specimens, whereas colour may be affected by factors such as age, food and climate (see references in Backeljau & De Bruyn, 1990).

Given the need for clear-cut species-specific characters that also allow the identification of juvenile specimens, and considering the current confusion in arionid taxonomy, several authors have started to explore the possibilities of protein electrophoresis as an additional source of systematic data. The present paper aims to (1) briefly review the literature with respect to the application of protein electrophoresis in arionid systematics (excluding the *A. ater* and *A. lusitanicus* species aggregates) and (2) present new electrophoretic comparisons of some poorly known species. Throughout this paper we will follow the original enzyme nomenclature as used by the different authors.

BRIEF LITERATURE REVIEW

The *Arion hortensis* complex

Following Davies' (1977, 1979) original suggestion that in NW Europe this complex consists of three species, viz. *A. hortensis* Férussac, 1819, *A. distinctus* Mabille, 1868 and *A. owenii* Davies, 1979, Backeljau (1985a, b) used polyacrylamide gel electrophoresis (PAGE) of albumen gland proteins (AGP) and isoelectric focusing (IEF) of non-specific digestive gland esterases (EST), to show that these three taxa correspond to biological species. The PAGE AGP profiles are indeed so characteristic and constant that they represent reliable taxonomic markers (Backeljau, 1989). Nevertheless, Backeljau *et al.* (1988) observed a difference between AGP profiles of *A. owenii* populations from Ireland and NW England, and those from S England. Yet, whether this variation reflects an intraspecific polymorphism or a taxonomic feature, is undecided.

Band counting analyses of the IEF patterns in the three species revealed a significant difference between intra- and interspecific EST similarities, suggesting that *A. owenii* may reproduce uniparentally in view of its nearly invariant EST patterns (Backeljau, 1985b). However, this is at variance with the results of Foltz *et al.*, (1982, 1984) who used starch gel electrophoresis (SGE) of allozymes to show that *A. owenii* is an outcrossing species. Indeed, Backeljau *et al.* (1994) subsequently noted that the specimens investigated by Backeljau (1985b) were probably highly inbred and therefore may have lost all EST variation.

Foltz *et al.* (1982, 1984) did not make interspecific comparisons. Hence, their allozyme data cannot be used as species markers. Such data were presented by Backeljau (1987) and Dolan & Fleming (1988). Using PAGE, Backeljau (1987) found that seven enzymes, i.e. α-glycerophosphate dehydrogenase (α-GPD), superoxide dismutase (SOD), malate dehydrogenase (MDH), glucosephosphate isomerase (GPI), amylase (AMY), lactate dehydrogenase (LDH) and phosphoglucomutase (PGM) allowed a differentiation between *A. hortensis* and the two other species. Yet, the distinction between *A. owenii* and *A. distinctus* was only possible with PGM. Dolan & Fleming (1988) on the contrary, found with SGE two enzymes by which the three species could be separated, i.e. PGM and enolase (ENO), while additional patterns for species identification were observed at aspartate aminotransferase (AAT), peptidases (PEP-1, PEP-2, PEP-3), malic enzyme (ME), MDH-3 and SOD.

The *Arion intermedius* complex

Recent allozyme studies showed that natural populations of *A. intermedius* Normand, 1852 consist of a number of homozygous multilocus genotypes or strains, defined by isocitrate dehydrogenase (IDH), MDH, LDH and EST (McCracken & Selander, 1980, Foltz *et al.*, 1982, Dolan & Fleming, 1988, Backeljau & De Bruyn, 1991, Backeljau *et al.*, 1992). The lack of heterozygotes, combined with (1) the absence of mating behaviour, (2) the production of viable offspring by specimens reared in isolation (Chichester & Getz, 1973, Davies, 1977) and (3) the invariant IEF EST patterns (Backeljau, 1985b), suggest that *A. intermedius* is a selfing species (McCracken & Selander, 1980, Selander & Ochman, 1983, Backeljau & De Bruyn, 1990, 1991). Yet, recently Garrido *et al.* (1995a) found spermatophores in Spanish *A. intermedius*. As spermatophores are only produced during coition (see references in Backeljau & De Bruyn, 1990), these findings suggest that at least in the Iberian peninsula *A. intermedius* may display mating behaviour and thus may have the capacity to outcross.

Currently, little is known about the ecogenetics and taxonomy of the strains in *A. intermedius*. Yet, Backeljau & De Bruyn (1991) used PAGE of allozymes to show that colour morphs in *A. intermedius* are not associated with particular electrophoretic strains, so that body colour is not a reliable taxonomic marker in this complex. Similarly, Backeljau *et al.* (1992) showed that *A. pascalianus* Mabille, 1868 (*sensu* Backhuys, 1975 and Rähle, 1992) is nothing but a dark, mottled colour morph of *A. intermedius*.

Finally, the subgeneric relationships of *A. intermedius* have caused some debate. IEF profiles of EST suggested a close association between *A. intermedius* and *A. owenii*, but Dolan & Fleming's (1988) allozyme study reported a closer relationship between *A. intermedius* and *A. circumscriptus* Johnston, 1828. Yet, Backeljau & De Bruyn (1990) reassessed all available electrophoretic (and other) data and found that if Dolan & Fleming's (1988) work was adapted for possible technical biases due to the possible different resolution of SGE compared to PAGE, there was more support for associating *A. intermedius* with the *A. hortensis* complex (subgenus *Kobeltia* Seibert, 1873), than with *A. circumscriptus* (subgenus *Carinarion* Hesse, 1926). Hence, until compelling evidence indicates otherwise we regard *A. intermedius* as a *Kobeltia* (Backeljau & De Bruyn, 1990).

The *Arion fasciatus* complex

Three morphologically very similar species belong to this highly controversial complex: *A. fasciatus* (Nilsson, 1823), *A. circumscriptus* Johnston, 1828 and *A. silvaticus* Lohmander, 1937. The species status of the three segregates was considered well-established since Chichester (1967) showed that each of them has a characteristic AGP profile on a PAGE gel. Further evidence was provided by the allozyme studies of McCracken & Selander (1980) and Foltz *et al.* (1982), who calculated that Nei's mean genetic identity between the species ranges from 0.62 - 0.69 in the U.S.A. to 0.74 in Ireland. Yet, at the same time these authors showed that each of the three species consists of one (*A. fasciatus*, *A. circumscriptus*) or two (*A. silvaticus* in Ireland) homozygous strains, apparently as a result of sustained selfing (e.g. Selander & Ochman, 1983). Subsequently, Backeljau *et al.* (1987) refined Chichester's (1967) work, by producing three distinct AGP profiles as resolved by IEF. A parallel IEF analysis of EST also showed a clear differentiation between *A. circumscriptus* and the two other species. The IEF profiles of AGP and EST were highly constant within species and over large geographic areas, thus confirming widespread selfing. Yet, Backeljau *et al.* (1987) remarked that the biological species concept is not applicable in such case. This opinion was recently further supported by a PAGE analysis of 18 enzyme loci showing that in NW Europe *A. fasciatus* consists of three, *A. circumscriptus* of two and *A. silvaticus* of 20 homozygous strains (Backeljau *et al.*, 1996, Jordaens *et al.*, unpublished data). Nei's mean genetic identity between these strains is 0.81 (cf. the corresponding values between genetic forms in the *A. subfuscus* complex) and a UPGMA tree shows a bootstrap supported cluster of all *A. fasciatus* and *A. silvaticus* strains, thus confirming the close relationship between these two complexes. Hence, it appears as if the morphological and genetic differentiation in the *A. fasciatus* complex is comparable to that of the selfing snail *Rumina decollata*, which also consists of different morphotypes and genetic strains (e.g. Selander & Ochman, 1983). This casts further doubt on the species status of members of the *A. fasciatus* complex, even though they can be electrophoretically separated by characteristic electromorphs at EST, PGM, alanine aminotransferase (ALAT) and fumarate hydratase (FUMH) (Backeljau *et al.*, 1996).

Interestingly, at least in *A. silvaticus* it appears that more than 50% of natural populations consist of two or more coexisting strains (Backeljau *et al.*, 1996). Moreover, unpublished data suggest that strains can be divided into narrow and broad niched ones (general purpose genotypes?). On the other hand we have data indicating that in other parts of Europe members of the *A. fasciatus* complex may more often outcross, thus suggesting a geographic or ecological component in the breeding biology of these slugs. Hence, an ecogenetic comparison of the *A. fasciatus* and the *A. intermedius* complexes seems highly relevant. Our current research is therefore focusing on these issues.

The *Arion subfuscus* complex

The taxonomy of this complex is extremely confused, particularly in the Iberian Peninsula, where a number of new species have been described recently (e.g. Garrido *et al.*, 1995b). Yet, at least in case of *A. urbiae* De Winter, 1986 and *A. anguloi* Martín & Gómez, 1988, both PAGE of allozymes (phosphogluconate dehydrogenase (PGD), leucine aminopeptidase (LAP), glucose dehydrogenase (GDH), diaphorase (DIA), AMY, PGM, SOD, α-GPD, FUM (= FUMH), IDH, GPT (= ALAT) and ME) and IEF of EST suggest that a single species is involved. This follows from the absence of specific protein profiles, the high genetic identity between both taxa (I = 0.947), the high EST similarities and a comparison of genetic variability measures (Backeljau *et al.*, 1994).

Arion subfuscus (Draparnaud, 1805) itself consists of a number of genetic entities that can be distinguished by SGE of allozymes. McCracken & Selander (1980) reported one outcrossing and one selfing (two strains) form in the U.S.A., while Foltz *et al.* (1982) found one monomorphic and one outcrossing form in the British Isles. The British forms hybridize and are genetically very similar (Nei's genetic identity is 0.816; cf. the *A. fasciatus* complex). Interestingly, the outcrossing British form is very similar to the outcrossing form in the U.S.A. (Nei's identity is 0.909), yet the monomorphic British form is very different from the selfing American form (Nei's identity is 0.467). Finally,

Backeljau *et al.* (1994) found with PAGE of allozymes that *A. subfuscus* on the European mainland consists of two genetic types (F and S) defined by a Fast and Slow electromorph at PGM. Both types show distinct PAGE profiles at AMY, α-GPD, DIA, FUM, IDH and LAP, and Nei's genetic identity between them is 0.421. Also their EST similarity obtained from IEF profiles is very low compared to other "intraspecific" values. Although the emerging picture of *A. subfuscus* is still far from complete, the current electrophoretic data clearly suggest that more than one species is involved.

TOWARDS AN ELECTROPHORETIC KEY

In order to further demonstrate the possible use of electrophoretic data in arionid taxonomy and species identification, we compared a number of recently described, but poorly known arionids by means of vertical PAGE of allozymes. The species involved are: *A. franciscoloi* Boato, Bodon & Giusti, 1983, *A. paularensis* Wiktor & Parejo, 1989, *A. wiktori* Parejo & Martín, 1990 and *Geomalacus malagensis* Wiktor & Norris, 1991. They were compared with previously analyzed specimens of *A. hortensis*, *A. distinctus*, *A. owenii*, *A. intermedius*, *A. fasciatus*, *A. circumscriptus*, *A. silvaticus*, *A. subfuscus*-S, *A. subfuscus*-F, *A. fagophilus* De Winter, 1986 and *Geomalacus maculosus* Allman, 1843. In this way we constructed a synoptic table of characteristic electromorphs (cf. Backeljau & De Winter, 1987). Sample preparation of individual digestive gland homogenates and PAGE procedures are described by Backeljau (1987, 1989). Electromorphs are labelled alphabetically according to their relative mobility with respect to the origin (A being the fastest = most anodal band). The results of the comparison are shown in Table 1.

Table 1. Electromorph characterization of 15 arionid species at four enzyme loci.

Species	PGD	AMY	SOD	AAT
A. hortensis	B	B	E	C
A. distinctus	B	A	A	C,D
A. owenii	B	A	A	C
A. intermedius	B	A	A	B
A. fagophilus	B	A	D	A
A. fasciatus	A	B	B	D
A. circumscriptus	A	B	B	D
A. silvaticus	A	B	B	D
A. subfuscus-F	B	F	A	?
A. subfuscus-S	B	C	A	?
A. paularensis	B	D,F	A	?
A. wiktori	B	D,F	A	?
A. franciscoloi	B	E	E	?
G. maculosus	C	G	A	?
G. malagensis	C	H,I	C	?

Although Table 1 does not yet allow many taxonomic conclusions, it shows some salient features, such as (1) the high similarity of the members of the *A. fasciatus* complex and their separation from other taxa, thus supporting the "subgenus" *Carinarion*, (2) the differentiation of the two *Geomalacus* species, (3) the suggestive supraspecific groupings implied by the PGD electromorphs (*Arion*, *Carinarion* and *Geomalacus*), and (4) the similarity between *A. paularensis* and *A. wiktori*, particularly for AMY at which both species are polymorphic for the same alleles.

Additional PAGE comparisons show that *A. paularensis* and *A. wiktori* also share three alleles at PGM, again supporting a close relationship between both taxa. The two *Geomalacus* species, on the other hand, consistently differ at IDH, PEP, DIA and PGM. *G. malagensis* shows a fixed pattern of three bands, while *G. maculosus* shows only one (Fig. 1A). It also appears that *G. malagensis* expresses two AMY loci (the more anodal of which is fixed for a single allele, the cathodal showing Mendelian segregation of two alleles) (Fig. 1B), whereas in *G. maculosus* only one weak zone of AMY activity can be detected. This suggests that the two *Geomalacus* species are probably not closely related.

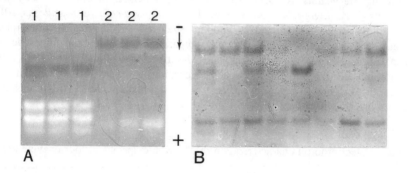

Figure 1.
Electrophoretic variation in *Geomalacus*. (A) Comparison between *G. malagensis* (1) and *G. maculosus* (2) at SOD (cathodal black bands) and PGM (anodal white bands); (B) AMY variation in *G. malagensis*, with Mendelian segregation in a monomeric enzyme with two alleles at a single locus (cathodal bands) and fixation of a single band coded by a putative second locus (anodal bands). The arrow indicates the direction of protein migration.

CONCLUSION

Protein electrophoresis has not only provided crucial data to evaluate arionid species boundaries, but has also uncovered unexpected, yet fundamental and controversial issues, such as (1) the geographical, ecogenetic and evolutionary significance of selfing versus outcrossing, (2) the wide distribution and coexistence of selfing homozygous multilocus genotypes and their eco-ethological correlates, (3) the taxonomic interpretation of multistrain complexes, and (4) the highly unreliable nature of colour features in arionid taxonomy.

ACKNOWLEDGEMENTS

We are indebted to L. Noble (University of Aberdeen) for commenting on the manuscript. This work was funded by F.J.B.R. grants 2.0004.91 and 2.0128.94.

REFERENCES

Backeljau, T (1985a) A preliminary account of species specific protein patterns in albumen gland extracts of *Arion hortensis* s.l. (Pulmonata, Arionidae). *Basteria.* **49**, 11-17.

Backeljau, T (1985b) Estimation of genic similarity within and between *Arion hortensis* s.l. and *A. intermedius* by means of isoelectric focused esterase patterns in hepatopancreas homogenates (Mollusca, Pulmonata: Arionidae). *Zeitschrift für zoologische Systematik und Evolutionsforschung.* **23**, 38-49.

Backeljau, T (1987) Electrophoretic distinction between *Arion hortensis, A. distinctus* and *A. owenii* (Mollusca: Pulmonata). *Zoologischer Anzeiger.* **219**, 33-39.

Backeljau, T (1989) Electrophoresis of albumen gland proteins as a tool to elucidate taxonomic problems in the genus *Arion* (Gastropoda, Pulmonata). *Journal of Medical and Applied Malacology.* **1**, 29-41.

Backeljau, T; Ahmadyar S Z; Selens M; Van Rompaey J; Verheyen W (1987) Comparative electrophoretic analyses of three European *Carinarion* species (Mollusca, Pulmonata, Arionidae). *Zoologica Scripta.* **16**, 209-222.

Backeljau, T; Breugelmans K; Leirs H; Rodríguez T; Sherbakov D; Sitnikova T; Timmermans J-M; Van Goethem J L; Verheyen E (1994) Application of isoelectric focusing in Molluscan systematics. *The Nautilus.* **Supplement 2**, 156-167.

Backeljau, T; Davies S M; De Bruyn L (1988) An albumen gland protein polymorphism in the terrestrial slug *Arion owenii. Biochemical Systematics and Ecology.* **16**, 425-429.

Backeljau, T; De Brito C P; Tristão da Cunha R M; Frias Martins A M; De Bruyn L (1992) Colour polymorphism and genetic strains in *Arion intermedius* from Flores, Azores (Mollusca: Pulmonata). *Biological Journal of the Linnean Society.* **46**, 131-143.

Backeljau, T; De Bruyn L (1990) On the infrageneric systematics of the genus *Arion* Férussac, 1819 (Mollusca, Pulmonata). *Bulletin de l'Institut Royal des Sciences Naturelles de Belgique, Biologie.* **60**, 35-68.

Backeljau, T; De Bruyn L (1991) Preliminary report on the genetic variability of *Arion intermedius* in Europe (Pulmonata). *Journal of Medical and Applied Malacology.* **3**, 19-29.

Backeljau, T; De Bruyn L; De Wolf H; Jordaens K; Van Dongen S; Winnepenninckx B (1996) Allozyme diversity in slugs of the *Carinarion* complex (Mollusca, Pulmonata). *Heredity.* in press.

Backeljau, T; De Winter A J (1987) An electrophoretic characterisation of three paratypes of *Arion fagophilus* De Winter, 1986, with notes on the subgeneric division of the genus *Arion* Férussac, 1819 (Mollusca, Pulmonata). *Zeitschrift für zoologische Systematik und Evolutionsforschung.* **25**, 169-180.

Backeljau, T; De Winter A J; Martín R; Rodríguez T; De Bruyn L (1994) Genital and allozyme similarity between *Arion urbiae* and *A. anguloi* (Mollusca: Pulmonata). *Zoological Journal of the Linnean Society.* **110**, 1-18.

Backhuys, W (1975) *Zoogeography and taxonomy of the land and freshwater molluscs of the Azores.* Amsterdam: Backhuys & Meesters, pp. 1-350.

Chichester, L F (1967) The zoogeography, ecology and taxonomy of arionid and limacid slugs introduced into Northeastern North America. Unpublished Ph.D.Thesis, University of Connecticut. pp. 138-157.

Chichester, L F; Getz L L (1973) The terrestrial slugs of Northeastern North America. *Sterkiana.* **51**, 11-42.

Davies, S M (1977) The *Arion hortensis* complex, with notes on *A. intermedius* Normand (Pulmonata: Arionidae). *Journal of Conchology.* **29**, 173-187.

Davies, S M (1979) Segregates of the *Arion hortensis* complex (Pulmonata: Arionidae), with the description of a new species, *Arion owenii. Journal of Conchology.* **30**, 123-127.

Dolan, S; Fleming C C (1988) Isoenzymes in the identification and systematics of terrestrial slugs of the *Arion hortensis* complex. *Biochemical Systematics and Ecology.* **16**, 195-198.

Foltz, D W; Ochman H; Jones J S; Evangelisti S M; Selander R K (1982) Genetic population structure and breeding systems in arionid slugs (Mollusca: Pulmonata). *Biological Journal of the Linnean Society.* **17**, 225-241.

Foltz, D W; Ochman H; Selander R K (1984) Genetic diversity and breeding systems in terrestrial slugs of the families Limacidae and Arionidae. *Malacologia.* **25**, 593-605.

Garrido, C; Castillejo J; Iglesias J (1995a) The spermatophore of *Arion intermedius* (Pulmonata: Arionidae). *Journal of Molluscan Studies.* **61**, 127-133.

Garrido, C; Castillejo J; Iglesias J (1995b) The *Arion subfuscus* complex in the eastern part of the Iberian peninsula, with redescription of *Arion subfuscus* (Draparnaud 1805) (Gastropoda: Pulmonata: Arionidae). *Archiv für Molluskenkunde.* **124**, 103-118.

McCracken, G F; Selander R K (1980) Self-fertilization and monogenic strains in natural populations of terrestrial slugs. *Proceedings of the National Academy of Sciences of the USA.* **77**, 684-688.

Rähle, W (1992) Nacktschnecken (Arionidae, Agriolimacidae und Limacidae) von Madeira und Porto Santo (Mittelatlantische Inseln) (Gastropoda: Pulmonata). *Malakologische Abhandlungen.* **16**, 13-24.

Selander, R K; Ochman H (1983) The genetic structure of populations as illustrated by molluscs. In: *Isozymes: Current Topics in Biological and Medical Research. Volume 10: Genetics and Evolution.* M C Rattazzi, J C Scandalios & G S Whitt (eds), New York: Alan R. Liss, pp. 93-123.

Session 2
Economic Impact

Session Organiser
and Chairman   J Oakley

CONSERVATION TILLAGE AND SLUGS IN THE U.S. CORN BELT

R B HAMMOND

Department of Entomology, Ohio Agricultural Research & Development Center, The Ohio State University, Wooster, OH 44691, USA

ABSTRACT

As conservation tillage practices have gained acceptance by corn and soybean growers in the midwest, slug problems have increased in the number of hectares being damaged and the with-in field severity. Four species of slugs are found, with *Deroceras reticulatum* being the most common and numerous. The other slugs, in order of prevalence, are *D. laeve, Arion subfuscus*, and *A. fasciatus*. Severe defoliation and stunting of corn by slugs have been observed, with corn having reached 5th-6th leaf stage being seriously damaged. Although severe defoliation by slugs occurs in soybean, the most serious damage is plant stand reduction. Stand loss occurs in two manners: 1) soybean emerges and reaches growth stage V1-V2 and then is consumed by slugs, and 2) the damage occurs prior to plant emergence from the seed furrow or from underneath the crop residue (thus, the plant never emerges). These two types of damage are quite different in terms of which slugs are causing the injury, how we will need to monitor for the slugs, and when best to treat. Efforts are also underway to develop control recommendations for midwestern field crops, including studies on molluscicides, application techniques, and cultural practices that might lead to efficient management.

INTRODUCTION

The corn belt region of the U.S. is one of the most productive farming regions in the world. Stretching from Ohio on the east to Nebraska on the west, the corn belt is the U.S.'s major producer of corn (*Zea mays* (L.), and soybean (*Glycine max* [L.] Merr.). Crop residue management systems have been gaining acceptance by growers in the corn belt with no-till hectares having gained the most. The top no-till states in the U.S. based on planted hectares are the corn belt states of Illinois (2.4 million ha), Indiana (1.8 million ha), Iowa (1.7 million ha), and Ohio (1.6 million ha). Based on percentage of total hectares, Ohio has the largest percentage (38%) of no-till hectares of these states. The corn belt has been, and continues to be, a leader in conservation tillage systems.

Conservation tillage is any tillage or planting system that allows for 30% or more of the soil surface to be covered with crop residue. These tillage systems include no-till where the soil is left undisturbed with planting accomplished in a narrow seedbed or slot; ridge-till where planting is completed in a seedbed prepared on ridges with sweeps, disk openers or row cleaners; and mulch-till where soil is disturbed prior to planting with tools such as chisels, field cultivators, or other implements leaving 30% or better residue cover (CTIC, 1995). Weed control in these systems is accomplished primarily with herbicides, and with the latter 2 systems, limited amounts of tillage. Another system often practiced is reduced-till where 15-30% of the soil surface is covered with residue. In the U.S., all these practices are known as crop residue management (CRM) systems (CTIC, 1995). Contrasted with these practices, conventional till is where less than 15% cover after planting, and usually involves plowing or intensive tillage. Growers adopt conservation tillage practices to reduce labor requirements and machinery wear, to increase fuel savings, to provide for higher soil moisture, to reduce soil erosion and compaction, to improve water infiltration, water quality, and soil tilth, to increase organic matter, and to provide for more wildlife.

CONCERNS WITH SLUGS ON CORN AND SOYBEAN

Slugs were associated with field corn as early as the late 1950s when Neiswander (1959) reported that corn fields planted into plowed alfalfa sod were being damaged by *Deroceras reticulatum* (Müller). At that time, the best method for control was early cultivation. The use of metaldehyde was considered unnecessary unless the slug population was unusually high. Later, Barry (1969) reported on damage by *D. reticulatum* in corn fields in conservation tillage fields in Ohio during 1968. Gregory & Musick (1976), who wrote extensively on invertebrate damage in conservation tillage crop, considered slugs the most serious non-insect problem in a review article of insect management in reduced tillage. They noted that methiocarb (Mesurol®) was the most promising slug control chemical on the horizon for growers.

Soybean began to be grown using conservation tillage methods in the late 1970s following the development of longer-residual, and more importantly, post-emergent herbicides. Hammond (1985) reported an association with slugs and soybean for the 1st time in the mid-1980s. It was noted that *D. reticulatum* was the primary slug causing injury in Ohio soybean fields planted using conservation tillage practices into corn residue. Hammond & Stinner (1987) reported that the amount of residue influences slug populations in corn and soybean with the largest numbers in no-till systems. They found more slugs when the previous crop was soybean, albeit reasons for this were unclear, suggesting that soybean residue might provide a more favorable habitat. They concluded that as more soybean was grown in conservation tillage systems in rotation with corn, the incidence of slug problems in both crops would increase.

As conservation tillage hectares have increased in the corn belt, so has the concern with slugs. Growers in the eastern corn belt who are experiencing the greatest amount of slug damage are questioning their continued use of these practices (Willson & Eisley, 1992). Growers in states immediately west and north of Ohio are beginning to experience slugs problems. States further to the west, now beginning to see more hectares of no-till crops, are concerned with the expansion of slug problems into their regions.

SLUG SPECIES

Although most earlier publications mentioned *D. reticulatum* as the damaging slug species, recent work has identified 4 species of concern in the corn belt. The first two slugs, *D. reticulatum* and *D. laeve* (Müller), are Agriolimacid slugs, while the remaining slugs, *A. subfuscus* (Draparnaud) and *Arion fasciatus* (Nilsson), are Arionid slugs. Of interest, while *A. fasciatus* is considered an important slug species in northeastern U.S., it has only been found in a few fields in the corn belt and then, in very low numbers. However, *A. subfuscus*, which is common in some fields, is only found in small numbers in the northeast.

Deroceras reticulatum, known as the gray garden slug in the U.S., is the most widespread and damaging slug found in field crops. This is an introduced species that is now widespread throughout North America. Corn and soybean are heavily defoliated by this species leading to significant yield reductions. *D. reticulatum* can also cause severe stand loss in both crops from feeding on the germinating seed within seed furrows that have not closed or underneath the crop residue. Observations indicate that most of the damage is from juveniles, with populations on corn reported as high as 10 to 30 juveniles per plant (Barry, 1969). All life stages have been found to overwinter; eggs that have overwintered or have been laid by adults in the spring produce the juveniles that cause the damage.

Deroceras laeve, known as the marsh slug, is the next slug in terms of damage. This is the only native species and is very widespread. Earlier literature usually had no mention of this species. When mentioned, smaller populations of *D. laeve* were reported compared with *D. reticulatum*. Similar differentials of their population size appear to currently exist in the corn belt.

Arion subfuscus, the dusky slug, is an introduced species. It is the predominant species within some fields in Ohio, with numbers in baited-beer traps often reaching >50 slugs within a 24 h period (unpublished data). *A. subfuscus* often causes growers greater concern than other slug species because of its larger size and bright coloration. Because of the limited amount of research on slugs in field crops, the damage potential of *A. subfuscus* is unclear. Literature suggests that *A. subfuscus* is mostly a woodland species and fungivorous (Chischester & Getz, 1969, 1973). However in Ohio, *A. subfuscus* appears capable of causing significant injury to young, germinating plants, especially when crops are planted in later spring and germination is occurring at the time of activity by juveniles (unpublished data). Its ability to feed on leaf tissue and cause significant defoliation is being examined.

SLUG DAMAGE

Slug injury to corn in the spring is different from that of other early season pests such as cutworms that cut the plant off at soil level or stalkborers that cause plants to wilt and die (Byers & Calvin, 1994). Although most corn injury occurs during the 2 to 4 leaf stage, we have observed corn in the 5 to 6 leaf stage being severely damaged. Corn damaged by slugs has leaf tissue missing between the veins, but with the lower epidermis intact. The leaves will have elongated holes and a ragged appearance as this feeding progresses and the plant grows. When injury is moderate to severe, the plant will live but grow at a reduced rate and result in lower yields. The plant can have no leaves remaining with only a stump left if injury is very severe. These plants do not recover and thus die, resulting in plant stand reductions.

The only economic injury level work specific to slugs on corn is by Byers & Calvin (1994), where corn was infested before seedling emergence and at the 2- and 4-leaf stage with 4 densities of juvenile *D. reticulatum*. Economic injury levels were established that ranged from 2 to 20% defoliation in a warm, wet season, and 39 to 59% in seasons that were not conducive to slug activity. Little effort has been done on slug-induced stand reduction.

Soybean injury from slugs is similar to that of other early soybean insect pests. Slugs will often begin by feeding on the cotyledons, defoliating the unifoliate leaves, and then progressing to feeding on the trifoliate leaves. We have also observed that slugs will feed on the tip of the leaves before it expands, resulting in the distal half of the leave missing; little further defoliation to the remaining leaf is seen. Severe defoliation can also cause the plants to become stumps resulting in plant stand reductions. An additional injury we are seeing in soybean is feeding on the germinating seed, the cotyledons, or young seedling while beneath the residue resulting in plant death and stand reductions. Soybean is usually planted after corn in the corn belt, nearer the time of heavy slug activity. If soybean begins to germinate and emerge at this time, stand reductions can be quite heavy resulting in large areas of a field having no plants. The concern with this type of injury is that it happens soon after planting, before a grower realizes that slugs are active and feeding. When plants fail to emerge, the damage has already occurred and it is too late to take therapeutic action.

Most of the information on pest-injury/yield-loss relationships on early season soybean comes from simulated defoliation studies (Hammond, 1989a, Hunt *et al*. 1994); no such studies on actual slug injury or simulated slug injury on soybean have been conducted (albeit that efforts are currently being undertaken). Hammond (1989a) found no consistent impact of early-season defoliation on yields. However, Hunt *et al*. (1994), simulating bean leaf beetle (*Cerotoma trifurcata*) injury to seedling soybean, found significant yield reductions up to 12% at defoliation levels of 68% because of a delay when the plant canopy reached the critical leaf area when 90% of the light is intercepted. Removal of cotyledons, although not impacting yield directly, resulted in less leaf area at growth stage V3 where the impact of defoliation was greater compared with plants where the cotyledons remained. They concluded that a given population of a pest defoliator would have a greater impact on plants without cotyledons than with cotyledons. Based on their study, while seedling defoliation by slugs may not be economical, it might reduce plant fitness by delaying canopy development and plant height. While the aforementioned studies had soybean grown in wide rows (usually 76.2 cm), slug injury occurs on plants grown in narrow rows because that is the common practice with conservation tilled soybean. Currently, there is no information available on the relationship between early season soybean defoliation, stand loss, and the use of narrow row widths.

Somewhat unique to slug injury in both crops is the interaction that occurs between defoliation and stand loss. Agronomic guidelines in most states give information as to the plant stands that are required to maintain yield or that necessitates replanting. However, because these stand losses are often not uniform throughout a field, a grower must determine the areas that require replanting. The slug population that exists in the field, along with the current and forecasted weather conditions, must be considered and a decision made as to the need to apply a molluscicide. It has been our experience that replanted fields often require treatment before replanting because the slug population still exists and slugs have continued to grow in physical size and density.

SAMPLING SLUGS IN CORN AND SOYBEAN

Research is attempting to develop sampling programs for slugs to determine the need for treatment. Numerous techniques have been examined, including various types of passive traps. We have found that once the crop has emerged, *in situ* sampling of slugs on the plants provides an excellent estimate of the species and their population size. Barry (1969) used this method when determining chemical efficacy, finding slug numbers ranging from 0 to 30 juveniles per corn plant. We have been using this method in both corn and soybean in numerous investigations, including population dynamic studies and molluscicide efficacy trials. The greatest benefit is that the numbers of slugs can be determined per individual plant. In corn, each plant becomes a sampling unit; each plant can be individually examined for slugs. Our experiences in soybean is to count the number of slugs observed on the plants and number of plants per unit area (for example per m^2) and then calculate the number per individual soybean plant. Conservation tilled soybean is usually planted in narrow rows, and it is much easier to count slugs per area rather than per individual plant. The difficulty with *in situ* sampling is the need to make the counts near or following dusk after slugs have emerged from underneath the residue and climbed onto the plants. This requires late evening or night field visits which might not be acceptable to those needing to sample for slugs. However, work is underway to determine if such data, i.e., number of slugs per plant, can be correlated to yield loss, thus providing growers with reliable economic injury levels.

SLUG MANAGEMENT

Cultural Control

Slug management is similar on both crops and should be first directed towards preventive methods to reduce the capacity of the slug population to increase in size or the ability of the slug to cause economic injury to the plant. Preventive methods are those cultural practices that growers can use to minimize the impact of slugs.

The primary cultural means of managing slugs has been and continues to be tillage. The problem with this approach is that it is contrary to the concepts of conservation tillage, and would remove the benefits. A consideration for growers in the corn belt are the various government programs that promote, and often necessitate, the use of conservation tillage practices on highly erodible land. The use of tillage is limited in numerous fields because of the requirement for residues in amounts of >30%. Where possible, light to moderate tillage is suggested in fields, or areas of fields, experiencing yearly slug problems when no other control measures are solving the problem.

Although there are no published studies on other cultural practices for slug management, numerous ones are being suggested for their ability in alleviating slug problems. The use of row cleaners in front of the planter which remove residue from over the seed zone is thought to impact slug damage. Although it is questionable if row cleaners have a significant impact on the slug population, they do allow for enhanced germination and quicker, more vigorous plant growth in the spring because of slightly warmer and drier soil. Increased plant growth allows the crop to outgrow slug injury. There is also thought that row cleaners may affect slugs' feeding behavior resulting in lesser defoliation. Although the use of row cleaners is becoming more common in corn production, its use is limited in soybean because of the use of narrow rows.

Vigorously growing plants that have more leaf area compared with other plants will usually be better able to tolerate slug injury. Thus, any practice that allows for quicker germination and growth, or the presence of more leaf area, will theoretically be less damaged. The use of starter fertilizer in corn has been suggested to give plants a quicker start. For crops that are being damaged in later spring, earlier planting dates are suggested to allow adequate time for the plants to produce sufficient leaf area to withstand slug injury. Contrasted with this approach, later plantings are suggested if the problems occur in mid-spring, where later planting would allow for warmer and drier soils that will enhance crop growth.

Another cultural practice that is being examined is the use of cover crops. Some growers using a legume cover crop with corn have reported lesser slug injury. In one example, observations taken in a corn field during 1995 indicated >20 medium size *D. reticulatum* per beer trap which would be considered a relatively large population (unpublished data). However, little to no slug injury was evident on the corn which already had about 5-6 leaves, normally an indication of a lack of slugs within the field.

Molluscicide Control

The only means of controlling an active slug infestation in either crop is the use of a molluscicide. While growers in some states had a special state registration need (known as a 24C) for a bait containing methiocarb mixed with cracked corn to control slugs on corn in the earlier 1980s, no such special label was available on soybean. However, a similar special needs label was later obtained by the mid-to-late 1980s for a cracked corn bait containing thiodicarb (Larvin®) rather than methiocarb for use on soybean (Hammond, 1989b). These labels for cracked corn/molluscicide baits were discontinued and are no longer useable by growers. For a number of years thereafter, no molluscicides were available for corn and soybean. Currently, corn and soybean growers in the corn belt have few choices of materials, and there is little information on how to best use them. All the materials currently available contain metaldehyde.

Commercially-prepared baits containing metaldehyde became readily available in the 1990s. These baits were found to be efficacious at low rates in field crops, 11.2 to 13.5 kg formulation/ha (10-12 lb formulation/acre), which made their use more economically justified. The 1st material made widely available was Deadline® Bullets which was a 4% metaldehyde bait sold by Valent® U.S.A. Corporation (Deadline® is a registered trademark of Pace International LP). The material consists of medium-to-large sized pellets that when applied at 11.2 kg formulation/ha places approximately 1 piece of material per 0.09 m² (1 ft²). Growers who used this material often did not achieve acceptable control; it was felt that the size of the pellets when applied at the lower rates prevented adequate coverage. A new formulation, Deadline® Granules, became available in 1994, that allowed much better coverage because of its smaller size particles. At 11.2 kg formulation/ha, coverage is approximately 9-10 pieces per 0.09 m² (1 ft²). Hammond conducted numerous efficacy trials with the granules in 1994 and 1995 (unpublished data) and found very good control with rates ranging as low as 5.6 kg formulation/ha (5 lb/acre). However, growers had problems with the material because of the consistency of the formulation (it was often too moist). More recently, a 3rd formulation was developed, Deadline® Minipellets, which contains particles that are more similar to the Bullet formulation in consistency but much smaller in size (B. Haddad, Valent Corp., personnel communication). This material was tested during the 1996 spring growing season.

A second product recently brought into the corn and soybean market in the corn belt was Prozap® Agri-brand Snail and Slug AG pelleted bait (Hacco, Inc.) containing 3.5% metaldehyde. Snail and Slug AG was found to be efficacious at rates of 11.2 to 16.8 kg formulation/ha (10-15 lb formulation/acre) in tests during 1995 (unpublished data).

Although commercially-prepared molluscicides are now available, there continues to be a need determine how best to use them. Hammond (1996) conducted studies of weekly applications of Deadline Granules in corn fields and found that the material needs to be applied near, or after, spring egg hatch for control of the juvenile population. When the molluscicide was applied earlier to adult slugs or simulating a prophylactic application at an earlier planting time, it did not control the subsequent juvenile population and thus, did not prevent later crop injury. Research is being conducted to determine the most appropriate equipment to use in applying the various molluscicides at low rates. Growers currently use spreaders intended to apply fertilizer at rates of 112 kg/ha (100 lb/acre) or more; applying molluscicide with even coverage at rates as low as 11.2 kg formulation/ha (10 lb/acre) has met with limited success. Various types of broadcast spreaders meant to apply material at low rates should be tested for even coverage of molluscicides.

Other Control Tactics

Growers are using other control tactics for slug management with various degrees of success. A method used on corn has been the application of liquid 28% nitrogen applied at night which is intended to kill the slugs on contact. The fertilizer causes slugs to emit copious amounts of slime which leads to dehydration and death. This method has met with limited success and is being further researched. In addition, growers have tried various liquid potash, lime mixtures, and other concoctions in an attempt to manage their slug problem. For those growers who appear to have had success, there has been as many that have had failures. Growers appear willing to try any home remedy because of the severe problems they are having in their crops. However, without scientifically-appropriate research that includes replications, checks, and statistical analyses are conducted, universities have been hesitant to recommend them.

FUTURE WORK AND NEEDS

With the continued adoption and further use of conservation tillage practices throughout the corn belt and the increasing slug problem both geographically and in the severity of damage that is occurring, the need for research on slugs and how to manage them is paramount. We need a much better understanding of the biology and life history of the slug species causing problems in field crops. Before IPM concepts are applied against slugs, we need to know the relationships between slug injury and yield losses on crops so that economic injury levels and thresholds can be determined. We need sampling techniques for use by growers, IPM specialists, and crop and pest consultants that are quick and easy to use in a slug management program.

More efforts are needed on control tactics including cultural practices that can serve as preventive measures against slugs and therapeutic tactics (i.e., molluscicides) that offer growers economically and effective control. This includes additional active ingredients, different formulations of currently available materials, and better application technology. Other possible control measures should be explored that will give growers consistent efficacy.

The challenge is to develop management programs for slugs that will allow for the continued acceptance and further adoption of conservation tillage practices.

REFERENCES

Barry, B D (1969) Evaluation of chemicals for control of slugs on field corn in Ohio. *Journal of Economic Entomology*. **62**, 1277-1279.

Byers, B D; Calvin D D (1994) Economic injury levels to field corn from slug (Stylomaatophora: Agriolimacidae) feeding. *Journal of Economic Entomology* **87**, 1345-1350.

Chichester, L F; Getz L L (1969) The zoogeography and ecology of Arionid and Limacid slugs introduced into Northeastern North America. *Malacologia* **7**, 313-346.

Chichester, L F; Getz L L (1973). The terrestrial slugs of Northeastern North America. *Sterkiana*, **51**, 11-42.

CTIC (1995) Conservation impact. Conservation Technology Information Center. **13**, 1-6.

Hammond, R B (1985) Slugs as a new pest of soybeans. *Journal of Kansas Entomology Society*. **58**, 364-366.

Hammond, R B; Stinner B R (1987) Seedcorn maggots (Diptera: Anthomyiidae) and slugs in conservation tillage in Ohio. *Journal of Economic Entomology*. **80**, 680-684.

Hammond, R B (1989a) Effects of leaf removal at soybean growth stage V1 on yield and other growth parameters. *Journal of the Kansas Entomological Society*. **62**, 96-102.

Hammond, R B (1989b) Soybean insect pest management and conservation tillage: options for the grower. *Proceedings, 1989 Southern Conservation Tillage Conference*, 3-5.

Hammond, R B; Smith J A; Beck T (1996) Timing of molluscicide applications for reliable control in no-tillage field crops. *Journal of Economic Entomology*. **89**, (in press).

Hunt, T E; Higley L G; Witkowski J F (1994) Soybean growth and yield after simulated bean leaf beetle injury to seedlings. *Agronomy Journal*. **86**, 140-146.

Gregory, W W; Musick G J (1976) Insect management in reduced tillage systems. *Bulletin of Entomology Society of America*. **22**, 302-304.

Neiswander, C R (1959) Slugs can injure young corn. *Ohio Agricultural Research and Development Center, The Ohio State University Research Bulletin*. May-June, pp 37.

Willson, H R; Eisley B R. (1992) Effects of tillage and prior crop on the incidence of five key pests on Ohio corn. *Journal of Economic Entomology*. **85**, 853-859.

THE USAGE OF MOLLUSCICIDES IN AGRICULTURE AND HORTICULTURE IN GREAT BRITAIN OVER THE LAST 30 YEARS

D G GARTHWAITE, M R THOMAS

Central Science Laboratory, MAFF, Sand Hutton, York, YO4 1LZ, UK

ABSTRACT

The work of the pesticide usage survey is outlined as is the methodology of sample selection, the range of surveys and data collection. Annually over 800,000 hectares of agricultural and horticultural crops are treated with molluscicides and over 250 tonnes of active substances are applied. Metaldehyde accounts for 55% of the total area treated, methiocarb 40% and thiodicarb 5%. Molluscicide use has increased almost 70 times since 1970-1974 with arable crops accounting for 99% of all usage. An estimated 4,800 tonnes of product are applied annually in Great Britain at a cost of approximately £10M.

INTRODUCTION

The pesticide usage survey group of the Central Science Laboratory (CSL), Harpenden has been conducting official surveys into pesticide use in England & Wales since 1965. The Scottish Office Agriculture, Environment and Fisheries Department (SOAEFD), began surveys in Scotland in the mid nineteen seventies and the programmes have been run in parallel since then. Early surveys were instigated in response to concerns over the use of organochlorines in horticultural crops, including hops and outdoor vegetables. Surveys of soft fruit and top fruit were added in 1966 and 1967, glasshouse crops in 1968, hardy nursery stock in 1971 and arable crops in 1974. A variable three to six year cycle of each of these commodity groups was continued until 1990 when a fixed cycle was introduced. This fixed cycle now includes a biennial survey of arable crops with all other surveys undertaken every four years. Ten commodity groups are now included as part of this regular rolling cycle.

The primary role of the survey group is to advise government on all aspects of pesticide usage and since 1985 there has been an obligation under the Food and Environment Protection Act to monitor the post-registration use of pesticides. Data are presented first to the Advisory Committee on Pesticides and then published for general sale and distribution to interested parties including other government departments, research establishments and pesticide manufacturers.

All aspects of pesticide use are covered by the surveys and data collected includes areas treated, growers' rates of application, methods and dates of application, reasons for use and any additional agronomic data which may influence applications.

This paper utilises the full molluscicide data set, archived at CSL, Harpenden, and aims to show the trends in usage over the last thirty years, the range of crops treated, the methods of application and the estimated cost to the agricultural and horticultural industry.

MATERIALS AND METHODS

Samples are selected for each individual commodity group within the cycle, for each crop within the specific commodity group and are drawn from the census returns of both England & Wales and Scotland to represent the area of crops grown in Great Britain. Samples are stratified according to the total area of crops grown within the old MAFF regions of England & Wales and land use regions of Scotland and by size group based on the total area of crops grown on each holding.

Data collected from each holding are raised by two factors to give an estimate of regional usage; the first factor being dependent on farm size group and region and the second dependent on crop area and region. Data are further adjusted by a third factor to give estimates of total pesticide usage related to the national cropping areas in Great Britain. The raising factors are based on the areas of crops grown during the period of each survey according to Agricultural Census Returns for both England & Wales and Scotland. Questionnaires for each survey consist of two forms and are completed during a personal interview with the grower by a trained member of the survey group. Form 1 summarises the areas of crops grown on each holding during the season of interest. Form 2 deals with all aspects of pesticide usage on the individual crops grown on a holding.

Data collected with regard to molluscicide use include dates of application, rates of application and methods of application. The latter area is of particular interest as more recent surveys include details of those molluscicides incorporated with seed at drilling or broadcast into the standing crop, with different environmental implications for both methods of application.

Estimates of the economic importance of molluscicide use have been made by contacting both manufacturers and suppliers in the agrochemical industry.

OVERALL USAGE OF MOLLUSCICIDES

Two molluscicide active substances, metaldehyde and methiocarb, have dominated usage over the last thirty years, however their importance has changed within each of the survey cycles (Table 1).

Currently over 800,000 hectares of agricultural and horticultural crops are treated with molluscicides. Metaldehyde accounts for approximately 55% of the total, methiocarb for 40%, thiodicarb (introduced between 1990-1993) for 5% and aluminium sulphate/copper sulphate/sodium tetraborate for less than 1% (Table 1).

Table 1. Estimated annual usage of molluscicides in Great Britain between 1970-1995 (treated ha).

	1970-1974	1975-1979	1980-1983	1984-1989	1990-1993	1994-1995
Metaldehyde	5,781	14,370	93,715	128,278	164,045	444,836
Methiocarb	6,278	44,345	590,562	404,212	158,655	322,045
Thiodicarb	.	.	.	.	98	44,933
Total	12,059	58,715	684,277	532,490	322,798	811,814

There has been a 67 fold increase in the area treated with molluscicides since the early nineteen seventies and annual usage rose from just over 12,000 treated hectares to 812,000 treated hectares in 1994-95. Much of the increase occurred during the early eighties with a 12-fold increase and a rise from just under 59,000 treated hectares annually over the period 1975-1979 to an annual estimate of 684,000 treated hectares over the period 1980-1983. Since then the treated area has increased by only 18%.

Similar increases are seen when the total weight applied is considered (Table 2). In the early surveys, up to and including 1982, the rate of application was assumed to be the full recommended rate, however in surveys subsequent to this date the growers' actual rates were used. There was a 38-fold increase in the use of molluscicides over the last twenty five years, from less than 7 tonnes of active substance used annually over the period 1970-1974 to over 250 tonnes in 1994-95. A nine-fold increase in the weight applied annually between 1975-1979 and 1980-1983 mirrors the changes in area treated. Since the early eighties the weight applied has increased by 19%. The relatively smaller increases in weight applied compared to area treated are due to the move to lower rates of application.

Table 2. Estimated annual usage of molluscicides in Great Britain between 1970-1995 (tonnes applied).

	1970-1974	1975-1979	1980-1983	1984-1989	1990-1993	1994-1995
Metaldehyde	5.2	12.9	81.8	87.3	85.8	190.5
Methiocarb	1.4	9.8	130	84.4	25.5	54.7
Thiodicarb	.	.	.	.	< 0.1	6.7
Total	6.6	22.7	211.8	171.7	111.3	251.9

In each cycle the area treated with methiocarb has remained consistently higher than metaldehyde until the current survey period, where usage of metaldehyde was 38% greater than methiocarb. In most years the weight of methiocarb applied has been less than metaldehyde because its products have lower rates of application and lower percentages of active substance per unit weight of product. In terms of quantity of product applied, usage within the current cycle is approximately 3,200 tonnes containing metaldehyde, 1,400 tonnes containing methiocarb and less than 200 tonnes containing thiodicarb. In order to calculate the quantity of product used it has been assumed that metaldehyde products contained 6% active substance, methiocarb products 4% and thiodicarb 4%, though these values do vary, with increasing numbers of newer products introduced containing lower rates of active substance.

Usage of the fourth molluscicide, aluminium sulphate/copper sulphate/sodium tetraborate, has been minimal and is not considered further.

USAGE OF MOLLUSCICIDES BY COMMODITY GROUP

Over the last thirty years there have been four main commodity groups treated with molluscicides, arable, fodder and forage, vegetable and soft fruit crops.

Table 3. Estimated annual usage of molluscicides on each of the major commodity groups in Great Britain between 1970-1995 (treated ha).

	1970-1974	1975-1979	1980-1983	1984-1989	1990-1993	1994-1995
Arable crops	6,862	41,890	666,655	518,698	291,948	800,366
Fodder crops	2,672	12,804	9,839	3,345	22,121	_[1]
Vegetables	894	1,344	3,678	5,781	5,289	9,055
Soft Fruit	780	1,334	2,021	3,067	2,008	2,393
Total	11,208	57,372	682,193	530,891	321,366	811,814

[1]not surveyed during this period

Usage on arable crops, and in particular wheat and oilseed rape, currently accounts for 99% of the total area treated with molluscicides. Since the late nineteen seventies there has been a 13% increase in the area of arable crops grown with a seven-fold increase in the area of oilseed rape grown during the same period. The total area treated with molluscicides declined by 56% from 1982 to 1992, however in 1994 there was a 20% increase compared with 1982. The increase in 1994 was due in part to the use of molluscicides on industrial linseed grown on an increased set aside area. The total molluscicide treated area in the 1994 arable survey comprised 55% metaldehyde, 40% methiocarb and 5% thiodicarb.

Use on fodder and forage crops was the second main area of molluscicide usage. The area grown has remained virtually unchanged over recent surveys but in the most recent survey in 1993, newly sown leys accounted for almost 90% of the total area treated. In 1989 usage on newly sown leys accounted for 63% of the total molluscicide treated area, and in 1982 for 72%. Usage in 1993 was almost seven times greater than in 1989. Metaldehyde comprised 83% of the total weight of molluscicides applied in 1993, with methiocarb accounting for a further 16% and thiodicarb making up the remainder.

The use of molluscicides in vegetable crops in 1995 was more than double that in 1981, even though the area grown in 1995 was 30% less. Usage in 1995 had increased by 71% since 1991 and by 57% since 1986. In the most recent survey in 1995, metaldehyde accounted for 63% of the total molluscicide treated area and methiocarb for 37%. Thiodicarb was used on less than one percent of the total treated area.

Molluscicide use on soft fruit was confined mainly to strawberries and blackcurrants. The treated area in 1995 was 18% higher than in 1980 even though the area grown had declined by 34% over the same period. The weight of molluscicide active substances applied increased by 53% between 1980 and 1995. Methiocarb accounted for 78% of the molluscicide treated area, metaldehyde being the only other molluscicide recorded.

Usage on other crops was minimal compared to those cited above.

USAGE OF MOLLUSCICIDES ON INDIVIDUAL CROPS

In 1994 usage on four crops accounted for 91% of the total area of molluscicide applications to arable crops, wheat, 61%, oilseed rape, 15%, winter barley, 9%, and potatoes 8%. However, in terms of areas grown, wheat comprised 38% of the total area of arable crops, oilseed rape 9%, winter barley, 13% and potatoes 3%. With the exception of winter barley all of the crops cited above show a proportionately higher incidence of application than would have been expected if all crops were treated on a *pro-rata* basis. This possibly indicates a greater perceived susceptibility to slug attack either because of the timing of drilling or the palatability of host crops. It is also possible that the increased area of oilseed rape grown since the mid-seventies, the move from spring to autumn drilling and the significant increase in use of "double-low" cultivars with low glucosinolate levels (Glen *et al.*, 1990) is responsible for the increase in molluscicide usage since that time.

The area of wheat grown increased by 8% between 1982 and 1994. Fifty six percent of the molluscicide treated area of wheat in 1994 received metaldehyde, 38% methiocarb and the remainder thiodicarb. Broadcasting accounted for 83% of molluscicide applications to wheat with the remainder being incorporated at drilling.

The area of oilseed rape grown in 1995 had more than doubled since 1982 and there had been a corresponding increase of 20% in the molluscicide treated area. Sixty three percent of the molluscicide treated area in 1994 was treated with metaldehyde, 29% with methiocarb and the remainder with thiodicarb. Ninety percent of molluscicides applied to oilseed rape were broadcast with the remainder being incorporated at drilling.

In contrast to the previous two crops, the area of winter barley grown decreased by 29% between 1982 and 1994. However, despite this the molluscicide treated area increased by 5%. In 1994 metaldehyde accounted for 59% of all applications, methiocarb for 39% and thiodicarb for the remainder. Eighty percent of all molluscicide applications to winter barley were broadcast.

Since 1982 the area of potatoes grown in Great Britain, both seed and ware, declined by 12%. Despite this decrease in area grown the area treated with molluscicides increased five-fold between 1982 and 1994. Methiocarb accounted for 72% of the molluscicide treated area, metaldehyde for 25% and thiodicarb for the remainder. In the most recent survey almost all molluscicides (99%) were broadcast onto the growing crop.

Usage of methiocarb as a seed treatment was confined to sugar beet within the arable crops commodity group where all of this crop (194,504 hectares) was treated. However, this treatment is targeted more specifically against pygmy beetle, wireworm and other soil pests than slug damage.

Fodder and forage crops as a commodity group is dominated by the area of grassland grown and the area of fodder crops alone has declined by 4% since 1982. In 1993, grassland less than five years old accounted for 13% of the total area of fodder and forage crops and almost 90% of the molluscicide treated area, entirely on newly sown leys. Metaldehyde accounted for 88% of the molluscicide treated area on such grassland, methiocarb for 12% and thiodicarb the remainder. Almost all molluscicides (98%) applied to new leys were broadcast onto the growing crop. The use of methiocarb seed treatments within fodder and forage crops was confined mainly to maize where 20% of the area grown received an application.

The area of vegetable crops grown had decreased by 30% since 1982, however, the area treated with molluscicides increased almost three-fold over the same period. Two crops accounted for almost three quarters of all molluscicide usage during 1995, Brussels sprouts and turnips/swedes. Since 1982 Brussels sprouts have consistently been the vegetable crop treated most intensively with molluscicides. In 1995 they accounted for 5% of the area of vegetable crops grown but for 56% of the molluscicide treated area. In the same year turnips and swedes comprised 3% of the vegetable area grown but 18% of the molluscicide treated area.

Metaldehyde comprised 59% of the total molluscicide treated area of Brussels sprouts in 1995, the only other molluscicide recorded being methiocarb. Metaldehyde was used on 89% of the molluscicide treated area of turnips and swedes, methiocarb again being used on the remainder. All molluscicides applied to Brussels sprouts and turnips/swedes were broadcast within the growing crop.

The area of soft fruit crops declined by 91% between 1980 and 1994, whilst molluscicide usage increased by 18% over the same period. In 1995 strawberry crops made up 82% of the soft fruit molluscicide treated area, blackcurrants grown for processing comprising a further 14%. In each survey since 1980 strawberries have been the main target for all molluscicide applications followed by blackcurrants grown for processing.

Methiocarb usage on strawberries accounted for 77% of all molluscicide applications to this crop, though some of this use may have been targeted against strawberry seed beetle. Metaldehyde was the only other molluscicide applied. On blackcurrants, methiocarb comprised 88% of the molluscicide treated area again with metaldehyde being the only other molluscicide recorded.

Within horticultural crops generally, increased molluscicide use may be in response to improved standards and quality control required for supermarket produce.

THE ECONOMIC IMPORTANCE OF MOLLUSCICIDES

The figures collected by the survey group are estimates based on a representative sample of farmers and growers. Using these figures and industry recommended prices it is possible to estimate the cost of molluscicide applications within each of the commodity groups. It is not possible to survey all crops each year and therefore usage during a year in which a commodity group was not sampled was extrapolated from the previous survey.

Usage of methiocarb on all crops during the current cycle of surveys, estimated at between £4.00-£5.00/kg product for almost 1,400 tonnes of product used annually, is equivalent to a cost of approximately £6M. Applications of metaldehyde at an approximate price of £1.00/kg product for an estimated annual usage of 3,200 tonnes of product is equivalent to a cost of just over £3M. Annual usage of thiodicarb is currently estimated at 200 tonnes of product, which at £4.00-£4.50/kg is equivalent to almost £1M. The latter figure may be an underestimate, however, as the product had only just been introduced and some surveys conducted before its introduction will not have accounted for its current market share.

Using the figures above it is estimated that the cost of mollusc control to the agricultural and horticultural industry in products alone is around £10M. The cost of application of these molluscicides is not included in the above estimates.

ACKNOWLEDGEMENTS

The authors would like to thank all the farmers and growers who have participated in pesticide usage surveys over the years. Thanks are also due to the survey group members who have diligently collected and processed the data. The help of Tim Green and David James of Rhone-Poulenc in producing the economic interpretation is greatly appreciated.

REFERENCES

Cutler, J R (1981) Review of Pesticide Usage in Agriculture, Horticulture and Animal Husbandry in Scotland 1975-1979. *Pesticide Usage Survey Report* **27**, Edinburgh: DAFS.

Garthwaite, D G; Thomas, M R (1996) Soft Fruit in Great Britain 1994. *Pesticide Usage Survey Report* **128**, London: MAFF.

Garthwaite, D G; Thomas, M R; Hart, M J (1995) Arable Farm Crops in Great Britain 1994. *Pesticide Usage Survey Report* **127**, London: MAFF.

Garthwaite, D G; Thomas, M R; Hart, M J; Wild, S W (1996) Outdoor Vegetable Crops in Great Britain 1995. *Pesticide Usage Survey Report* **134**, London: MAFF.

Glen, D M; Jones H; Fieldsend, J K (1990) Damage to oilseed rape seedlings by the field slug *Deroceras reticulatum* in relation to glucosinolate concentration. *Annals of Applied Biology.* **117**, 197-207.

Sly, J M A (1977) Review of Usage of Pesticides in Agriculture and Horticulture in England and Wales 1965-1974. *Pesticide Usage Survey Report* **8**, London: MAFF.

Sly, J M A (1981) Review of Usage of Pesticides in Agriculture and Horticulture in England and Wales 1975-1979. *Pesticide Usage Survey Report* **23**, London: MAFF.

Thomas, M R; Garthwaite, D G (1994) Grassland and fodder crops in Great Britain. *Pesticide Usage Survey Report* **119**, London: MAFF.

PRODUCING AND MARKETING A MOLLUSCICIDE

I U HEIM, R A BLUM, A W LÖLIGER

Lonza AG, Münchensteinerstrasse 38, CH-4002 Basel, Switzerland

ABSTRACT

Lonza as the sole relevant manufacturer of the molluscicide metaldehyde finds itself in a network of interactions with other counterparts partly determining and partly contributing to the decision making within the market of this molluscicide. Registration authorities, press and concerned groups are considered to have a strong and decisive influence on all the parties involved. The manufacturer of metaldehyde containing formulations and independent research institutes provide relevant contributions to the framework influencing all the participants, as well. Lonza interacts with all partners by a strong scientific and technical presence.

INTRODUCTION

Metaldehyde has been produced by Lonza for more than 75 years. First, as solid fuel tablets which were highly appreciated by all-weather proof hikers and campers but others like hairdressers and physicians did also make use of the tablets. In the 80s, META[1] tablets were more and more replaced by portable gas containers and Lonza abandoned the production of tablets.

The molluscicidal activity had been detected in the early 30s both in South Africa and Europe, but it was not until 1960 that metaldehyde became commercially important as a molluscicide. Initially, several manufacturers supplied this active but ceased due to technical and commercial reasons. Lonza maintained its metaldehyde (META) production as the process is part of a chain of synthesis which allows the use of bi-products for further processes.

Within the last 35 years, the production of META has permanently been subjected to improvements with respect to safety and protection of the environment. Concurrently, the environment for the commercialisation and application has changed considerably within the same period. Lonza - not only concerned in the improvement of its production - has also invested in the optimisation of the use of META as a molluscicide. This included assistance to the manufacturer to provide the applying customer with an efficient, economical, environmentally safe and easy-to-apply formulation.

---

[1] META = Registered trademark of Lonza AG, Basel, Switzerland

Lonza not producing any formulated product has found itself as an element in a framework constituting „the market" (Fig 1.). Optimising META as a molluscicide, therefore, has been and will be subjected to interactions of various parties, which *contribute* to or *determine* this process of advancement. The relevance of these interactions for Lonza as a manufacturer of the a.i. META is the subject of this presentation.

Fig 1. The framework of the market of a molluscicide and its determining and contributing elements

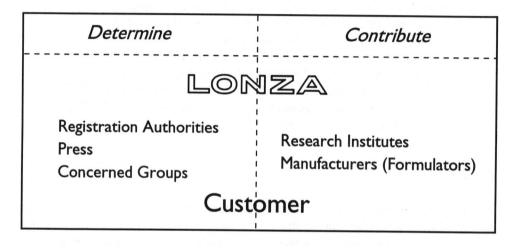

CHARACTERISATION OF THE INTERACTIONS BETWEEN THE ELEMENTS IN THE FRAMEWORK OF THE META MARKET

Customer

The applying customer or end user is the key factor pushing for the advancement of the final product. His requirements in terms of quality, quantity and price have changed significantly due to the scaling-up of the agricultural production illustrated by the expansion of monocultures, the rapid mechanisation and the extensification of soil cultivation, new crop varieties, the intensification of traditional production and the overall lean economic situation in agriculture. Finally, the farmer's input/output ratio has significantly changed. As a result he has *asked* for a more economic means to control the slugs or snails in a wider range of crops and under increasingly varying ambient conditions on the one hand. On the other, the customer has also *contributed* information on newly surfacing slug or snail pest problems and encouraged the development of formulations for new applications.

Manufacturer of formulated products

The manufacturers of META containing products find themselves also exposed to interactions with all other counterparts, especially to the requirements of the enduser, which let them to develop a range of formulations for different applications (Tab. 1). With research institutes the manufacturers collaborate and support their activities. The registration authorities, the press and concerned groups do also influence the commercial and technical activities of the manufacturers, whose activities of development and innovation are *a contribution* to the framework.

Table 1. Available formulations of META-Metaldehyde. In broad acre crops and horticulture, where mainly baits are used, improvements have been achieved with respect to attractiveness, pellet stability (rainfastness) and user safety (inclusion of repellents). Improvements in pellet size and consistency allowed the application to be fully mechanised. Sprays are used in horticulture and arboriculture. The paste is mainly applied in citrus against snails. The availability of the formulations differ from country to country.

Description	Characteristics	Mode of application
Mini pellet	Dry processed pellet	Bait
Quality pellet	Dry processed pellet with particular ingredient or technique (patented)	Bait
Quality pellet	Wet processed pellet	Bait
Granular	Inert carrier coated with a.i.	Bait
Wetable Powder	Suspensions	Spray
Paste	Suspensions	Special equipment

Research institutes

Research institutes and laboratories play a significant role both in generating information for a better understanding of slugs and snails as pests as well as in the improvement of META formulations.

A great effort is being done in several countries to determine the factors relevant for slug and snail damage. Cultivation methods, climatic factors like temperature and precipitation, soil conditions and tillage methods, the different susceptibility of crop varieties are among the factors taken into consideration in the studies. The identification of new foci of pests (cropwise and geographical) and new pest species has been done by institutes. Further, monitoring pest development and advising treatment strategies provide important agronomic and commercial information.

Less applied but more basic research has been done on the mode of action of molluscicides and their impact on the environment and non-target organisms. The results of these studies have had an important impact on improvement of the efficacy of the formulations and have also led to the development of particular and patented META formulations. Further, it can be assumed, that the results also have been considered by the registration authorities while defining the requirements for registration. Finally, efficacy tests under field and semi-field conditions carried out by independent institutes are an important element in quality control of formulations.

It is, therefore, obvious that independent research institutes are the most important *contributing* element within the framework while generating knowledge with scientific, agronomic and commercial relevance.

Registration authorities

This body is the most *determining* element within the framework by setting the limits and requirements for the approval or the use of a formulation. These demands must be met by those seeking a registration, which are both the producer of the a.i. and the formulator.

Press

World-wide, professional publications spread the relevant information mostly in a very concise and easy-to-understand form to the customer. Although, the relevant press claim to be independent, the articles can lack balance. However, the image of independence gives the press a *determining* influence within the framework.

Concerned groups

Groups specifically concerned with environmental and safety issues with respect to molluscicides play a significant role in only a few selective geographical areas. However, the issue of the poisoning of hedgehogs by dead slugs, is an example which shows how concerned groups can *determine* activities of authorities and the commercial counterparts.

THE ACTIVITIES OF LONZA AS AN A.I. PRODUCER WITHIN THIS FRAMEWORK

Customer

Lonza has only a very limited direct contact to the end user of META products. However, the need to find a solution for a new or particular problem in the field is often forwarded to Lonza, who has, consequently, initiated trials directly. Also, new applications in aquatic systems are being developed in collaboration with end-users, while local research institutes are involved, as well.

Regarding established markets and applications, Lonza gains information with practical relevance directly through market studies. To the greater extent, the manufacturer of META formulations reports to Lonza on the needs of the end-user.

Manufacturer of formulated products

Lonza is permanently in contact on technical and commercial grounds with the manufacturers: Especially, quality control (efficacy, stability, META content, packaging, labelling) is performed and quality improvements or advancements are in some cases initiated and always fully supported. New potential applications are developed together while Lonza funds the respective evaluations studies. The registration of high quality formulations is especially supported.

Research Institutes

Regular contacts are cultivated between researchers and Lonza supporting a wide range of activities on both the scientific level and the level of application. This generated wider knowledge of physiological and agricultural implications of the use of META as a molluscicide. New pest problems have been identified and new applications realised. Standard tests on efficacy of treating with META products are regularly carried out on a contract basis which, indirectly, supports other activities of the involved institutes.

It can be concluded, that the interaction between Lonza and scientific institutes are of mutual benefit and play also an important role in strengthening the commercial activity.

Registration Authorities

Lonza is fully exposed to the moves of the authorities as the manufacturer of the a.i. META. The increasing number of tests required to maintain the registration for META produce costs well in excess of 1 million Swiss francs per year.

Communications between Lonza and the registration authorities are limited due to the fact that commercial aspects do not play a role in the activities of the authorities. The industry has also to consider commercial aspects together with environmental issues and user safety. Lonza's position is distinctive due to the lack of the production of an end product. As the authorities disseminate their information mainly among approval holders or seekers, Lonza has to acquire the major part of information via third parties. Some direct formal communication, however, exists. With regard to the pending re-registration of metaldehyde in Europe direct contacts are more and more important.

As present demands and conditions for the registration significantly influence Lonza's commercial and technical decisions relevant for all the counterparts, expected requirements in future for the registration of a META containing product are closely followed world-wide.

Press

The manufacturers of final products interact more actively with the professional press. Lonza follows the scientific publication activity closely considering it as the valuable source of information for the advancement of existing formulations or the development of new applications. Lonza gathers scientific information on all aspects of slug & snail research world-wide, adapts the relevant part for the manufacturer and end user and disseminates it through a newsletter. This information is also forwarded to the manufacturer who is responsible for its diseminations to end-users. Additionally, Lonza maintains a presence in the professional media through selective advertisements. Finally, Lonza supports the publication in scientifically reviewed journals of studies performed in collaboration with research institutes.

Concerned Groups

To avoid engagement on particular issues Lonza permanently analyses the risks and establishes measures to minimise the genuine risks. Especially, the reduction of accidental poisoning by children, pets and wildlife has been of great concern. For instance, Lonza supported a detailed study on the risk of hedgehogs being poisoned by slugs killed by META. Additionally, Lonza currently implements the inclusion of repellent compounds into pellets by the manufacturer and supports studies with the aim to optimise this safety measure.

CONCLUDING REMARKS

To maintain the marketability of META Lonza has to invest significantly into scientific activities of third parties or of its own. The company runs a rearing facility under controlled conditions to have an independent supply of slug material for laboratory studies. Additionally, an infrastructure has been built up allowing physical and chemical tests for the refinement of formulations and their quality control. Concurrently, Lonza encourages and supports basic research by third parties for a better understanding of physiological and behavioural factors determining the efficacy of META. It is obvious that producing and marketing of META is not simply printing a price list and filling bags but meeting appropriately the requirements of all counterparts involved.

ASSESSMENT OF THE MOLLUSCICIDAL ACTIVITY OF A COPPER COMPLEX COMPOUND

P R DAVIS, J J VAN SCHAGEN, M A WIDMER AND T J CRAVEN

Entomology Branch, Agriculture Western Australia, Baron-Hay Court, South Perth, Western Australia, 6151

ABSTRACT

The molluscicidal activity of a new patented copper complex, formulated as a spray, was trialled against slugs and snails in the laboratory where it caused significant mortality of *Deroceras reticulatum*. The product was tested, in both spray and granular formulations, against *Helix aspersa* in a vineyard and a citrus orchard where it was compared to methiocarb baits and copper oxychloride spray. The copper complex spray demonstrated a high level of repellency to *H. aspersa* for extended periods in the field and proved superior to the other products tested in reducing snail numbers in the vines/trees. The product has potential in the management of pest molluscs especially as part of integrated pest management programs.

INTRODUCTION

Despite terrestrial molluscs being significant agricultural pests world-wide, there are surprisingly few chemicals which are effective in their control; especially when compared to the number of insecticides available to control pest insects. Only two chemicals are in common use as commercial molluscicides in Australia: metaldehyde and methiocarb. Other chemicals with demonstrated molluscicidal activity include copper sulphate, copper oxychloride, carbaryl, mexacarbate and thiodicarb.

The major use of molluscicides currently is via incorporation into baits. Until recently, only methiocarb, in Australia, was registered for use as a molluscicide in spray formulation. In commercial citrus orchards, Bordeaux mixture (copper sulphate/lime) is used as a schedule fungicide with the added advantage of providing some snail control.

Clearly there is plenty of scope for new molluscicides; especially those which are effective when applied as sprays. Copper salts are known to have molluscicidal qualities (Godan, 1983) and copper sulphate, in particular, has been used extensively in the control of pest molluscs. Ryder and Bowen (1977) reported that copper sulphate is a contact molluscicide being absorbed through the foot epithelium of the slug, *Agriolimax reticulatus*. In aquatic snails, copper affects osmoregulation causing death (Cheng and Sullivan, 1977). Metallic copper bands have been used as repellent barriers on trees.

Recently, 'Four-S', a patented copper complex compound (containing mainly copper silicate and henceforth called copper silicate) has been registered in Australia for the management of snails and slugs. The trials reported here describe initial assessments of the molluscicidal activity of the registered liquid formulation of this product. Some trials were also undertaken using a granular formulation of copper silicate.

MATERIALS AND METHODS

The copper silicate liquid formulation used in all trials was a ready-to-use product, as supplied by the manufacturer, containing 2.8 g/L copper. This is a buffered solution having a pH in the range 3.2 to 3.5.

<u>Laboratory trials:</u>

Snail Immersion Trials: Common garden snails, (*Helix aspersa*), were collected from park land and introduced into the laboratory where they were provided with water and chicken pellets *ad lib*. The snails were allowed to acclimatise to the laboratory conditions for 4 days before healthy, mature specimens were selected for the trial.

Exposure to the copper silicate was by immersion in a beaker of the product. Snails were suspended in mid-air until they were fully extended out of their shells, with eye-stalks also extended, before being immersed for 2 seconds. Following immersion, the snails were placed on clean paper towelling to drain off the excess fluid. Following treatment, ten snails were placed in each plastic tray (31cm x 62cm x 12cm) which had a double layer of damp paper towelling on the base and which was covered with fly-wire. This was then three-quarters covered by black plastic to increase the humidity and therefore activity of the snails.

Each cage was provided with water, sliced butternut pumpkin and chicken pellets *ad lib*. The cages were cleaned on a regular basis and food was replaced as required. They were kept in the laboratory where ambient temperature varied between 15 and 25°C. There were 10 replicates of the treatment and 3 replicates of the 'control' (untreated) group.

Slug Trial: Mature slugs of two species , *Milax gagates* and *Deroceras reticulatum*, were kept in the laboratory for 3 days before healthy specimens were chosen for the trial. Twelve slugs were placed on paper towelling and sprayed with the copper silicate as supplied using a fine spray from a plastic hand-sprayer. The slugs were left on the treated paper for 1 minute then placed in cages. The cages were plastic trays (31cm x 62cm x 12cm) with a double layer of damp paper towelling on the base and completely covered with fly-wire and black plastic sheeting. Ventilation was provided by 6 holes (6 mm in diameter) in the plastic at one end of the trays. Water and a slice of butternut pumpkin were provided *ad lib* and replaced as required. The treatments were replicated 10 times and there were 3 'control' groups.

Dead snails/slugs were counted in both laboratory trials with death defined as failure to respond to stimulation, by a dissecting needle, of the underside of the foot.

<u>Field trials</u>

Vineyard trial: The molluscicidal activity and protection afforded by several molluscicides was assessed at Banara Wines, Benara Rd., Caversham. Twelve rows, each typically containing 19 vines, were divided into 3 blocks of 4 treatments. At the time of treatment, the vines were dormant and pruned with some buds starting to burst. Assessment of snail activity was made by counting all snails on each vine and within a 100mm radius of the vine butt; both prior to treatment and at various intervals post-treatment.

A 'whipper-snipper' was used to remove vegetation from around each vine so that the only contact with the ground was the vine butt. Therefore, any snails leaving or re-entering the vine were forced to cross the treatments. All support posts were treated in the same manner, effectively isolating the entire row.

Two treatments were applied on September 10, 1993. The copper silicate was applied to 'run-off' onto the vine butts to the major fork and to the ground within 100mm of the base of the vines, using a back-pack sprayer. A total of 41.5L was used on 57 vines and 30 posts. It was estimated that twice the volume was applied to vines as to posts, giving an average of 576ml per vine and 288ml to each post. Conditions were ideal with dew still present at the time of application, no wind and the weather for the rest of the day being clear, dry and warm (25°C). 'Mesurol®' snail bait pellets (20g/kg methiocarb) were applied to the bases (within a 200mm radius) of vines in three rows; 4.3kg was used in total. It was estimated that only 2/3 of the amount of bait was required for posts as vines, giving averages of 56g/vine and 37g/post. The third treatment, copper silicate granules ('Four-S Granules'), was applied on September 13, 1993. A pre-measured vial was used to put 50g of granules around the base (within a 200mm radius) of each vine and post.

Citrus orchard trial: Fifty mandarin trees in 2 rows were chosen in an orchard which was heavily infested with *H. aspersa*. The trees were skirt-pruned and the vegetation under the canopy slashed with a 'whipper-snipper' leaving the trunk as the sole access to the trees for the snails.

Snails were counted on the trunks and lower branches and on the ground within a 150mm radius of the trunks. The trees were then ranked in order of the number of snails counted; from highest to lowest. The top 5 trees were selected and the treatments randomly assigned to each of these 5 trees. This process was repeated with each group of 5 trees working down the order of ranking. There were 4 treatments and an untreated 'control' with 10 replications of each.

A total of 5.5kg of the copper silicate was sprayed on the butts of 10 trees. Copper silicate granules, 900g, were applied to the base of another 10 trees while Copper oxychloride (@5g/L) was applied to a further 10 trees. Four litres was used on the ten trees. Two hundred grams of Mesurol (20g/kg methiocarb) bait was applied to the base of another set of ten trees. The weather was fine and warm during and following the applications of the treatments.

In both field trials, counts were made pre-treatment and at various times post-treatment. Dead snails/shells were removed pre-treatment so that estimates of the number of snails killed by individual treatments could be obtained. Dead snails found within 150mm of the bases of the trees/vines were counted post-treatment. Once counted, the dead snails were removed, thereby avoiding the chances of counting a dead snail on more than one occasion. Death was defined as no response to stimulation of the foot area.

Results for both field trials were analysed using analysis of variance (ANOVA) techniques and Fisher's least significance difference (LSD) at $P \leq 0.05$.

RESULTS

Laboratory trials

Snail Immersion trial: Observations indicated that the snails were obviously distressed by the treatment. They immediately produced viscous, white mucus followed by bubbles and then (within 30 seconds) a very viscous, bright yellow mucus.

While mortality amongst the treated snails was low, amounting to only 3% after 8 days, behavioural effects were evident. Treated snails were markedly less active than the untreated 'control' group. Many treated snails remained immobile and sealed the aperture to their shells with mucus, similar to aestivation but not identical as the mucus dried transparent. After 8 days the effects on the treated snails appeared to be diminishing so the trial was terminated.

Slug spray trial: When sprayed, the slugs were obviously discomforted and produced a viscous greeny/blue mucus. A brief inspection 3 hours after the treatment revealed that the slugs were overtly affected by the spray. Mortality was significantly higher (t-test, $P<0.05$) in the treatments where 58.3% mortality was recorded after 13 days compared to only 8.33% in the 'control' group. It is interesting to note that most of this mortality occurred within the first days following treatment; 41.7% mortality being recorded in the treatment group within 24 hours.

There was also a clear distinction in the susceptibility of the 2 species to the treatment. The more mobile and faster moving *D. reticulatum* were particularly vulnerable with 85% mortality while only 31.7% of the *M. gagates* succumbed. This difference was significant (t-test, $P<0.01$).

The symptoms of death also varied. All the *M. gagates* which died or were obviously affected had their penises everted. Only some of the *D. reticulatum* had swollen genital orifices while most showed no such external signs.

Field trials

Vineyard trial: The pre-treatment assessment of this trial shows the 'control' plots to have a significantly lower initial snail population than both of the copper silicate treatments but that the methiocarb bait treatment is not significantly different from either.

Post-treatment, snail populations fell in all plots and it is believed this was caused, in part at least, by inter-row cultivations done by the vigneron during the trial period. Despite the fall in the snail population in the 'control' plot, the copper silicate spray treatments had significantly lower snail populations at all post-treatment sampling events (Table 1 and Figure 1). Its performance was also significantly superior to the methiocarb bait treatment at both 8 and 15 days post-treatment but not after 33 days. The granular formulation of copper silicate did not perform as well but was significantly better than the 'control' for the first 2 weeks and was better, or not significantly different from, the methiocarb bait treatment on all occasions. Methiocarb baits were not distinguishable from the 'controls' until the last sample event when snail numbers in all plots were low.

Table 1: Mean number (n = 57) of live snails on grape vines treated with molluscicides.

Time (days) post-treatment	Control	Methiocarb baits	Copper silicate granules	Copper silicate spray
-2	4.21a	5.48ab	6.11b	5.71b
8	2.03c	2.05c	1.44b	0.44a
15	2.04c	1.68bc	1.51b	0.52a
33	0.96b	0.37a	0.62ab	0.24a

Note: Treatment means, for the same date (across rows), followed by the same letter are not significantly different from each other as established by ANOVA with the means separated by Fisher's least significant difference test (P<0.05). Means followed by a different letter are significantly different at P<0.05. Comparisons for significance between dates are not possible with this table.

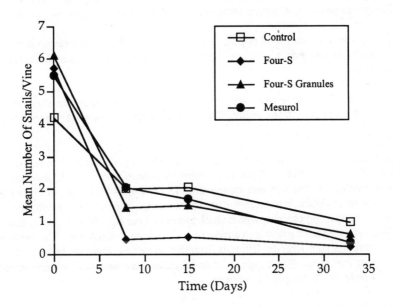

Figure 1: Mean number of live snails per vine following treatment with molluscicides.

The greatest numbers of dead snails (Table 2) on the 3 sampling events were in the methiocarb bait plots and this was significantly higher than the 'controls' and other treatment plots on all occasions. The copper silicate spray treatment only had a significantly higher number of dead snails from the 'control' 8 days after treatment but was not significantly different on subsequent sampling dates. It is possible snails died outside the search area and mortality could be underestimated in this trial.

Table 2: Mean number (n = 57) of dead snails on grape vines treated with molluscicides.

Time (days) post-treatment	Control	Methiocarb baits	Copper silicate granules	Copper silicate spray
8	0.53a	2.53c	0.76ab	1.26b
15	0.45a	4.86b	0.9a	1.34a
33	0.04a	0.39b	0.02a	0.00a

Note: Treatment means, for the same date (across rows), followed by the same letter are not significantly different from each other as established by ANOVA with the means separated by Fisher's least significant difference test ($P<0.05$). Means followed by a different letter are significantly different at $P<0.05$. Comparisons for significance between dates are not possible with this table.

At the time of treatment, some buds had burst on the vines and these were deliberately sprayed with copper silicate to test for phytotoxicity. Subsequently, no obvious phytotoxicity was observed although some occasional 'browning' of leaf margins was noted.

Orchard trial: Pre-treatment snail populations were uniform between treatments due to the method of treatment assignment and consequently there was no significant difference between any of the plots.

Many snails were observed to froth and drop to the base of the trees soon after the application of the copper silicate spray. This was the only treatment to produce this type of effect.

Following treatment, snail numbers in trees treated with copper silicate were significantly lower (Table 3, Figure 2) than in trees for all other treatments and 'controls' on every sampling date (except methiocarb baits on one sample date 15 days post-treatment) extending to 71 days after treatment (when assessments were terminated). The next best performing treatment was copper oxychloride where snail numbers were significantly lower than the 'control' and copper silicate granule treatments on all sampling occasions. The methiocarb bait treatments were not significantly different from the copper oxychloride spray treatments up to 57 days post-treatment. The copper silicate granules were statistically indistinguishable from the 'controls' on all sampling occasions.

Only the methiocarb bait treatment caused noticeable mortality of snails and this was significantly higher than the 'control' and all other treatments on all sampling occasions (Table 4, Figure 3). As could be expected, most of the dead snails were found in the first 3 weeks post-treatment.

Table 3: Mean numbers (n = 10) of live snails per mandarin tree after treatment with
molluscicides.

Time (days) post-treatment	Control	Copper oxychloride spray	Methiocarb bait	Copper silicate granules	Copper silicate spray
-1	308.6a	292.2a	282.9a	307.2a	297.7a
8	436.1c	234.9b	300.4b	402.3c	78.4a
15	396.9c	196.5b	161.2ab	318.7c	70.0a
22	454.2c	248.4b	182.7b	376.7c	70.6a
29	497.8c	209.8b	220.0b	454.9c	61.8a
43	362.9c	154.6b	166.9b	289.2c	53.3a
57	215.6d	106.1b	123.8bc	177.4cd	30.8a
71	243.7c	119.1b	207.4c	262.3c	38.3a

Note: Treatment means, for the same date (across rows), followed by the same letter are not significantly
different from each other as established by ANOVA with the means separated by Fisher's least significant
difference test (P<0.05). Means followed by a different letter are significantly different at P<0.05.
Comparisons for significance between dates are not possible with this table.

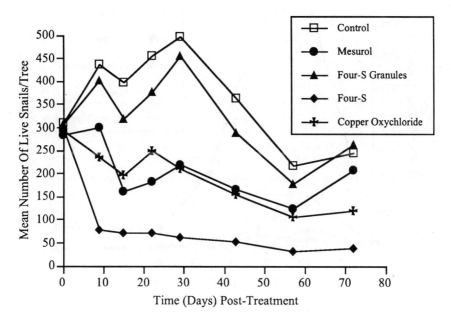

Figure 2: Mean number of live snails per mandarin tree following treatment with
molluscicides.

Table 4: Mean numbers (n = 10) of dead snails per mandarin tree following treatment with molluscicides.

Time (days) post-treatment	Control	Copper oxychloride spray	Methiocarb bait	Copper silicate granules	Copper silicate spray
8	3.2a	3.7a	159.5b	10.5a	31.6a
15	1.1a	6.9a	143.9b	0.6a	11.6a
22	0.5a	0.6a	47.2b	0.3a	3.5a
29	0.0a	0.6a	11.5b	0.5a	0.6a
43	0.1a	0.1a	13.9b	0.2a	0.2a
57	0.0a	0.2a	6.0b	0.2a	0.9a

Note: Treatment means, for the same date (across rows), followed by the same letter are not significantly different from each other as established by ANOVA with the means separated by Fisher's least significant difference test (P<0.05). Means followed by a different letter are significantly different at P<0.05. Comparisons for significance between dates are not possible with this table.

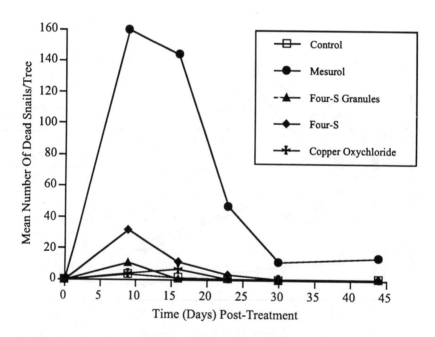

Figure 3: Mean number of dead snails per mandarin tree following treatment with molluscicides.

DISCUSSION

The laboratory trials indicated that the liquid formulation of copper silicate was repellent to adult specimens of *H. aspersa* but only caused a low level of mortality. Of the 2 species of slugs tested, *D. reticulatum* suffered significant mortality but the other species, *M. gagates*, did not. The former species is the more mobile and active but whether this was a factor in the differential mortality is not clear.

From our observations it appeared that the adult snails, in particular, were able to survive the treatment through the production of abundant mucus enabling them to simply 'slough-off' the copper silicate. Obviously size and exposed body surface area are factors and hence juvenile *H. aspersa* (unpublished data) and slugs were more susceptible. Loss of body water is likely to be a significant factor in the mode of action and those molluscs killed in the trials appeared dehydrated. As treated snails and slugs were kept in semi-closed trays, with a consequent high humidity, and were given free access to water, conditions favoured recovery from the treatment. Under more harsh post-treatment conditions, higher mortality rates could be expected.

Since the laboratory trials demonstrated low mortality to snails but repellency was obvious, the two field trials were designed specifically to assess the potential of copper silicate, in 2 formulations, to protect vines/trees against snails via its repellent effect.

The granular formulation did not demonstrate significant activity in these trials but the treatments were limited to spot applications of the product where snails may well have been able to avoid contact with the material. It is possible general broadcasting of the granules over an entire area may be more effective.

In both these trials, the copper silicate spray demonstrated significant repellency over an extended period of time resulting in reduced snail populations in vines/trees for periods in excess of 10 weeks. The reduction in snail populations in both vines and trees was superior to reductions produced by spot treatments with standard methiocarb baits.

The degree and longevity of the effect of copper silicate spray was somewhat surprising in the 'orchard trial' as the snail population at the time of treatment was so high that the snails completely covered the surface of the butt of the trees, often being 2 deep. Under these circumstances, the amount of spray actually being able to deposit on the trunks must have been dramatically reduced as most would have been deposited on the snail shells. It is therefore likely that better performances can be expected where the butt surfaces can be thoroughly treated.

These trials did not demonstrate sufficient control of mollusc populations via a direct lethal effect to warrant commercial development along these lines. Lethal effects on juvenile *H. aspersa* and some slug species deserve further investigation. It is likely that the toxic effect of copper silicate spray could be improved by applying it when the molluscs are active. Application early on a damp morning followed by sunny weather would be optimal.

The manufacturer claims that the copper silicate remains active in the field for extended periods because the copper silicate is soluble in water only under acid conditions and once the formulation has dried after application it is not dissolved by rain or irrigation. The results from the vineyard and orchard trials support the claim of an extended effect of copper silicate under field conditions.

The use of metaldehyde and methiocarb baiting methods for the protection of plants and crops can have undesirable side effects on dogs, fowls, native birds and the non-target invertebrate soil fauna. Bieri et al (1989) reported on the toxicity of methiocarb snail baits to earthworms while Purvis and Bannon (1992) reported both short and long term effects of these baits on predatory carabid beetles. Effects of pesticides on earthworms is a growing concern among producers of fruits and vegetables aiming to reduce their use of pesticides. Bowen and Mendis (1995) review some of the recent advancements in pest mollusc control and express the need for greater innovation in developing integrated management strategies for pest molluscs. Our assessment of the spray formulation of copper silicate demonstrated significant repellency to molluscs in a series of laboratory and field trials. In addition, this repellency is apparent for extended periods in the field. The product is well suited to integrated pest management programs aimed at the reduction of the use of pesticides.

REFERENCES

Bieri, M; Schweizer, H; Christensen, K; Daniel, O (1989) The effect of metaldehyde and methiocarb slug pellets on *Lumbricus terrestris* Linne. In: *Slugs and Snails in World Agriculture*, BCPC Monograph 41, Henderson, I F (ed.) British Crop Protection Council, pp. 237-244.

Bowen, I D; Mendis, V W (1995) Towards an integrated management of slug and snail pests. *Pesticide Outlook.* **6**, 12-17.

Cheng T C; J T Sullivan (1977): Alterations in the osmoregulation of the pulmonate gastropod *Biomphalaria glabrata* due to copper. *Journal of Invertebrate Pathology* **29**, 101-104.

Godan D (1983): *Pest Slugs and Snails*. Springer-Verlag. 445pp.

Purvis, G; Bannon, J W (1992) Non-target effects of repeated methiocarb slug pellet application on carabid beetle (Coleoptera: Carabidae) activity in winter-sown cereals. *Annals of Applied Biology* **121**, 401-422.

Ryder, T A and I D Bowen (1977): The slug foot as a site of uptake of copper molluscicide. *Journal of Invertebrate Pathology* **30**, 381-386.

Session 3
Physiology and Function

Session Organiser
and Chairman   Dr N W Runham

SLUGS AS TARGET OR NON-TARGET ORGANISMS FOR ENVIRONMENTAL CHEMICALS

R TRIEBSKORN
Zoological Institute, Dept. of Physiological Ecology, University of Tübingen, Auf der Morgenstelle 28, D-72076 Tübingen, Germany

I F HENDERSON, A MARTIN
IACR-Rothamsted, Harpenden, Herts, AL5 2JQ, UK

H-R KÖHLER
Zoological Institute, Dept.of Cell Biology, University of Tübingen, Auf der Morgenstelle 28, D-72076 Tübingen, Germany

ABSTRACT

The ability of slugs *(Deroceras reticulatum)* to survive exposure to various chemical stressors such as molluscicides (carbamates, metaldehyde, iron-chelates) or other environmental pollutants (heavy metals, pesticides) was investigated using x-ray analyses, autoradiography, filter electron microscopy, enzyme histochemistry, enzymatics, electron microscopy, and stress protein analyses. Uptake pathways, target sites, modes of action and general responses in the animals were studied, and the results (including aspects of resorption and metabolism of the toxins as well as their cellular and biochemical effects) interpreted in terms of both molluscicidal efficiency and of the consequences to slugs´ecological role as soil decomposers.

INTRODUCTION

Slugs, particularly *Deroceras reticulatum,* are important pests in agriculture and horticulture in many parts of the world (Barker, 1989, Beer, 1989, Byers & Calvin, 1994, Ferguson, 1994). As with many other pest species, their pest status arose mainly from Man´s activities favouring high population densities of slugs, for example by creating monocultures of favoured food, by cultivation methods which provide refuges from desiccation and predation, and by providing glasshouses in which slugs can escape from both harsh winters and dry summers. In nature however, slugs, like earthworms, are ecologically important decomposers of plant litter in woodland and other soils (e.g. Swift *et al.*, 1979). Slugs are therefore of interest not only to agriculturists and the pesticide industry but also to ecotoxicologists as monitor species for environmental pollution. Both disciplines are interested in similar endpoints of toxicity: 1) Mortality, as the best (for pest control) or the worst case (for ecotoxicologists) of toxic impact; 2) Cell damage as evidence of molluscicidal effects or as an early warning marker for sublethal environmental pollution; and 3) The organisms´ defense mechanisms, as barriers to be overcome in molluscicide development or as a strategy for survival in a contaminated environment.

This paper investigates the toxicology of slugs in both situations: 1) Slugs as target organisms for molluscicides (methiocarb, cloethocarb, metaldehyde or iron chelates), and 2) Slugs as non-target organisms exposed to environmental pollutants (heavy metals, γ-HCH (lindane) or pentachlorophenol (PCP). Responses at the organ, cell and biochemical levels and their energetic implications are discussed.

MATERIALS AND METHODS

Test animals

Adult *Deroceras reticulatum*, either laboratory-reared and between four and six months old or field collected and between 400 and 600 mg weight, were used in all the experiments.

Application of substances

All the molluscicides were administered orally. The compounds used were 1) Commercially available slug pellets [*Mesurol* - 4% of 4-(methylthio)-3,5-xylyl-methyl-carbamate and *Spiess Urania 2000* - 4% metaldehyde]; 2) Dried wheat germ agar containing either 2% or 0.1% cloethocarb [phenol-2-(2-chloro-1-methoxyethoxy)-methylcarbamate]; 3) Agarose gel containing either 23μg or 63 μg metaldehyde per animal directly injected into the buccal cavity; and 4) Whole wheat flour (5mg) containing either 30μg or 500μg per animal of either iron butan [tris(1-oxo-1,2-diazabutan-2-oxido)Fe(III)] or iron octan [tris(1-oxo-1,2-diazaoctan-2-oxido)Fe(III)](Clark *et al.*, 1995).

Heavy metals at various concentrations (59, 358 or 751 μg Cd/g dry wt; 405, 2456 or 3751 μg Pb/g dry wt; 1874, 9305 or 14915 μg Zn/g dry wt) or γ-HCH (58, 116 or 231 ng/g dry wt) or PCP (29, 87 or 867 ng/g dry wt) were also given orally by feeding the animals lettuce leaves soaked with aqueous solutions of each toxin. The slugs fed with heavy metals were also exposed to 1.4, 5.8 or 14.1 μg Cd/g dry wt, 103.1, 164.6 or 292.3 μg Pb/g dry wt or 136.4, 86.9, or 490.9 μg Zn/g dry wt in the substrate.

For all tests, control animals were kept on uncontaminated substrate and fed uncontaminated food.

Tracing methods

X-ray analysis: Slugs were fed a standard food (Standen, 1951) containing 20% ThO_2 and 2% cloethocarb and x-rayed at fixed intervals after feeding began (method as in Triebskorn & Florschütz, 1993).

Autoradiography: Slugs were fed either ^{14}C-labeled cloethocarb (1 μCi) or metaldehyde (3 μCi) and dissected at fixed intervals after feeding began (method as in Triebskorn *et al.*, 1990).

Filter electron microscopy: Slugs were fed with low (30μg or 0.6%) or high (500μg or 10%) concentrations of both iron octan and iron butan, and dissected and fixed 16 hours after feeding began. Samples of hepatopancreas and skin were fixed in 2% glutardialdehyde in 0.01 M cacodylate buffer (pH 7.4) (2h at 4°C), then 1% aqueous OsO_4 (2h at 4°C), then embedded in Spurr resin. Ultrathin sections were analyzed for iron according to the method described by Triebskorn *et al.* (in press). Similarly prepared sections of hepatopancreas of metal-treated slugs were analyzed for zinc according to the method described by Triebskorn & Köhler (in press).

AAS (atomic absorption spectroscopy) measurements: The metal content of slugs was determined according to the method described by Köhler *et al.* (1995), and Triebskorn & Köhler (in press).

Structural investigations

For light and electron microscopy slugs were dissected at fixed intervals after feeding began and fixed in 2% glutardialdehyde in 0.01 M cacodylate buffer (pH 7.4). The samples were either routinely processed for electron microscopy (e.g. Triebskorn, 1989; Triebskorn & Künast, 1990) or for paraffin histology (e.g. Triebskorn, 1995)

Enzyme histochemical tests and enzymatics

For enzyme histochemical tests tissues were either frozen in liquid nitrogen or fixed for 1 h in 2% glutardialdehyde. Fixed material was embedded, without dehydration, in *HistoResin.* Non-specific esterases (UE), alkaline phosphatase (AP), acid phosphatase (AcP), Ca-ATPase (ATP) and neotetrazoliumreductase (NTR) were investigated according to the methods described by Triebskorn (1991, 1995). In parallel, enzyme activities of AP, AcP and arylhydrocarbonhydroxylase (AHH) were determined photometrically according to Triebskorn (1991).

Stress protein analyses

In animals exposed to heavy metals, the stress protein (hsp70) was analyzed by a standardized Western blotting technique and subsequent image analysis according to Köhler *et al.* (1992, 1994) and Zanger *et al.* (1996).

RESULTS

Contact, absorption, and storage

Slugs encounter toxic materials either by contact or during feeding. If they crawl over surfaces recently sprayed with pesticides, on toxic slug baits or in heavily polluted substrates the skin is the first point of contact. However, the primary target for toxic baits and the point of first contact for pollutants ingested with contaminated food is the digestive tract. For stomach poisons such as molluscicidal baits the foregut and crop can be the main site of absorption. X-ray studies of animals fed with 2% cloethocarb showed that the transport of the food pulp along this part of the gut was inhibited, and the molluscicide retained there for at least 15h. Direct cellular effects on the foregut and crop and the enhanced opportunity for toxins to be absorbed there could therefore both be important factors in the toxicity of such compounds.

Autoradiography of animals fed with radiolabelled cloethocarb or metaldehyde demonstrated that after rapid absorption in the foregut and crop the molluscicides were transported in the haemolymph to other organs such as the hepatopancreas, stomach, and skin. In the hepatopancreas, radioactivity was found mainly in the basophilic cells 1h to 10h after feeding and in the connective tissue (including macrophages) adjacent to the midgut gland tubules after 16h.

However, the sites and pathways of absorption and distribution of different compounds are dependant on their individual properties, such as lipophilicity. For example, 16h after feeding, the iron residue from chelate molluscicides was found in the resorptive cells of the hepatopancreas rather then in the basophilic cells, in a particular type of secondary lysosome.

The ability of slugs to store pollutants, particularly in the hepatopancreas, makes them useful indicators of environmental contamination, especially heavy metals. Experiments in which slugs were fed various metals at concentrations similar to those encountered in the field resulted in accumulations of up to 71µg Cd, 1169µg Pb or 437µg Zn per g dry wt of slug tissue.

Responses in different organs

Generally, in all the organs examined, certain cells were damaged or seriously disrupted directly by the toxin immediately after its application. Such seriously affected cells were usually localized in "hot spots" in a particular part of the organ, and cells elsewhere in the organ usually maintained their function or showed only specific and limited responses to toxic impact.

Crop: With all the test materials lipid storage was reduced. The epithelial cells in control animals contained many lipid droplets, those from treated animals did not. In the molluscicide-treated animals, the activity of enzymes involved in transport processes (ATPases, alkaline phosphatases) was increased. When the toxic dose exceeded a certain limit a second stage of symptoms was observed where the epithelial cells were damaged and showed pathological symptoms.

Skin: Enhanced production of mucus and increased mucopolysaccharide synthesis was observed (this occurs in the stomach and the intestine to a lesser extent as well). Depending on the duration of exposure and concentration of toxin all types of mucocytes initially showed ultrastructural responses related to increased mucus production. Above a certain threshold, mucus cells (and other epithelial cells) showed signs of pathology. Metaldehyde caused an increase in ATPase activity especially at the bases of the mucus cells.

Hepatopancreas: In response to molluscicides, heavy metals, PCP and lindane, the resorptive cells ceased food absorption from the lumen and this was associated with an activation of intra- and/or extracellular digestion: Fusion of vacuoles and lysosomes increased and large vacuoles and secondary lysosomes dominated the cells. The lysosomes were not mainly near the cell apices as in controls, and after molluscicide exposure the pattern of acid phosphatases changed. Glycogen storage was generally reduced, microvilli shortened, and the number of pinocytotic vesicles or channels drastically reduced. The activity of non-specific esterases especially in the lumen, and of alkaline phosphatases at the base of the tubules was increased.

In the basophilic cells organic and inorganic pollutants (which were both located within these cells by tracing techniques) caused structural modification of organelles (especially endoplasmic reticulum), which was correlated with an induction of enzymes involved in biotransformation (NTR, AHH) and in cellular transport (ATP, AP). Storage products in the basophilic cells were significantly reduced.

Stress protein expression

The stress protein (hsp70) level increased by up to ten times compared to control animals in slugs treated with all the compounds except γ-HCH. A dose-response relationship was observed with all the heavy metal and PCP treatments, with increased proteotoxicity induced even by low concentrations of toxins. When very high levels were applied however, the relationship broke down and the hsp70 level declined in acutely poisoned slugs, as reported in other soil invertebrates (Guven et al., 1994, Eckwert et al., in press). This effect might explain the low hsp70 levels in slugs treated with γ-HCH.

DISCUSSION

The results obtained after exposure to both organic and inorganic chemicals suggest that there are three main groups of cellular response: **1) Responses indicating cell damage** such as cellular pathology, a decrease in enzyme activies or a reduction in the level of stress protein (both the latter probably caused by a breakdown in the protein synthesis apparatus). These reactions are widespread in target cells (and in non-target cells as well with acute poisoning) and are the desired effects of molluscicides. Slugs might survive chronic exposure as long as cell damage remains below a critical level and is confined to restricted portions of their organs which can then still function adequately. Such conditions are frequently encountered in sublethally polluted environments. **2) Responses associated with defence or repair mechanisms.** The most obvious of these is enhanced mucus secretion by the skin, although it occurs in the stomach, intestine and parts of the genital tract (Triebskorn & Ebert, 1989). The copious mucus could serve to dilute the toxin and sometimes, for example in the digestive tract where the pH is 4 - 5, even detoxify it if it is pH sensitive (Triebskorn, unpublished). Exposure to organic toxins often induces biotransformation enzymes, and exposure to metals can lead to the formation of spherites (e.g. Hopkin, 1989) or increased expression of metal binding proteins or metallothioneins (Dallinger et al., 1989; Berger et al., 1995). Another major response in this group is the expression of stress proteins which indicate proteotoxicity and can refold malfolded proteins to some extent in affected cells (e.g. Hightower, 1991). All these responses begin or are enhanced in order to get rid of the toxin or to limit damage, and in the event of long-term chronic exposure slugs can keep them activated over long periods; they are therefore useful as "biomarkers" for chemical pollution. To molluscicides, these reactions are obstacles to be overcome to improve efficacy. The final group of reactions are **3) responses associated with energy production.** All the processes in group 2 are energetically expensive. For example, even under unpolluted conditions, a limpet expends about a quarter of its daily energy budget in production of pedal mucus for adhesion and locomotion (Davies et al., 1990). A large amount of extra energy is needed to fuel the repair and defence mechanisms described above, and it may be provided by intensified intracellular digestion in the hepatopancreas which leads to the increased fusion of vacuoles, increased activity of enzymes involved in intra- and extracellular digestion, depletion of storage products in the crop and hepatopancreas, and increased activities of enzymes involved in transport.

It may be that the responses in the three groups described above are interdependent, and the limiting factor for tolerance of chemical stress is the energy available to maintain the defence mechanisms. Fig. 1 summarizes this hypothesis and illustrates how the responses in group 1 (represented by "cellular pathology") and group 2 (represented by "stress proteins") may be affected differently by the finite "energy" available in both chronic and acute toxic conditions.

In the environment, chronic exposure conditions could have a significant effect on an entire ecosystem, since in order to compensate for a condition of permanent chemical stress many organisms may have to keep repair and defence mechanisms continually activated, and invest large amounts of energy into limiting cell damage. This energy expenditure will build up a considerable deficit and so the animals' requirement for energy from feeding is consequently increased and any short-term food shortages which would be otherwise tolerable may not be survived. Energy for reproductive processes may also be limited and chronic exposure to pollutants often results in reduced reproductive success (summarized by Kammenga, 1996). The balance of the community of decomposers in polluted soils may ultimately be significantly altered, and chemical pollution or extended intensive chemical crop protection treatments provide a strong selective pressure for resistant genotypes which, in the long term, bodes ill for chemical pest control.

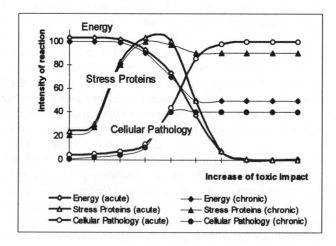

Fig.1: Hypothetical response to acute and chronic toxic exposure of stress proteins (hsp70) and cellular pathology in relation to energy budget.

"Increased toxic impact" means longer exposure to moderate toxin concentration in chronic exposure; but under acute conditions high concentrations can only be tolerated for a short time. The timescales for the two conditions are therefore different which is not reflected in the plot.

ACKNOWLEDGEMENTS

The authors are grateful to LONZA Ltd, Basle/Visp, Switzerland, BASF AG, Ludwigshafen/Limburgerhof, Germany, the German Ministry of Education, Science, Research and Technology (BMBF), and the UK Biotechnology and Biological Sciences Research Council (BBSRC) for financial support of these studies.

REFERENCES

Barker, G M (1989) Slug problems in New Zealand pastoral agriculture. In: *Slugs and Snails in World Agriculture.* I Henderson (ed.). BCPC Monograph No. 41, pp. 59-68.

Beer, G J (1989) Levels and economics of slug damage in potato crops 1987 and 1988. In: *Slugs and Snails in World Agriculture.* I Henderson (ed.). BCPC Monograph No. 41, pp. 101-106.

Berger, B; Dallinger, R; Thomaser, A (1995) Quantification of metallothionein as a biomarker for cadmium exposure in terrestrial gastropods. *Environmental Toxicology and Chemistry.* **14(5),** 781-791.

Byers, R A; Calvin, D D (1994) Economic injury levels to field corn from slug (Stylommatophora: Agriolimacidae) feeding. *Journal of Economic Entomology.* **87(5),** 345-350.

Clark, S J; Coward, N P; Dawson, G W; Henderson, I F; Martin A P (1995) Metal chelate molluscicides: The redistribution of iron diazoalkanolates from the gut lumen of the slug, *Deroceras reticulatum* (Müller) (Pulmonata: Limacidae). *Pesticide Science.* **44,** 381-388.

Dallinger, R; Janssen, H H; Bauer-Hilty, A; Berger, B (1989) Characterization of an inducible cadmium-binding protein from hepatopancreas of metal-exposed slugs (Arionidae, Mollusca). *Comparative Biochemistry and Physiology.* **92C(2),** 355-360.

Davies, M S; Hawkins, S J; Jones, H D (1990) Mucus production and physiological energetics in *Patella vulgata* L.. *Journal of Molluscan Studies.* **56,** 499-503.

Eckwert, H; Alberti, G; Köhler, H-R (in press) The induction of stress proteins (hsp) in *Oniscus asellus* (Isopoda) as a molecular marker of multiple heavy metal exposure: I. Principles and toxicological assessment. *Ecotoxicology.*

Ferguson, A W (1994) Pests and Plant injury on lupins in the south of England. *Crop Protection.* **13 (3),** 201-210.

Guven, K; Duce, J A; de Pomerai, D I (1994) Evaluation of a stress-inducible transgenic nematode strain for rapid aquatic toxicity testing. *Aquatic Toxicology.* **29,** 119-137.

Hightower, L E (1991) Heat shock, stress proteins, chaperones, and proteotoxicity (Meeting review). *Cell.* **66,** 191-197.

Hopkin, S P (1989) *Ecophysiology of metals in terrestrial invertebrates.* Elsevier Applied Science, London, New York, 366pp.

Kammenga, J E (1996) *Manual of BIOPRINT. Biochemical fingerprint techniques as versatile tools for the risk assessment of chemicals in terrestrial invertebrates.* Ministry of the Environment, NERI, Denmark.

Köhler, H-R; Triebskorn, R; Stöcker, W; Kloetzel, P-M; Alberti, G (1992) The 70 kD heat shock protein (hsp 70) in soil invertebrates: A possible tool for monitoring environmental toxicants. *Archives of Environmental Contamination and Toxicology.* **22,** 334-338.

Köhler, H-R; Rahman, B; Rahmann, H (1994) Assessment of stress situations in the Grey Garden Slug, *Deroceras reticulatum,* caused by heavy metal intoxication: Semi-quantification of the 70 kD stress protein (Hsp70). *Verhandlungen der Deutschen Zoologischen Gesellschaft.* **87,** 228.

Köhler, H-R; Körtje, K-H; Alberti, G (1992) Content, absorption quantities and intracellular storage sites of heavy metals in Diplopoda (Arthropoda). *BioMetals.* **8,** 37-46.

Standen, O D (1951) Some observations upon the maintenance of *Australorbis glabratus* in the laboratory. *Annals of Tropical Medicine and Parasitology.* **45,** 80-83.

Swift, M J; Heal, O W; Anderson, J M (1979) *Decomposition in terrestrial ecosystems.* Blackwell Scientific; Oxford, London, Edinburgh, Melbourne. 372 pp..

Triebskorn, R (1989) Ultrastructural changes in the digestive tract of *Deroceras reticulatum* (Müller) induced by a carbamate molluscicide and by metaldehyde. *Malacologia* **31(1),** 141-156.

Triebskorn, R; Ebert, D (1989) The importance of mucus production in slugs' reaction to molluscicides and the impact of molluscicides on the mucus producing system. In: *Slugs and Snails in World Agriculture.* I Henderson (ed.). BCPC Monograph No. 41, pp. 373-379.

Triebskorn, R; Künast, C; Huber, R; Brem, G (1989) The tracing of a ^{14}C-labeled carbamate molluscicide through the digestive system of *Deroceras reticulatum. Pesticide Science.* **28,** 321-330.

Triebskorn, R; Künast, C (1990) Ultrastructural changes in the digestive system of *Deroceras reticulatum* (Mollusca; Gastropoda) induced by lethal and sublethal concentrations of the carbamate molluscicide Cloethocarb. *Malacologia* **32(1),** 89-106.

Triebskorn, R (1991) The impact of molluscicides on enzyme activity in the hepatopancreas of *Deroceras reticulatum* (Müller). *Malacologia.* **33(1-2),** 255-272.

Triebskorn, R; Köhler, H-R (1992) Plasticity of the endoplasmatic reticulum in three cell types of slugs poisoned by molluscicides. *Protoplasma.* **169,** 120-129.

Triebskorn, R; Florschütz, A (1993) Transport of uncontaminated and molluscicide-containing food in the digestive tract of the slug *Deroceras reticulatum* (Müller). *Journal of Molluscan Studies.* **59,** 35-42.

Triebskorn, R (1995) Tracing molluscicides and cellular reactions induced by them in slugs' tissues. In: *Cell Biology in Environmental Toxicology.* M P Cajaraville (ed.). pp. 193-220.

Triebskorn, R; Köhler, H-R (in press) The impact of heavy metals on the grey garden slug, *Deroceras reticulatum* (Müller): Metal storage, cellular effects and semi-quantitative evaluation of metal toxicity. *Environmental Pollution.*

Triebskorn, R; Higgins, R P; Körtje, K-H; Storch, V (in press). Localization of iron stores and iron binding sites in blood cells of five different priapulid species by energy-filtering electron microscopy (EFTEM). *Journal of Trace and Microprobe Techniques.*

Zanger, M; Alberti, G; Kuhn, M; Köhler, H-R (1996) The stress-70 protein family in diplopods: induction and characterization. *Journal of Comparative Physiology B.* **165,** 622-627.

SLUG CHEMICAL ECOLOGY: ELECTROPHYSIOLOGICAL AND BEHAVIOURAL STUDIES

C J DODDS,[1] M G FORD,[2] I F HENDERSON,[1] L D LEAKE,[2] A P MARTIN,[1] J A PICKETT,[1] L J WADHAMS,[1] P WATSON[2]

[1] IACR-Rothamsted, Harpenden, Hertfordshire, AL5 2JQ, UK

[2] School of Biological Sciences, University of Portsmouth, King Henry Building, King Henry I Street, Portsmouth, Hampshire, PO1 2DY, UK

ABSTRACT

The chemical ecology of terrestrial molluscs such as the arable crop pest, the grey field slug (*Deroceras reticulatum*), is mediated by volatile semiochemicals. Little evidence has been found for pheromones, but a range of interactions associated with food location has been reported at the behavioural level. Identification of host kairomone components has been achieved by linking chemical analysis with behavioural bioassays. However, recently developed techniques employing neurophysiological recordings from the olfactory epithelium on the optic tentacles of slugs are showing considerable promise for identification of kairomones and other semiochemicals which act as repellents or interfere with responses to kairomones. Attempts are being made to link these recording techniques directly to chemical analytical systems. The impact of such approaches on understanding mollusc chemical ecology and in underpinning new control strategies is described.

INTRODUCTION

Terrestrial molluscs (Gastropoda: Pulmonata), i.e. slugs and snails, can be important crop pests, for example the grey field slug, *Deroceras reticulatum* (Limacidae), and certain species of fresh-water snail are vectors of schistosomiasis in human beings (Purchon, 1977). Although molluscs employ essential biochemical and behavioural processes that set them apart from other organisms, these have not yet been exploited to provide selective methods of chemical control. As with any organism, these molluscs interact with chemicals both as toxicants and as signalling agents (semiochemicals). It is in the latter category that considerable promise lies for selectivity in chemical control of pest species, particularly slugs.

Semiochemicals

Various types of semiochemicals have been investigated for slugs, but no clear evidence has been found for use of sex pheromones, aggregation pheromones or alarm pheromones (Burke, Stephenson & Pickett, unpublished). However, as volatile food-plant attractants and non-volatile phagostimulants (feeding stimulants) are known to be important for molluscan survival in general, and for *D. reticulatum* in particular (Stephenson, 1979), these have been investigated in depth. The most attractive volatiles, as judged by statistically significant levels of odour trail following, were from the leaves of lettuce and dandelion (*Lactuca sativa*, *Taraxacum officinale*, Asteraceae = Compositae) and the roots of carrot (*Daucus carota*, Apiaceae = Umbelliferae). The activity of these volatiles was almost entirely accounted for by their content of fatty acid metabolites, particularly the leafy alcohol (Z)-3-hexen-1-ol (Pickett & Stephenson, 1980).

The main damage caused by *D. reticulatum* to cereals is by seed hollowing, which prevents germination. For seed treatment, semiochemicals with repellent or antifeedant activity, and with negligible bird and mammal toxicity, were sought by investigating the activities of extracts of 60 plant species and various known arthropod antifeedants. These were deposited via ethereal solution onto wheat seeds and assessed in a choice test between treated and untreated seeds (Scott *et al.*, 1977). Eleven of these extracts were shown to be active and further assessment was made in a no-choice test. In this assay, two samples showed high activity, a root extract of horse-radish (*Armoracia rusticans*, Brassicaceae = Cruciferae) and a leaf extract of scented geranium (*Pelargonium graveolens*, Geraniaceae).

The activity of the *A. rusticans* extract was accounted for by its content of 2-phenylethyl isothiocyanate, a glucosinolate catabolite widely distributed in the Brassicaceae. The active component of the *P. graveolens* extract was found to be the monoterpenoid alcohol, geraniol, but its use in the field was unpromising because of high instability to aerial oxidation. Thirty related compounds, mostly monoterpenoids, were then tested, but only the bicyclic monoterpenoid ketone, (+)-fenchone (Figure 1), showed activity. The (-) enantiomer (Figure 1) was inactive, indicating that the effect was behavioural rather than biocidal. (+)-Fenchone was effective in protecting seeds in box trials, but the high volatility and poor persistence of this compound frustrated attempts at field deployment. A wide range of physical and chemical approaches to producing slow release formulations was investigated, but without commercial success (Scott *et al.*, 1984; Airey *et al.*, 1989). However, because of the promising activity of these plant-derived semiochemicals against slugs, it was decided to investigate the neurophysiological response to (+)-fenchone and related compounds in detail.

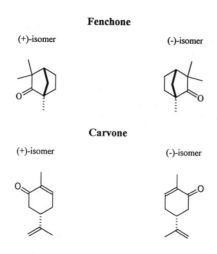

Fenchone

(+)-isomer   (-)-isomer

Carvone

(+)-isomer   (-)-isomer

Fig. 1. The Chemical Structure of Fenchone and Carvone isomers

MATERIALS AND METHODS

Olfactory nerve preparation

Electrophysiological recordings from the olfactory epithelium on the optic tentacles of *D. reticulatum* were obtained using methods described by Garraway *et al.* (1991, 1992). The signal processing was originally accomplished by converting nerve signals recorded on magnetic tape to a digital form suitable for storage and processing by a computer (Dempster, 1993). This has now been improved by the acquisition of a PC-based analogue to digital recorder board that is able to take digital samples and examine them at very high resolutions and even more recently by a Cambridge Electronic Design (CED) which has the advantage of built-in routines for event counting, spike sorting, power spectra and burst analysis.

Feeding bioassay

Individual oven-dried wheat flour pellets treated with 1.5% sucrose, plant extracts or solvent were presented to slugs which had previously been fed on Chinese cabbage (*Brassica chinensis*) for 24 h, then starved for a further 24 h. After 24 h, the mean percentage increase or decrease in feeding attributable to each treatment was calculated, compared to the control pellets (Clark *et al.*, in preparation).

Chemicals

Authentic chemicals were obtained commercially (Aldrich) and diluted in hexane for electrophysiological and behavioural assays.

Isolation of plant volatiles

Plants were harvested with minimum damage and placed, undried, in a glass vessel. Energy was applied to the vessel in the form of microwave radiation and nitrogen passed through the system, so that volatiles released were trapped in the solvent (hexane) (Craveiro *et al.*, 1989).

Gas chromatography coupled-mass spectrometry (GC-MS)

Coupled gas chromatography-mass spectrometry was performed on a Hewlett-Packard 5880A gas chromatograph with an HP-1 capillary column (50m × 0.32 mm i.d., 0.25 μm film thickness) directly coupled to a VG 70-250 mass spectrometer (electron impact, 70eV, 250°C). Sample was introduced via a cold (30°C) on-column injector. The gas chromatograph was temperature programmed with an initial 5 min at 30°C, then a rise of 5°C/min to a final isothermal period at 250°C.

RESULTS AND DISCUSSION

In the electrophysiological assay on fenchone, already shown to have behavioural activity against *D. reticulatum*, the (+) isomer gave an immediate burst of spike activity, whereas with the (-) isomer, the frequency of bursting was much lower (Figure 2). Furthermore, at low stimulus concentrations of (+)-fenchone, e.g. 0.001%, bursts of spikes could be measured, whereas none were detected for (-)-fenchone (Figure 3). However, at higher concentrations, e.g. 1-100%, the burst duration for the two isomers was similar. Fenchone is a monoterpenoid ketone and is related structurally to carvone (Figure 1) and again, (+)-carvone showed an immediate burst of spikes, whereas this was slow to develop with (-)-carvone (Figure 4). The burst duration was also shorter for the (+) isomer. The clearly demonstrated differences in neurophysiological activity with the fenchone isomers, and the similar situation with the carvone isomers, indicated that a real semiochemical effect was being observed rather than the pharmacological effect of an irritant terpenoid. It was therefore concluded that since (+)-fenchone and (+)-carvone are common components of plants in the Apiaceae (= Umbelliferae), other members of this family should be investigated in order to identify compounds more suitable for use in crop protection than (+)-fenchone itself.

Extracts of plants from 33 genera in the Apiaceae were investigated in the feeding bioassay. Percentage change in feeding varied from +20% for the standard sucrose treatment, *ca* -10% for ground elder (*Aegopodium podagraria*) and greater burnet saxifrage (*Pimpinella major*), *ca* -30% for wild celery (*Apium graveolens*) and cow parsley (*Anthriscus sylvestris*) and *ca* -70% for rock samphire (*Crithmum maritimum*) and hemlock (*Conium maculatum*). These plant extracts were then investigated using the electrophysiological assay. Some differences in the order of activity were observed, with *D. carota* eliciting the weakest response and *A. graveolens* giving an activity of 50 (based on the total number of spikes recorded), whereas hemlock, again the most active, gave over 150 spikes, closely followed by coriander (*Coriandrum sativum*) and parsley (*Petroselinum crispum*), which all showed relatively high activity in the behavioural studies. Conversely, *C. maritimum* gave a very low response, similar to that from *D. carota*.

The extracts of *C. sativum* and *C. maculatum* were investigated by gas chromatography coupled-mass spectrometry (Pickett, 1990). For *C. sativum*, the major components were tentatively identified as decenal, 2-decenal, 2-undecenal and 2-dodecenal. These compounds were screened in the neurophysiological and behavioural assays and preliminary findings suggest that they all possess some antifeedant properties. Confirmation of identity was by peak enhancement in GC, using the same conditions as for GC-MS. For *C. maculatum*, one compound proved to be particularly active. The chemistry associated with the identification of this material will be described elsewhere. Other compounds identified from *C. maculatum* were quantified and investigated in the electrophysiological and feeding bioassays, at concentrations comparable with those in the original plant extract. These included monoterpenes and a sesquiterpene, β-caryophyllene, although some, e.g. limonene, had only weak electrophysiological activity. In the feeding assay, activity ranged from -20% for the weakest response to -70% for the total plant extract, with the identified compound also approaching this level. Thus, it was concluded that most of the activity for the *C. maculatum* extract had been accommodated by the GC-MS identification work and subsequent confirmation by GC peak enhancement with authentic compounds. The compounds ensuing from the work on *C. maculatum* will be investigated in field simulation trials and, for the most effective, in the field. Previous investigations have studied the chemical composition of various species of Apiaceae, for example *C. sativum* (Pino *et al.*, 1993; Potter *et al.*, 1993). However, we believe that the aldehydes 2-decenal, 2-undecenal and 2-dodecenal are referenced as components here for the first time.

A novel source of biologically active material discovered during the course of this work arises from predatory beetles, particularly those in the family Carabidae, such as *Pterostichus melanarius*. This known slug predator provided an extract which gave a strong

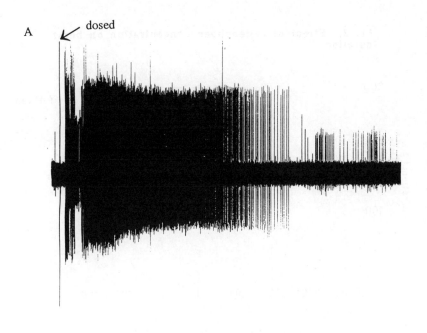

25 secs

Fig 2.　ETG responses to the isomers (A) (+)fenchone and (B) (-)fenchone

Fig 3. Effect of (+)fenchone concentration on burst duration

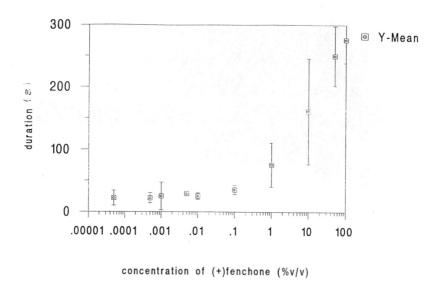

concentration of (+)fenchone (%v/v)

Effect of (-)fenchone concentration on burst duration

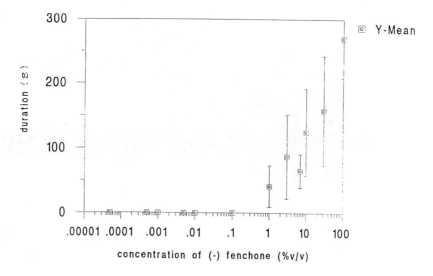

concentration of (-) fenchone (%v/v)

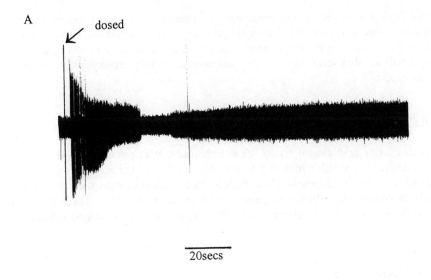

A

dosed

20secs

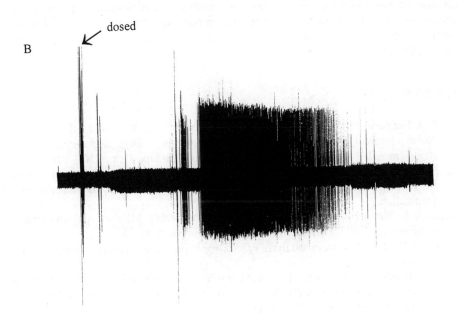

B

dosed

Fig 4. ETG responses to the isomers (A) (+)carvone and (B) (-)carvone

electrophysiological response in the tentacular preparation, whereas a similarly produced extract of the phytophagous carabid, *Zabrus tenebrioides*, gave no detectable response (Dodds *et al.*, submitted). Active components will be identified subsequently and investigations are being extended to slug predators in the coleopteran family Staphylinidae, including *Staphylinus olens*.

CONCLUSIONS

It is seen that, with new electrophysiological techniques, it is possible to identify novel biologically active compounds which may be useful as antifeedants or repellants against slugs for future agricultural development. Nonetheless, further identification work is necessary, particularly in connection with the interaction with predatory beetles. The highly active compounds from *C. maculatum* offer particular promise and will be developed and described in detail.

ACKNOWLEDGEMENTS

IACR receives grant-aided support from the Biotechnology and Biological Sciences Research Council of the United Kingdom. This work was in part supported by the United Kingdom Ministry of Agriculture, Fisheries and Food, the Perry Foundation and the Home Grown Cereals Authority.

REFERENCES

Airey, W J; Henderson I F; Pickett J A; Scott G C; Stephenson J W; Woodcock C M (1989) Novel chemical approaches to mollusc control. In: *Slugs and Snails in World Agriculture*, I F Henderson (ed.). BCPC Monograph No. 41, pp. 301–307.

Clark, S J; Dodds C J; Henderson I F; Martin A P. A bioassay for the screening of materials influencing feeding behaviour in the grey field slug *Deroceras reticulatum*, in preparation.

Craveiro, A A; Matos F J A; Alencar J W; Plumel M M (1989) Microwave oven extraction of an essential oil. *Flavour and Fragrance Journal* **4**, 43-44.

Dempster, J (1993) *Computer analysis of electrophysiological signals*, Academic Press, London.

Dodds, C J; Henderson I F; Watson P. Induction of activity in the olfactory nerve of the slug *Deroceras reticulatum* (Müller) in response to volatiles emitted by carabid beetles. *Journal of Molluscan Studies*, submitted.

Garraway, R; Leake L D; Ford M G; Pickett J A (1991) The action of a range of volatile compounds on a tentacular preparation of the field slug, *Deroceras reticulatum* (Müll.). *Pesticide Science* **33**, 240-242.

Garraway, R; Leake l D; Ford M G; Pickett J A (1992) The development of a chemoreceptive neurophysiological assay for the field slug, *Deroceras reticulatum* (Müll.). *Pesticide Science* **34**, 97-98.

Pickett, J A (1990) Gas chromatography-mass spectrometry in insect pheromone identification: three extreme case histories. In: *Chromatography and Isolation of Insect Hormones and Pheromones*, A R McCaffery & I D Wilson (eds), pp. 299-309.

Plenum Press.

Pickett, J A; Stephenson J W (1980) Plant volatiles and components influencing behaviour of the field slug, *Deroceras reticulatum* (Müll). *Journal of Chemical Ecology* **6**, 435–444.

Pino, J; Borges P; Roncal E (1993) Compositional differences of coriander fruit oils from various origins. *Die Nahrung* **37**, 119-122.

Potter, T L; Fagerson I S; Craker L E (1993) Composition of Vietnamese coriander leaf oil. *Acta Horticulturae* **344**, 305-311.

Purchon, R D (1977) *The Biology of the Mollusca*. 2nd Edition. 560 pp. Pergamon Press.

Scott, G C; Griffiths D C; Stephenson J W (1977) A laboratory method for testing seed treatments for the control of slugs in cereals. *Proceedings of the British Crop Protection Conference: Pests and Diseases*, 129-134.

Scott, G C; Pickett J A; Smith M C; Woodcock C M; Harris P G W; Hammon R P; Koetecha H D (1984) Seed treatments for controlling slugs in winter wheat. *Proceedings of the British Crop Protection Conference: Pests and Diseases*, 133–138.

Stephenson, J W (1979) The functioning of the sense organs associated with feeding behaviour in *Deroceras reticulatum* (Müll). *Journal of Molluscan Studies* **45**, 167-171.

CAN DIFFERENT pH ENVIRONMENTS IN SLUG DIGESTIVE TRACTS BE EXPLOITED TO IMPROVE THE EFFICACY OF MOLLUSCICIDE BAITS?

C R KELLY

Huntingdon Life Sciences Ltd, PO Box 2, Huntingdon, Cambs, PE18 6ES

S GREENWOOD

The Department of Child Health, St. Mary's Hospital, Hathersage Rd, Manchester, M13 0JH

S E R BAILEY

School of Biological Sciences, Manchester University, Oxford Rd, Manchester, M13 9PT

ABSTRACT

The pH of the intact digestive tract of a major pest species of slug was investigated using a technique developed in medical research to measure intracellular spaces. Single recording microelectrodes filled with a liquid ion exchange resin sensitive to protons were inserted into a number of different regions of the gut of freshly dissected *Deroceras reticulatum* (Müller). Results indicated a clear pH gradient in the presence of digesting food, with pH values ranging from 5.7 in the crop to 7.9 in the late intestine. In the light of these findings we discuss the potential for reformulating molluscicides to exploit this feature of molluscan physiology. If the environment of the digestive tract in a larger population of *D. reticulatum* and other pest slug species are shown to exhibit similar characteristics, it may be possible to improve molluscicide bait formulations by use of slow release formulations sensitive to pH.

INTRODUCTION

Terrestrial slugs are soil-borne generalists, found across large parts of the world. In the UK and northern Europe, an abundant source of fresh food is provided as emerging autumn sown arable crops, just as conditions at the soil surface become clement for surface activity and procreation. Hence slugs are widely regarded as one of the most serious crop pests in arable agriculture. This aspect of terrestrial gastropod biology creates a complex pest control problem; how to effectively deliver sufficient amounts of toxin to enough small targets in an unevenly distributed population, occupying a large bulk of soil, in order to exert some measure of pest control? Presently on an agricultural scale, use of molluscicidal baits is favoured.

Two active ingredients (methiocarb and metaldehyde) have dominated the market for nearly 3 decades in various bait forms. Advances in molluscicide technology have been relatively small compared to the number of new and selective chemicals available for other pests and diseases. Briggs and Henderson (1987) suggested this was due in part to a lack of candidate molluscicides that were non-repellent yet toxic enough to ensure a lethal dose. No new commercially available molluscicidal agent has been shown to improve upon either of the two main molluscicidal compounds and make a significant increase to slug control in the field.

The inherent problem with slug baits is that the pest must be first attracted, initiate a meal and consume a lethal dose (Bailey and Wedgwood, 1991). Even when actively crawling, slugs will only respond to food 50 to 60 % of the time (Hunter and Symonds, 1970). Therefore it is essential that once a meal is initiated it is effective. Numerous authors have called for improvements in the performance of the existing compounds by reformulation (Wright and Williams, 1980, Henderson and Parker, 1986). As yet no major advances have been announced.

One of the commonest properties exploited in drug delivery in humans and other vertebrates is the pH difference in the various parts of the digestive tract. This study attempts to demonstrate that the pH of the gut of the important pest species the Field Slug (*Deroceras reticulatum*, Müller) varies enough to be exploited similarly in molluscicide formulations. Using a technique developed to measure intracellular pH of mammalian epithelial cells, we attempted to measure the pH of various regions of the digestive system with minimal disturbance to these tissues in the field slug. Walker (1969) investigated the pH of the guts of two species, *D. reticulatum* and *Arion ater*, using a Micro-spear combination glass/reference electrode and showed small differences according to location and conditions. Bourne, Jones and Bowen, (1991) pointed out that since then there has been little work published on the physiology of molluscan gut, with attention centred on the digestive gland and its function.

METHODS AND MATERIALS

Test organisms

Mature field slugs (>0.7g) collected from Jodrell Bank Experimental Grounds (Cheshire, UK) were kept in plastic holding boxes (17 x 11.5 x 6 cm) on moist peat in groups of ten. Holding boxes were placed in conditions of 10 hours of subdued light at $12 \pm 1°C$ and 14 hours darkness at $7 \pm 1°C$ in incubators. Fresh lettuce and carrots were provided *ad libitum* and any dead or sick animals were removed.

Preparation and dissection of slugs

An individual, selected randomly from the holding box, was allowed to crawl along a large plastic spatula to gain cohesion of the 'foot'. The slug was then slowly lowered into an insulated flask containing liquid nitrogen for a few minutes to ensure death. Care was required so that the epidermis did not crack and damage the internal organs. Once frozen solid the slug was transferred to a dissecting dish and bathed in slug Ringer solution at ambient temperature (g per litre: NaCl 7.6; KCl 0.335; $CaCl_2$ 1.0; Jullien *et al.*, 1955a, cited by Lockwood, 1961). Once thawed, the animal was pinned at the head and tail before dissection under a binocular microscope (x 40). The exposed gut was cleared of connecting tissue allowing access for the microelectrode.

Manufacture of microelectrodes.

Single barrelled pH sensitive microelectrodes were manufactured from borosilicate "Kwikfil" glass capillaries (Clark Electromedical), using a vertical electrode puller (Narashige). Final tapers were approximately 1cm in length and tip diameters approximately 0.1μm. The glass was silanised by exposure to hexamethyldisilazane (Fluka) vapour by inverting the electrodes over a beaker containing the solution for 2-6 hours, thus rendering the tip hydrophobic. A small drop of proton selective ion exchange resin (Fluka proton cocktail A, which gives a linear response over the pH range 4-9; Schulthess, Shijo, Pham, Ammann and Simon, 1981) was introduced into the electrode shank and allowed to fill the tip overnight.

The shank of the electrode was back-filled with phosphate buffer (millimoles per litre: $KH_2(PO_4)$ 40; $Na(OH)$ 23; NaCl 15) at pH 7.0. The microelectrode was clamped in a micromanipulator (Leitz) and the electrical potentials were measured using a high impedance electrometer (WP1 model FD223).

The voltage response of the pH-sensitive microelectrode placed in the gut is the sum of the Nernst potential generated by the asymmetry of the hydrogen ion activities across the sensor membrane and the electrical potential difference (pd) across the epithelial cell layer of the gut. The transepithelial pd was measured directly in separate experiments using conventional "Ling-Gerard" microelectrodes (Fein, 1977). These pd sensing electrodes were pulled from "Kwik-fil" borosilicate glass capillaries and had resistances of 40 MΩ when filled with 1.5M KCl solution. Recordings from the pH-sensitive microelectrodes could then be corrected for the contribution of the transepithelial pd measured from the corresponding area of the slug gut.

All the electrical potential differences were measured with respect to a reference "Ling-Gerard" microelectrode placed in the Ringer solution bathing the dissected slug.

Calibration of the electrodes.

Before calibration the test electrode tip was ground for 2-6 seconds to reduce the electrical resistance at the tip. The impedance was measured and found to be R $\approx$ 4 x 10^{10} Ω, equivalent to a tip less than 1 μm .

Test microelectrodes were calibrated in a solution of buffer at pH 7, against a reference electrode (1.5 M KCl) in the same buffer. The reading was zeroed on a chart recorder and stability recorded for a few minutes. Both electrodes were then placed in a buffer solution at pH 9.2 and the response monitored. This was repeated with a pH 4 buffer solution, checking that the response returned to 0 mV at pH 7 between buffers. Measurement could proceed provided that the slope was close to the theoretical value of 58 mV / decade and that the response was fast with a stable plateau ($\pm$ 2mV) and returned to the baseline, otherwise the electrode was rejected.

Gut pH measurement with liquid ion exchange microelectrodes.

The dissected slug in slug Ringer was placed in a Faraday cage, screened with brass chains on an antivibration table. A binocular microscope (Olympus CK 2) at 400 times magnification was used for observation. The equipment was set up with the reference electrode in the Ringer bathing the slug. Zero mV was fixed on the chart recorder with the recording electrode in the slug Ringer. Measurements were then taken, as soon as possible, from the exposed regions of the gut using the test electrode. The micromanipulator was used to direct the tip of the test electrode, under the microscope, through the gut wall. It was occasionally necessary to make a tiny incision with a scalpel for tougher tissues. The highly coloured nature of ingested lettuce and carrot facilitated identification of the gut tissue. Progressive dissection allowed measurements of the gut contents *in situ*. Electrical potentials were recorded for the contents of the crop, stomach and intestine, and converted to equivalent pH values. Slug Ringer was replaced as required to maintain a clean bath for the reference electrode. The baseline was checked between measurements. Readings from each slug were taken over a maximum of 2 hours. As the microelectrode tip was sub-microscopic, criteria were set to judge the integrity of a measurement:

i) the recording achieved a stable value for at least 20 seconds ($\pm$ 2.5 mV).

ii) the reading returned to 0 mV ($\pm$ 2.5 mV) when the recording electrode was removed from the tissue.

Although the tip was not visible a binocular microscope was used to observe the insertion of the electrode. The shank of the electrode and the depression of the gut wall during the initial stab were used to estimate the position of the tip

Measurements of potential difference (pd) were taken from the crop, the stomach, the early and late intestine for three slugs.

Figure 1 Diagram of the whole of the digestive tract of *D. reticulatum* (after Walker, 1969)

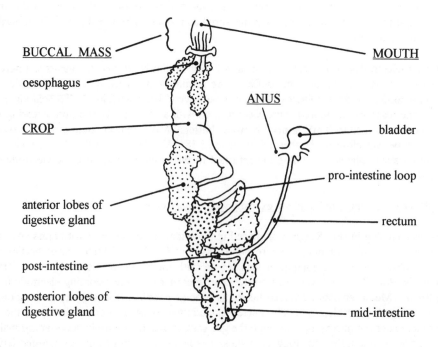

RESULTS

Calibration of the microelectrodes.

Calibration was completed before any measurements were taken. Stability of the readings and speed of the response were all acceptable. The slope of the line had a mean of 58.18 mV over the pH range 4 to 9.2 (n=6) with a standard error of 0.25 mV. This was acceptable and close to the theoretical value of 58 mV / decade and represents a variation of 0.0043 pH units between electrodes, a negligible error.

Gut pH of *Deroceras reticulatum*.

Using this technique pH measurements taken in various regions of the gut of six slugs were recorded with a reasonable degree of accuracy. Results from two other slugs were rejected due to resin displacement at the tip and an infestation of gut parasites. The three main regions of investigation were the crop, the stomach and the intestine (Figure A after Walker, 1969). These were further split into localities. Data (mV), was converted to equivalent pH units by subtracting the recorded value (mV) multiplied by the gradient of the linear response for each recording electrode from pH 7 (values > pH 7 were negative).

Table 1 presents the pH of the gut for each of eight localities from the crop, just after the oesophagus, to the late intestine. The locality late stomach was the junction of the pro-intestinal loop and the duct from the anterior digestive gland. The number of recordings at each locality is not the same for every slug due either to damage caused in dissection or to microelectrode problems whilst recording.

Table 1: pH values measured in various regions of the gut of the Field Slug (*D. reticulatum*).

Gut Region	Mean pH	Locality	pH	± SE	No	pH due to pd
Crop	5.71 (± 0.10)	Early	5.71	0.20	4	+ 0.085 (± 0.025)
		Mid	5.75	0.11	2	
		Late	5.67	0.11	2	
Stomach	6.14 (± 0.22)	Early	5.93	0.32	3	- 0.068 (1 Value)
		Late	6.46	0.06	2	
Intestine	7.39 (± 0.18)	Early	7.05	0.30	6	- 0.053 (± 0.011)
		Mid	7.23	0.34	4	
		Late	7.93	0.16	5	

Transepithelial potential difference (pd)

The transepithelial pd measured directly was ~ +5, ~ -4, and ~ -3mV in the crop, stomach and intestine respectively. The response of the pH sensing electrode which can be attributed to the pd is expressed in pH units in Table 1. The transepithelial pd contributed an equivalent of less than 0.1 pH units and the difference in the pH between regions of the gut are therefore genuine. Pd readings were stable (less than ± 2.5 mV) for much longer than the 20 seconds criteria adopted and quickly returned to zero when the tip was removed. This suggests leakage around the puncture hole was very slight therefore electrical potential was stable over the recording period. Even when the tip of the microelectrode was inserted through a small incision into the gut contents the recording stayed stable and returned to zero after removal.

DISCUSSION

Gut pH of *Deroceras reticulatum.*

The results of this study demonstrate clear and definite differences in the pH environments, in the presence of digesting food, in various localities of the gut of *D. reticulatum*. The procedure was shown to be accurate and precise, with negligible interference from transepithelial pd, at the tip of the recording electrode. The data indicate that the gut pH shifts from initially weakly acidic to mildly basic in the intestine. Extremes differed by 2.26 pH units, a biologically important difference. Walker (1969) found that extracted juice from the crop and intestine of *D. reticulatum* had pH values of 6.34 ± 0.05 and 7.1 ± 0.1 respectively. He repeated these measurements, with a microspear electrode at various regions of the gut, and found that pH varied according to conditions. A value of pH 5.8 or 7.7 was found in the crop with and without food present. In digestive glands under similar conditions the values were pH 6.3 and 7.0. He was unable to measure the pH in the intestine of the field slug but found the crop and mid intestine of recently fed *Arion ater* to be 5.3 and 7.0 respectively. This, he said contradicted previous work that had shown the pH to be 6.0 ± 0.1 throughout the digestive tract of *A. ater*. The results of the present study agree with Walker's (1969) findings and give more detailed data for various other locations from the stomach and along the intestine. The present study and Walker's (1969) *in situ* measurements for the crop lumen are notably similar.

Significance of findings.

These results aid our understanding of molluscan gut function. Triebskorn and Schweizer (1990), stated that after involuntary dosing of *D. reticulatum* with 63 mg of the molluscicide metaldehyde in alginate gel, the content of the crop lost its initial neutrality to become acidic after 10 hours at 5°C or 1.5 hours at 10 and 15 °C. The results from this present study and those of Walker (1969) show that this is not an unusual reaction in the crop to the presence of food (alginate). Walker (1969) proposed that the majority of crop juices were secreted by the digestive gland. Bourne et al. (1991) state that they found no excretory cells or mucocytes in the crop wall of *D. reticulatum*. The pH response in the crop lumen observed by Triebskorn and Schweizer (1990) was independent of any local effects of metaldehyde on the crop wall because the secretions are introduced from lower in the digestive canal. The opening of the duct from the digestive gland into the gut lumen is between the stomach and the pro-intestinal loop (See figure 1).

Toxic effects on the molluscan gut, due to ingested molluscicides, affect feeding mechanisms and are proposed as the reason for non-ingestion of a lethal dose (Wright and Williams, 1980; Wedgwood and Bailey, 1988; Bailey et al., 1989). Yet formulation chemists can control release of active ingredients by using pH sensitive polymers that dissolve in a selected pH environment once in the gut. If the results of this study were confirmed in a wider population and perhaps in other pest species of slug, this feature of molluscan physiology could be used to mask and deliver molluscicides however repellent. Theoretically this would overcome sublethal effects of the poison and ensure ingestion of a targeted lethal dose. It would be interesting to repeat the present study of gut pH in the presence of either methiocarb or metaldehyde molluscicidal pellets.

The problem however needs careful consideration. A number of extra factors need to be considered with molluscicides. The baits must be attractive and soft enough for the slugs to begin to feed on the pellet. However, the integrity of the coating on the active ingredient must not deteriorate whilst in the field, perhaps for several weeks; so the coating must be waterproof. Once a meal is initiated the coating must be able to withstand the physical action of the radula. In medicine, premature breakdown of enteric coatings in the mouth, due to the action of saliva, is not usually a problem because most tablets or capsules are swallowed immediately and the coating remains intact to prevent release in the gastric environment (Dreher, 1975). *D. reticulatum* has a large buccal cavity which will fill with food from a number of rasps before the food is passed into the crop (Henderson, 1969). The coating must stay intact through this process or paralysis of the mouthparts and rapid disruption of the neural mechanism involved in controlling the process of clearing the buccal cavity of food would certainly reduce feeding efficiency (Wedgwood and Bailey, 1988).

A material that could be applied as a coating that would dissolve in the lower pH of the stomach exists. However to really ensure sufficient ingestion of toxic pellet, an enteric coating that would survive the crop and release toxin once in the gut, would be ideal. This would theoretically mean that a slug may only be paralysed several hours after completing a meal.

Unfortunately this kind of complex formulation, designed with perhaps two or three reactive coatings, would be costly to manufacture. In the pharmaceutical industry this cost may be easier to justify than in the plant protection industry. However this may be considerably less costly than trying to develop a new molecule. Even if actually possible to produce with improved efficacy, would such a formulation be likely to be registered and released? Ecotoxicological considerations, like implications to non-target invertebrates, environmental fate and persistence would further increase the costs of releasing such a formulation. The development of a new formulation is therefore not a simple option and it is not surprising that in today's highly regulated industry, the ability to release improved, targeted formulations may be hampered by factors other than available technology.

ACKNOWLEDGMENTS

This work was carried out as part of a PhD research project at the University of Manchester, funded by an SERC Total Technology Case Award and was sponsored by Bayer plc, Crop Protection Business Group.

REFERENCES

Bailey, S E R; Wedgwood, M A (1991) Complementary video and acoustic recordings of foraging by two pest species of slug on non-toxic and molluscicidal baits. *Annals of Applied Biology.* **119**, 163-176.

Bailey, S E R; Cordon, S; Hutchinson, S (1989) Why don't slugs eat more bait? A behavioural study of early meal termination produced by methiocarb and metaldehyde baits in *Deroceras caruanae*. In *Slugs and snails in world agriculture*. I F Henderson (ed.). B.C.P.C. Mono No. 41, pp. 385-390.

Bourne, N B; Jones, G W; Bowen, I D (1991) Endocytosis in the crop of the slug, *Deroceras reticulatum* (**Müller**) and the effects of the ingested molluscicides, metaldehyde and methiocarb. *Journal of Molluscan Studies.* **57**, 71-80.

Briggs, G G; Henderson, I F (1987) Some factors affecting the toxicity of poisons to the slug *Deroceras reticulatum*. *Crop protection.* **6**, 341-346.

Dreher, D (1975) Film coatings on acrylic resin basis for dosage forms with controlled drug release. *Pharma International.* **1/2**, pp 6.

Fein, H (1977) *An introduction to Microelectrode Technique and Instrumentation.* 1st. Ed. W-P Instruments Inc., U S A, 24 p.

Henderson, I F (1969) A laboratory method for assessing the toxicity of stomach poisons to slugs. *Annals of Applied Biology.* **63**, 167-171.

Henderson, I F and Parker, K A (1986) Problems in developing chemical control of slugs. *Aspects of Applied Biology.* **13**, 341-347.

Hunter, P J; Symonds, B V (1970) The distribution of bait pellets for slug control. *Annals of Applied Biology.* **65**: 1-7.

Lockwood, A P M (1961) Ringer solutions and some notes on the physiological basis of their ionic composition. *Comparative Biochemistry and Physiology* **2**, 241-289.

Schulthess, P; Shijo, Y; Pham, HV; Pretsch, E; Amman, D; Simon, W (1981) A hydrogen ion-selective liquid-membrane electrode based on Tri-n-Dodecylamine as a neutral carrier. *Anal. Chim. Acta.* **131**, 111-115.

Triebskorn, R; Schweizer, H (1990) The impact of the molluscicide metaldehyde on the mucus cells in the digestive tract of the grey garden slug *Deroceras reticulatum*. In: *A N P P - 2nd International Conference on Pests in Agriculture*. Versailles, 1990. **1(3)**, pp. 183-190.

Walker, G (1969) Studies on the digestion of the slug *Agriolimax reticulatus*. *Ph.D. Thesis, University College of North Wales, Bangor*.

Wedgwood, M A; Bailey, S E R (1988) The inhibitory effects of the molluscicide metaldehyde on feeding, locomotion and faecal elimination of three pest species of slug. *Annals of applied Biology.* **112**, 439-537.

Wright, A A; Williams R (1980) The effect of molluscicides on the consumption of bait by slugs. *Journal of Molluscan Studies* **46**, 265 -281.

INTERNAL DEFENCE SYSTEM IN THE LAND SLUG *INCILARIA* SPP.

E FURUTA

Department of Anatomy, Dokkyo University School of Medicine, Mibu, Tochigi, 321-02, Japan

K YAMAGUCHI

The Laboratory of Medical Sciences, Dokkyo University School of Medicine, Mibu, Tochigi, 321-02, Japan

ABSTRACT

Three morphologically distinct haemolymph cell-types (blood cell-types), Type I, II and III, were found in haemolymph from the terrestrial slugs, *Incilaria fruhstorferi* and *I. bilineata*. In addition, vast numbers of platelet-like structures were present in collected haemolymph. Type I cells were macrophage-like, Type II cells were lymphocyte-like and Type III cells were fibroblast-like, morphologically. Among these haemolymph cells, only Type I cells were able to recognize and phagocytose foreign materials, in *in vitro* and *in vivo* experiments. Platelet-like structures with long fine fibers attached to foreign materials such as latex beads (0.8 μm in diameter). The latex beads were often seen deep in aggregates of the platelet-like structures. The enhancement factor(s) of phagocytosis by Type I cells existed in the haemolymph and body surface mucus. The body surface mucus was easily soluble in water (WSF). WSF agglutinated human type A and B blood cells and this activity was specifically inhibited by a low concentration of N-acetyl D-galactosamine (GalNAc). The WSF contained three 15kD C-type lectins, named Incilarin A, B and C, which consist of 150, 149 and 156 amino acid residues containing signal peptides of 17 amino acid residues at their N-termini. Studies on haemopoiesis by us revealed that in *Incilaria fruhstorferi* injected with yeast particles, there is no special haemopoietic organ: blood cell proliferation occurs throughout the body, throughout the connective tissue and the vascular system.

INTRODUCTION

All metazoan animals need an internal defence system to protect themselves against foreign organisms that succeed in getting past the external defence. It is well known that the internal defence mechanism in vertebrates consists of non-specific humoral and cellular factors, and specific cellular and humoral ones. Vertebrates are able to respond against invading microorganisms first by non-specific elements, such as lysozyme, complements, interferon, lysin, transferrin, and then by typical immune reactions of a clonal nature. Moreover, macrophages are involved in part or all of these sequential defence processes both as phagocytes and antigen-presenting cells in some way, and T- and B-lymphocytes are engaged in only the later stage of specific defence reactions.

However, invertebrates including the molluscs do not possess classical immune recognition molecules of vertebrates such as immunoglobulins, T-cells, MHC or antigen receptors. But they still manage to normally keep their internal body fluids sterile. They possess the capacity to distinguish not only between self and non-self, but also between non-self materials which differ in chemical properties (Bayne, 1990, Cooper *et al.*, 1992). Phagocytosis is considered to be the primary clearance mechanism in molluscs.

HAEMOLYMPH CELL PROFILES IN *INCILARIA* SPP

Type I cells (macrophage-like cells)

In *Incilaria fruhstorferi* which is a native, the largest slug in Japan, and *Incilaria bilineata*, Type I cells are the most common cells in the haemolymph. When first collected they are round with short pseudopodia (Fig. 1a). However, with time, they become more extensive, spreading in all directions over the surface of substratum by extending their pseudopodia (lamellipodia) with supporting ribs (Fig. 1b, c) and show random locomotion. Thus, Type I cells are considered to be the only cells which are capable of movement, just like an amoeba or mammalian macrophages. These cells measure approximately 20-30 μm in diameter with a kidney-shaped or lobulated nucleus measuring about 5 μm. The cytoplasm contains numerous mitochondria, rough endoplasmic reticulum, multivesicular bodies, residual bodies, Golgi apparatus and glycogen deposits (Fig. 2a, b).

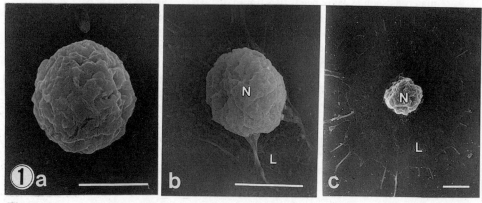

Fig. 1. SEM micrographs of Type I cells (macrophages) in contact with plastic dishes.
a, A cell immediately after harvest from the mantle cavity. b, A cell with lamellipodia with supporting ribs when incubated with SH7 culture medium (Furuta & Shimozawa, 1983) for 30 min. c, A cell flattened and spread over the surface with extending lamellipodia. L, lamellipodia; N, nucleus; Scales 5 μm.

Type II cells (lymphocyte-like cells)

Type II cells are smaller than Type I cells. They are oval, about 5 μm in diameter, with a central round nucleus and the nucleus to cytoplasm ratio is higher. Like Type I cells, Type II cells also adhere to the substratum but they remain spherical and do not form or only slightly form pseudopodia. They contain scattered free ribosomes and mitochondria surrounding the nucleus (Fig. 3).

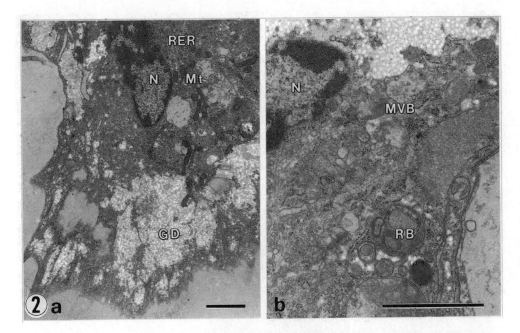

Fig. 2. TEM micrographs of spreading Type I cells incubated with the medium for 30 min. This cell possesses a kidney-shaped nucleus (N), small amounts of rough endoplasmic reticulum (RER), few mitochondria (Mt), well-developed Golgi apparatus, multivesicular bodies (MVB), residual bodies (RB), and many glycogen-like deposits (GD). **a**, low magnification; **b**, high magnification. Scales 2 μm.

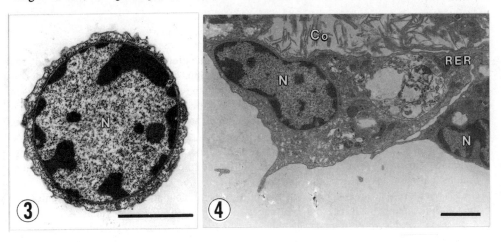

Fig. 3. TEM micrograph of Type II cell (lymphocyte-like).

This cell contains a relatively large round nucleus (N), free ribosomes and scanty cytoplasm. Scale 2 μm.

Fig. 4. TEM micrograph of Type III cell (fibroblast-like).

This cell possesses an oval nucleus (N), small amounts of rough endoplasmic reticulum (RER) and a few mitochondria, and collagen-like fibers (Co) are located outside of the cell. Scale 2 μm.

Type III cells (fibroblast-like cells)

Type III cells, measuring approximately 75×15 μm, are spindle-shaped cells. They contain microfibrils (12-15 nm in diameter), residual bodies, and collagen-like fibers present outside of the cytoplasmic membrane (Fig. 4).

Besides these cell types, numerous platelet-like structures (PLS) are observed that possess long fine fibers (Fig. 7). Foreign materials, such as sheep red blood cells and latex beads, are often seen deep in aggregates of them (Fig. 6). PLS seem to be derived from Type I cells and play a role in the clumping of foreign materials (Furuta *et al.*, 1987, 1990, 1991, Yamaguchi *et al.*, 1988).

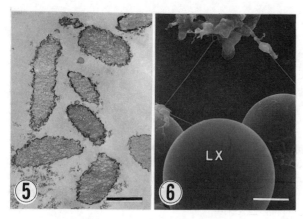

Fig. 5. TEM micrograph of platelet-like structures (PLS).
PLS are numerous in feshly collected haemolymph. Scale 2 μm.

Fig. 6. SEM micrograph of PLS.
PLS attach to latex beads (LX, φ 15.8 μm) with long fine fibers. Scale 5 μm.

ASPECTS OF CELLULAR DEFENCE REACTIONS

Invertebrates, like vertebrates, are protected against invading microorganisms by an internal defence system. In terrestrial gastropods, several functions have been attributed to blood cells which occur in the haemolymph (Bayne, 1973, Furuta *et al.*, 1987, 1989, 1990, Yamaguchi *et al.*, 1988, Adema *et al.*, 1992). In order to obtain much information on the early phases of phagocytosis, various types of foreign materials, biotic as well as abiotic, such as sheep blood cells, yeasts, Indian ink or latex beads of various sizes, were introduced into the haemocoel of the slugs, *Incilaria fruhstorferi* and *I. bilineata,* and Type I cells phagocytosed foreign materials. However, no cells phagocytosed different foreign material to the same extent. Smaller particles are more readily phagocytosed (Fig. 7). The phagocytic process of these cells is composed of three phases: attachment, filopodial elongation and actual internalization by veil-like membrane processes (Furuta *et al.*, 1987, 1990). Phagocytosis is a dynamic process in which foreign materials are attached to the phagocytic cell membrane in preparation for ingestion. In other words, phagocytosis is usually carried out with immunorecognition, ingestion and clearance from the body.

When the foreign particles are introduced into an animal body those that are too large to be phagocytosed by a single cell are apparently encapsulated by large numbers of host cells. A number of experimental studies on freshwater snails have been made of encapsulation of parasites but on terrestrial gastropods, few investigations have been made.

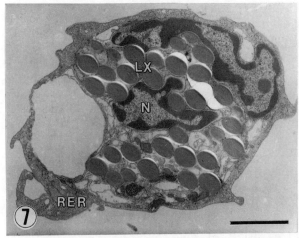

Fig. 7. TEM micrograph showing phagocytosis by Type I cell.

The cell was fixed 20 h after injection of latex beads (LX, φ 0.79 μm). N, nucleus; LX, latex beads; RER, rough endoplasmic reticulum. Scale 2 μm.

For the study of the formation and morphology of these capsules, gastropods were injected with particles (Cheng & Rifkin, 1970, Pan, 1965, Sauerländer, 1976) or with biologically inert objects. In other experiments, heterologous tissues were implanted into the gastropods (Sminia, 1981, Yamaguchi *et al.*, 1996). The results show that the encapsulation process is basically the same in gastropods. In the first phase of encapsulation, infiltration of phagocytic cells is observed immediately after the introduction of a foreign material in the region in which it is located. And then the material is surrounded by a loose layer of phagocytic cells. Gradually, the innermost cell layers become flattened and closely packed. The outer layers become separated by intercellular spaces filled with tiny connective tissue fibrils. The intracellular substances are synthesized by the cells which are transformed into fibroblasts (Sminia, 1981). Encapsulation is an important mechanism to isolate invading large foreign materials.

THE ROLE OF BODY SURFACE MUCUS AND HAEMOLYMPH (SERUM)

In molluscs, introduced foreign materials are phagocytosed or encapsulated by phagocytic cells from the blood and connective tissue. In vertebrates, recognition of foreign material by phagocytic cells is mediated by specific serum proteins, the immunoglobulins. In invertebrates, no substance comparable to vertebrate immunoglobulin is contained in the haemolymph. However, invertebrates do possess several classes of biologically active humoral factors that functionally mimic antibodies. Substances with agglutinating, opsonizing and lytic activities have been found in a number of invertebrates (Renwrantz & Mohr, 1978, Anderson & Good, 1976, Van der Knaap *et al.*, 1983, Furuta *et al.*, 1991, 1995, Tripp, 1992a, b, Blanco *et al.*, 1995).

Using body surface mucus and haemolymph of the terrestrial slug, we studied whether these elements may have opsonic activity when tested against relatively small particles.

The body surface mucus from *Incilaria fruhstorferi* is easily soluble in water and this water soluble fraction (WSF) agglutinates human A and B red blood cells (HRBC). And specifically, WSF causes haemolysis of B type HRBC 12 h after agglutination (Fig 8). The phagocytosis level was enhanced when HRBC was treated with the WSF (Fig. 9). The mucus-

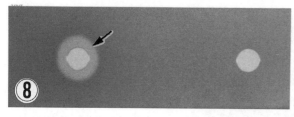

Fig. 8. Haemolysis by water soluble mucous fraction (WSF) on blood agarose plate.

WSF agglutinates B type human blood cells and causes haemolysis of the erythrocytes. Left, non-treated mucus; right, 56°C, 30 min treated mucus. Haemolytic circle (arrow) is present around the left well.

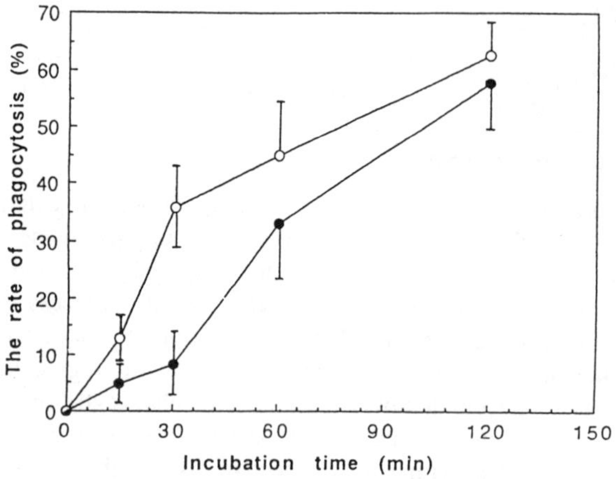

Fig. 9. Opsonizing effect of WSF-treatment on phagocytosis by Type I cells.

Sheep red blood cells (SRBC) were added to the haemolymph cell suspension and then kept for 15-120 min at 25°C. After methanol fixation and Giemsa staining, phagocytic ratios were calculated.

-○-, WSF-treated; -●-, non-treated. Vertical bars show standard deviation.

Table 1. Inhibition by various sugars of hemagglutination by WSF (diluted with 128 volumes of PBS).

Sugar	Concentration of sugar (mM)								
	100	50	25	12	6	3	1.5	0.7	0.3
NeuNAc	+	+	+	+	+	+	+	+	+
GlcNAc	−	+	+	+	+	+	+	+	+
GalNAc	−	−	−	−	−	−	−	−	−
D-galactose	+	+	+	+	+	+	+	+	+
L-fucose	+	+	+	+	+	+	+	+	+
D-glucose	+	+	+	+	+	+	+	+	+

Table 2. Inhibition by various sugars of hemagglutination by hemolymph (diluted with 5 volumes of PBS).

Sugar	Concentration of sugar (mM)								
	100	50	25	12	6	3	1.5	0.7	0.3
NeuNAc	−	−	−	−	−	−	−	+	+
GlcNAc	−	−	−	−	−	−	±	+	+
GalNAc	−	−	−	−	−	−	−	+	+
D-galactose	+	+	+	+	+	+	+	+	+
L-fucose	+	+	+	+	+	+	+	+	+
D-glucose	+	+	+	+	+	+	+	+	+

+, hemagglutination was observed; −, hemagglutination was not observed.　　Human A type RBC induced haemagglutination was specifically inhibited by a low concentration of N-acetyl D-galactosamine (GalNAc)(Table 1). On one hand, haemolymph of the land slug agglutinated A and B type HRBC, but GalNAc, N-acetyl glucosamine (GlucNAc) and N-acetyl Neuraminic acid (NeuNac) effectively inhibited the haemagglutination (Table 2)(Furuta et al., 1991). The WSF contained three 15kD C-type lectins, named Incilarin A, B and C. The cDNAs of these three C-types lectins were amplified by the PCR method and deduced amino acid sequences were determined. Incilarin A, B, and C consisted of 150, 145 and 156 amino acid residues, which contained signal peptides of 17 amino acid residues at N-termini (Yuasa et al., 1996).

ACTIVATION OF HAEMOPOIETIC REGION AGAINST INVADING FOREIGN MATERIALS

Molluscs infected with trematodes or other organisms show increases in the number of circulating blood cells that function in defence responses against parasitic infection (Cheng & Guida, 1980, Furuta et al., 1994). In fact, in Incilaria fruhstorferi injected with yeast particles, the number of circulating blood cells increased within 1 h to about 5-7 times of the original number. According to Furuta et al. (1994), studies of haemopoiesis by immunohistochemical methods under light and electron microscopes in vivo and in vitro, revealed that, in the slug there is no special haemopoietic organ; blood cell proliferation occurs throughout the body, throughout the connective tissue and vascular system. The function of this redistribution of blood cells may be interpreted in terms of enhancing the effectiveness of the cellular defence reactions.

REFERENCES

Adema, C E; Harris R A; Van Deuteron-Mulder E C (1992) A comparative study of hemocytes from six different snails: Morphology and functional aspects. Journal of Invertebrate Pathology. 59, 24-32.

Anderson, R S; Good R A (1976) Opsonic involvement in phagocytosis by mollusk hemocytes. Journal of Invertebrate Pathology. 27, 57-64.

Bayne, C J (1973) Molluscan internal defense mechanisms: The fate of C[14] labelled bacteria in the land snail Helix pomatia (L.). Journal of Comparative Physiology. 86, 17-25.

Bayne, C J (1990) Phagocytosis and non-self recognition in invertebrates. Bioscience. 40, 723-731.

Blanco, G A C; Alvarez E; Amor A; Majos S (1995) Phagocytosis of yeast by coelomocytes of the sipunculan worm Themiste petricola: Opsonization by plasma components. Journal of Invertebrate Pathology. 66, 39-45.

Cheng, T C; Rifkin E (1970) Cellular reaction in marine molluscs in response to helminth parasitism. In: A Symposium on Diseases of Fishes and Shellfishes. S F Snieszko (ed.). American Fisheries Society of Special Publication. No. 5, 443-496.

Cheng, T C; Guida V G (1980) Hemocytes of Bulinus truncatus rohlfsi (Mollusca: Gastropoda). Journal of Invertebrate Pathology. 35, 158-167.

Cooper, E L; Rinkevich B; Uhlenbruch G; Valembois P(1992) Invertebrate immunity: Another viewpoint. Scandinavian Journal of Immunology. 35, 247-266.

Furuta, E; Shimozawa A (1983) Primary culture of cells from the foot and mantle of the slug, Incilaria fruhstorferi Collinge. Zoological Magazine. 92, 290-296.

Furuta, E; Yamaguchi K; Shimozawa A (1987) Phagocytosis by hemolymph cells of the land slug, *Incilaria fruhstorferi* Collinge (Gastropoda: Pulmonata). *Anatomischer Anzeiger*. **163**, 89-99.

Furuta, E; Yamaguchi K; Shimozawa A (1989) Hemolymph coagulation in the land slug, *Incilaria fruhstorferi* Collinge. *Zoological Science*. **6**, 1204.

Furuta, E; Yamaguchi K; Shimozawa A (1990) Hemolymph cells and the platelet-like structures of the land slug, *Incilaria bilineata* (Gastropoda: Pulmonata). *Anatomischer Anzeiger*. **170**, 99-109.

Furuta, E; Yamaguchi K; Shimozawa A (1991) Current aspect of comparative immunology. *Dokkyo Journal of Medical Science*. **18**, 1-15.

Furuta, E; Yamaguchi K; Shimozawa A (1994) Blood cell-producing site in the land slug, *Incilaria fruhstorferi*. *Acta Anatomica Nipponica*. **69**, 751-764.

Furuta, E; Takagi T; Yamaguchi K; Shimozawa A (1995) *Incilaria* mucus agglutinated human erythrocytes. *The Journal of Experimental Zoology*. **271**, 340-347.

Pan, C T (1965) Studies on the host-parasite relationship between *Schistosoma mansoni* and the snail *Australorbis glabratus*. *American Journal of Tropical Medicine and Hygiene*. **14**, 931-976.

Renwrantz, L; Mohr W (1978) Opsonizing effects of serum and albumin gland extracts on the elimination of human erythrocytes from circulation of *Helix pomatia*. *Journal of Invertebrate Pathology*. **31**, 164-174.

Sauerländer, R (1976) Histologische Veränderungen bei experimentell mit *Angiostrongylus vasorum* order *Angiostrongylus cantonensis* (Nematoda) infizierten Achatschnecken (*Achatina fulica*). *Zeitschrift für Parasitenkunde*. **49**, 263-280.

Sminia, T(1981) Gastropods. In *Invertebrate Blood Cells*. N A Ratcliffe & A F Rowley (eds), Academic Press, London, Vol 1, pp 190-232.

Tripp, M R (1992a) Phagocytosis by hemocytes of the hard clam, *Mercenaria mercenaria*. *Journal of Invertebrate Pathology*. **59**, 222-227.

Tripp, M R (1992b) Agglutinins in the hemolymph of the hard clam, *Mercenaria mercenaria*. *Journal of Invertebrate Pathology*. **59**, 228-234.

Van der Knaap, W P W; Sminia T; Schutte R; Boerrigter-Barendsen L H (1983) Cytophilic receptors for foreignness and some factors which influence phagocytosis by invertebrate leukocytes: *In vitro* phagocytosis by amoebocytes of the snail *Lymnaea stagnalis*. *Immunology*. **48**, 377-383.

Yamaguchi, K; Furuta, E; Shimozawa, A (1988) Morphological and functional studies on hemolymph cells of land slug, *Incilaria bilineata, in vivo* and *in vitro*. In *Invertebrate and Fish Tissue Culture*. Y Kuroda, E Kurstak & K Maramorosch (eds) Japan Scientific Societies Press, Tokyo/Springer-Verlag, Berlin, pp247-250.

Yamaguchi, K; Furuta E (1996) Allo- or xenograft recognition in the land slug, *Incilaria fruhstorferi*. *Developmental & Comparative Immunology*. **20**, I-XIII.

Yuasa, H J; Takagi T; Furuta E (1996) Primary structure of three-C-type lectins from mucus of the land slug *Incilaria fruhstorferi*. *Developmental & Comparative Immunology*. **20**, I-XIII.

MONOCLONAL ANTIBODIES AGAINST *DEROCERAS RETICULATUM* AND *ARION ATER* EGGS FOR USE IN PREDATION STUDIES

V W MENDIS, I D BOWEN, J E LIDDELL AND W O C SYMONDSON

School of Pure & Applied Biology and School of Molecular & Medical Biosciences, University of Wales Cardiff, PO Box 915, Cardiff CF1 3TL

ABSTRACT

Monoclonal antibodies were developed against the eggs of *Deroceras reticulatum* and *Arion ater*. Positive antigen-antibody reactions for the *D. reticulatum* monoclonal antibody were associated with all the stages (0-15 day) of eggs, neonates and also a whole body extract of *D. reticulatum*, but not with mucus nor with any other mollusc eggs tested except *D. caruanae*. The *A. ater* monoclonal antibody produced positive antigen-antibody reactions with all the stages (0-20 days) of *A. ater* eggs but not with the neonates, whole body (pre-gravid stage) or mucus nor with the eggs of other non-*Arion* mollusc species. The antigen-antibody reactions were determined by qualitative indirect enzyme linked immunosorbent assays (ELISA). The use of these monoclonal antibodies as diagnostic probes for gut content analyses of potential natural predators of mollusc eggs is discussed.

INTRODUCTION

Severe economic damage by slugs to agricultural and horticultural crops in the British Isles is mainly caused by *Deroceras, Arion* and *Tandonia* species (Godan, 1983; Port & Port, 1986; Glen, 1989). Although there are many methods of controlling slugs, chemically (Bowen & Jones, 1985; Port & Port, 1986; Glen & Orsman, 1986; Martin, 1993) and biologically (Symondson & Liddell, 1993a; Wilson *et al.*, 1994; Symondson *et al.*, 1996), very little consideration has been given to the eggs of the above mentioned species. Ryder & Bowen (1977) pointed out that copper can be taken up by the eggs and initially retained in the perivitelline membrane, but the effect of copper on slug egg hatchability was not investigated. There are no other reported chemical control methods or predatory arthropods to control slug eggs except phorid fly larvae, *Megaselia aequalis* (Robinson & Foote, 1968). Therefore, identification of appropriate arthropod predators of slug eggs will be an obvious advantage for integrated slug control programmes in low-input agriculture and also for organic farming (Bowen *et al.*, 1993).

A major barrier to obtaining such information is that many arthropods ingest only liquids from their prey, which are unrecognisable by direct gut examination (Cohen, 1990). Additionally, the small size and secretive nature of many arthropod predators make direct field observations on feeding behavior extremely difficult. One way of overcoming the problem of confirming predator diets was the development of immunoassays with polyclonal antisera for possible prey species (Tod, 1973; Allen & Hagley, 1982; Dennison & Hodkinson, 1983; Symondson & Liddell, 1993b). However, due to their relatively poor specificity, polyclonal antisera can often give positive results against many other closely related prey species consumed by the same predator (Dempster, 1960; Greenstone, 1979). It is essential, therefore, to use more specific and sensitive methods of identifying prey species in predator gut samples. Symondson & Liddell (1993a) have developed a genus-specific monoclonal antibody for the detection of predation upon arionid slugs. Recently, monoclonal antibodies (MAbs) have been used to determine species (Symondson & Liddell, 1996), stage (Hagler et al., 1991) and even instars (Greenstone & Morgan, 1989) of prey.

With the development of hybridoma technology (Köhler & Milstein, 1975), antibody producing cells can be obtained by fusing spleen cells from an immunized mouse with myeloma cells (Liddell & Cryer, 1991). The resulting hybridoma lines are cultured and cloned until a monoclonal cell line is produced generating antibodies specific for a target antigen site. The specificity of monoclonal antibodies (MAbs), raised against a particular epitope of prey protein, largely avoids other problems associated with polyclonal antisera. These MAbs can be produced in unlimited quantities over time as a uniform reagent. These characteristics are not achievable with conventional polyclonal antisera. Producing monoclonal antibodies against D. reticulatum and A. ater eggs would be very useful for the identification of the predators of those two species.

To evaluate these specific and highly sensitive MAbs against target antigens, it is essential to follow a sensitive and reliable assay system. Greenstone & Morgan, (1989); Hagler et al., (1992); Symondson et al. (1995); and Symondson & Liddell, (1996) used a combination of MAbs and enzyme-linked immunosorbent assays (ELISA) which fulfill those conditions. ELISA offers the advantages of quantitative sensitivity, rapidity and simplicity at relatively low cost.

This paper reports how anti-egg MAbs were produced and tested for their specificity. These MAbs were tested against six stages of eggs, neonate and adult slugs for both D. reticulatum and A. ater to determine their stage specificity. Additionally, fourteen other species of mollusc eggs were tested, to establish the genus and species specificity of these MAbs using ELISA tests.

MATERIALS AND METHODS

Antigen preparation

Ten batches of 0-1 day old Deroceras reticulatum and A. ater eggs, from a laboratory culture, were separately pooled for homogenisation. The eggs were washed three times with distilled water (to remove any mucus and sand particles attached to them) and weighed. The eggs were prepared using a glass homogeniser. Homogenate was diluted 1:20 (w/v), with phosphate buffered saline (PBS), pH 7.4, and centrifuged at 8000 g. for 15 min. 1 ml aliquots of the resulting supernatants were stored at -20°C. Other samples such as eggs of different genera, species and stages used for antibody specificity tests, were obtained from laboratory cultures maintained at the University of Wales Cardiff. 1:20 stock solutions were prepared following the same procedure.

Antibody production

Five 10-12 week-old female Balb/C mice were immunized by subcutaneous injection of 200 µl of a 1:1 emulsion of Freund's complete adjuvant and 100 µg of egg protein in PBS. Three weeks after the first dose, the mice received a secondary immunisation of egg protein, using Freund's incomplete adjuvant (1:1). A final booster injection, containing 200 µg of egg protein in PBS without adjuvant, was given intravenously 2-3 weeks after the second injection and three days prior to the fusion.

Splenocytes (from the immunised mouse) were fused with myeloma cells in log growth phase using polyethylene glycol (MW 4000). The methods employed for fusion, hybrid selection, cloning and ascites production followed the protocols described in detail in Liddell & Cryer (1991).

Once identified, a clone which is producing antibodies against the antigen of interest can be grown from a single cell to produce monoclonal antibodies. These monoclonal cell lines were injected into pristane-primed Balb/C mice for production of ascitic fluid, which contains a high level (approx. 5 mg/ml) of specific antibody.

Hybridoma supernatant screening

Supernatant screening was performed at room temperature using the indirect ELISA procedure (Symondson et al., 1995). Egg antigen at a concentration of 50 µg/ml in PBS was incubated overnight in 96 well assay plates (Falcon Pro-Bind) at a volume of 100 µl/well. Wells were washed three times with PBS-Tween 20 (0.05%) (PBS/T) to remove the unbound antigen. 100 µl of the supernatants from each hybridoma line were added and incubated for two hours. Each plate included a PBS blank, a positive control (polyclonal antiserum taken from the same mouse as the original spleen cells) and a negative control (normal mouse serum). The unbound antibody was removed by washing three times with PBS/T. 100 µl of enzyme-labeled anti-mouse Ig conjugate to horseradish peroxidase (HRP) at the predetermined dilution (1:10,000) was added and incubated for one hour. Unbound conjugate was again removed by three washes with PBS/T. 100 µl of enzyme substrate OPD (o-phenylenediamine dihydrochloride) per well were added for 20-30 min until the colour developed sufficiently. 50 µl of 2.5 M sulphuric acid per well were used to stop the reaction. Optical densities were measured at 492 nm in a spectrophotometer designed to read microtitre plates.

Antibody specificity

Egg and slug stock solutions were further diluted 1:1000 in **PBS** for ELISA testing. samples tested were; different stages (0-1 day, 3, 5, 7, 10 and 12 days) of eggs, neonate and adult slugs of *Deroceras reticulatum*; different stages (0-1 day, 3, 5, 10, 15 and 20 days) of eggs, neonate and juvenile slugs of *Arion ater* and eggs (various ages pooled together) of *D. caruanae, Limax maximus, A. ater ater, A. ater rufus, A. hortensis, A. circumscriptus, A. distinctus, A. intermedius, Geomalacus maculosus, Tandonia budapestensis, T. sowerbyi, Helix aspersa, Cepaea nemoralis* and *Pomacea caniculata*. Ten replicates of each development stage were tested in duplicate. The MAbs DrE-1E9 and AaE-2E3 were tested following the ELISA procedure described above.

RESULTS

Antibody Production & Supernatant Screening

Initial screening tests revealed monoclonal antibodies against *D. reticulatum* and *A. ater* eggs in about 60% of the fusion plate supernatants. Following repeated cloning and screening against *D. reticulatum* and *A. ater* egg antigens, two cell lines (DrE-1E9 and AaE-2E3) were selected for their high specificity and their rapid growth rates and stability.

Antibody Specificity

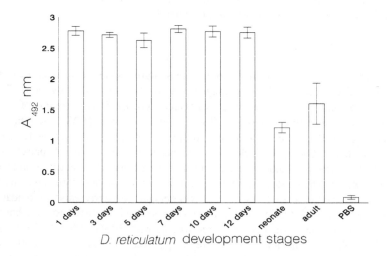

FIGURE 1. Absorption values at 492 nm (A_{492}) obtained in an ELISA to test reactivity of the MAb DrE-1E9 with different life stages of *D. reticulatum*. Mean values are given for 10 replicates, while bars indicate ± SE.

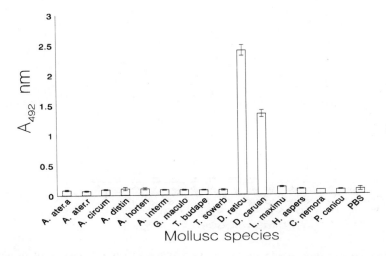

FIGURE 2. Absorption values at 492 nm (A_{492}) obtained in an ELISA to test reactivity of the MAb DrE-1E9 against eggs of different mollusc species. Mean values are given for 10 replicates, while bars indicate ± SE.

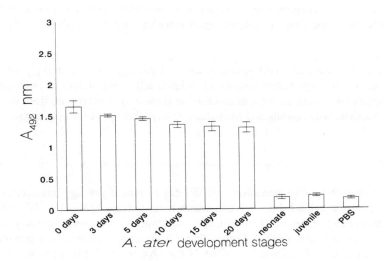

FIGURE 3. Absorption values at 492 nm (A_{492}) obtained in an ELISA to test reactivity of the MAb AaE-2E3 with different life stages of *A. ater*. Mean values are given for 10 replicates, while bars indicate ± SE.

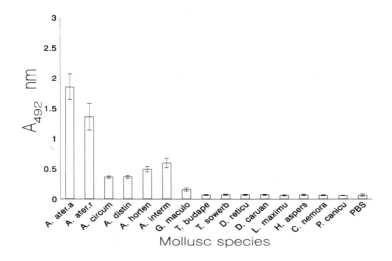

FIGURE 4. Absorption values at 492 nm (A_{492}) obtained in an ELISA to test reactivity of the MAb AaE-2E3 against eggs of different mollusc species. Mean values are given for 10 replicates, while bars indicate ± SE.

The MAb DrE-1E9 which was raised against *D. reticulatum* 0-1 day old egg antigen reacted strongly with all *D. reticulatum* egg stages tested (Figure 1). It was also found to be cross reactive with *D. caruanae* eggs, but not with any other species of eggs tested (Figure 2). Furthermore there was a moderate reaction with neonate and adult slugs of *D. reticulatum*.

The MAb AaE-2E3 which was raised against *A. ater* 0-1 day eggs reacted strongly with all the stages of *A. ater* eggs tested (Figure 3). Additionally it was found to be cross reactive at different levels with all the species of *Arion* (Figure 4). Furthermore there was no significant reactivity with neonate and adult slugs, or any other species of eggs tested.

DISCUSSION

Fusion of splenocytes from mice immunised with 0-1 day *D. reticulatum* egg homogenate and myeloma cells resulted a hybrid cell line which produced the MAb DrE-1E9 that reacted with all the egg stages of *D. reticulatum*. Additionally, at a lower intensity, it reacted with neonate and adult slugs, indicating that it is detecting a protein, or a protein site, shared by all stages of development. One of the objectives of this study was to develop an egg specific MAb for *D. reticulatum*. The positive antibody response associated with mature adult slugs was not surprising, since a gravid slug may contain eggs or egg proteins. The reactivity to adult slugs does not necessarily reduce the value of the MAb for detecting predation upon *D. reticulatum* eggs, as many predators can be assumed to be too small to attack an adult slug. The reaction of the MAb DrE-1E9 with

D. caruanae means that it is capable of detecting predation on *Deroceras* spp. eggs in field samples (V.W. Mendis, Ph.D. Thesis in preparation).

The MAb AaE-2E3 was also produced from 0-1 day old *A. ater* egg antigens. All the stages of *A. ater* eggs reacted strongly, showing that there is a similar epitope present throughout egg development as well as in the eggs of other species within the genus *Arion*. The main objective of producing an egg-specific MAb was fully achieved, since the neonate and juvenile slugs did not react with the MAb AaE-2E3. Reactions with fully grown egg-bearing adults was to be expected (V.W. Mendis, Ph.D. Thesis in preparation). Furthermore the usefulness of the MAb AaE-2E3 is enhanced by its reactivity with the eggs of other *Arion* species as it can be used more widely as a MAb to identify predation on the eggs of the genus *Arion*.

Further characterisation of these two MAbs is progressing in order to establish their reactivity with semi-digested slug egg material from predator gut samples. They can then be used as tools for routine identification of insect and other predators of *Deroceras* and *Arion* eggs. MAbs produced in this study provide a means of investigating the impact of predators on populations of *Deroceras* sp. and *Arion* sp. in the field.

REFERENCES

Allen, W R; Hagley, E A C (1982) Evolution of immunoelectroosmophoresis on cellulose polyacetate for assessing predation on Lepidoptera (Torticidae) by Coleoptera (Carabidae) species. *Canadian Entomologist*. **114**, 1047-1054.

Bowen, I D; Jones, G W (1985) Getting pesticides into cells. *Industrial Biotechnology*. **5**, 29-32.

Bowen, I D; Mendis, V W; Symondson, W O C; Liddell, J E; Leclair, S (1993) The integrated management of terrestrial slug pests. *Proceedings of the Second Malacological Convention*, Los Banos, The Philippines. 26-36.

Cohen, A C (1990) Feeding adaptation of some predaceous Hemiptera. *Annals of Entomological Society of America*. **83**, 1215-1223.

Dempster, J P (1960) A quantitative study of the predators on the eggs and larvae of the broom beetle, *Phytodecta olivacea* Forester, using the precipitin test. *Journal of Animal Ecology*. **29**, 149-167.

Dennison, D F; Hodkinson, I D (1983) Structure of the predatory community in a woodland soil ecosystem. I. Prey selection. *Pedobiologia*. **25**, 109-115.

Glen, D M (1989) Understanding and predicting slug problems in cereals. In: *Slugs and Snails in World Agriculture*. I Henderson (ed.). BCP Monograph c / 41, pp. 253-262.

Glen, D M; Orsman, I A (1986) Comparison of molluscicides based on metaldehyde, methiocarb or aluminium sulphate. *Crop Protection*. **5**, 371-375.

Godan, D (1983) *Pest Slugs and Snails - Biology and control* (translated by S.Gruber), Springer-Verlag, Berlin.

Greenstone, M H (1979) A passive haemagglutination inhibition assay for the identification of stomach contents of invertebrate predators. *Journal of Applied Ecology.* **14**, 457-464.

Greenstone, M H; Morgan, C E (1989) Predation on *Heliothis zea* (Lepidoptera: Noctuidae): An instar-specific ELISA assay for stomach analysis. *Annals of Entomological Society of America.* **82**, 45-49.

Hagler, J R; Cohen, A C; Enriquez, F J; Bradley-Dunlop, D (1991) An egg specific monoclonal antibody to *Lygus hesperus. Biological Control.* **1**, 75-80.

Hagler, J R; Cohen, A C; Bradley-Dunlop, D; Enriquez, F J (1992) Field evaluation of predation on *Lygus hesperus* (Hemiptera: Miridae) using a stage specific monoclonal antibody. *Environmental Entomology.* **21**, 896-900.

Köhler, G; Milstein, C (1975) Continuous cultures of fused cells secreting antibody of predefined specificity. *Nature.* **256**, 495-497.

Liddell, J E; Cryer, A (1991) *A Practical Guide to Monoclonal Antibodies.* John Wiley & Sons, Chichester.

Martin, T J (1993) The ecological effects of arable cropping including the non-target effects of pesticides with special reference to methiocarb pellets used for slug control. *Pflanzenschutz-Nachrichten Bayer.* **46**, 49-102.

Port, C M; Port, G R (1986) The biology and behaviour of slugs in relation to crop damage and control. *Agricultural Zoology Reviews.* **1**, 255-299.

Robinson, W H; Foote, B A (1968) Biology and immature stages of *Megaselia aequalis*, a phorid predator of slug eggs. *Annals of Entomological Society of America.* **61**, 1587-1594.

Ryder, T A; Bowen, I D (1977) The use of x-ray microanalysis to demonstrate the uptake of the molluscicide copper sulphate by slug eggs. *Histochemistry*, **52**, 55-60.

Symondson, W O C; Liddell, J E (1993a) A monoclonal antibody for the detection of arionid slug remains in carabid predators. *Biological Control.* **3**, 207-214.

Symondson, W O C; Liddell, J E (1993b) The development and characterisation of an anti-haemolymph antiserum for the detection of mollusc remains within carabid beetles. *Biocontrol Science & Technology.* **3**, 261-275.

Symondson, W O C; Mendis, V W; Liddell, J E (1995) Monoclonal antibodies for the identification of slugs and their eggs. *EPPO Bulletin.* **25**, 377-382.

Symondson, W O C; Liddell, J E (1996) A species-specific monoclonal antibody system for detecting the remains of Field Slugs, *Deroceras reticulatum* (Müller) (Mollusca: Pulmonata), in carabid beetles (Coleoptera: Carabidae). *Biocontrol Science & Technology.* **6**, 91-99.

Symondson, W O C; Glen, D M; Wiltshire, C W; Langdon, C J; Liddell, J E (1996) Effects of cultivation techniques and methods of straw disposal on predation by *Pterostichus melanarius* (Coleoptera: Carabidae) upon slugs (Gastropoda: Pulmonata) in an arable field. *Journal of Applied Ecology.* **33**, (in press).

Tod, M E (1973) Notes on beetle predators of molluscs. *The Entomologist.* **106**, 196-201.

Wilson, M J; Glen, D M; George, S K; Pearce, J D; Wiltshire, C W (1994) Biological control of slugs in winter wheat using the rhabditid nematode *Phasmarhabditis hermaphrodita. Annals of Applied Biology.* **125**, 377-390.

Session 4

Behaviour and Ecology

Session Organiser
and Chairman   Dr S E R Bailey

THE EFFECTS OF MANAGEMENT OF ROTATIONAL SET-ASIDE ON ABUNDANCE AND DISPERSION OF SLUGS

A BOLTON, L D INCOLL, S G COMPTON, AND C WRIGHT

Department of Biology, University of Leeds, Leeds, LS2 9JT, UK

ABSTRACT

The effects of four methods of managing rotational set-aside on number and dispersion of slugs were determined by both relative and absolute monitoring techniques. The set-aside methods included the sowing of mustard and of ryegrass and natural regeneration of cereal stubble with and without cultivation. Numbers were highest on uncultivated plots during establishment of vegetative cover, but highest in ryegrass plots after establishment. Dispersion tended to be random, rather than aggregated. *Deroceras reticulatum* was the most common species, on all types of set-aside, both during establishment and through the following season.

INTRODUCTION

In 1988, the rising cost of storage and disposal of foodstuffs, brought about by overproduction by European farmers, prompted the introduction of a policy of removing viable agricultural land from production (Floyd, 1992). This began on a voluntary basis, but was soon superseded, in 1992, by a compulsory system of removing a proportion of land from production (MAFF, 1992). Although the original aims were financially driven (Floyd, 1992), it soon became apparent that there were potential benefits to wildlife of applying different methods of managing set-aside land. These methods, prescribed annually (e.g. MAFF, 1995a, b), are still in development, and have changed from the original inception to the present (MAFF, 1992, 1995a, b) as the effects of set-aside have been ascertained (Clarke, 1992, 1993, 1994, 1995).

Land can currently be removed from production in two ways, either on an annual basis (rotational set-aside), or for a longer period of up to 5 years (guaranteed set-aside). The main aim of the methods of managing set-aside prescribed by MAFF is to produce 'green cover'. This can be achieved in a variety of ways, including natural regeneration, the sowing of grasses, or 'wild bird' cover (a mixture of two crop groups) or mustard (MAFF, 1995b). Under current EC rules, the vegetation on rotational set-aside land must be cut short, between 15 July and 15 August, or destroyed by 31 August. After mowing, all cuttings must be left *in situ*. The cover may be cut more regularly than this, and a series of exemptions exist which further complicate management options.

Slugs are well-known pests of agricultural systems (Port & Port, 1986), and previous work has shown that the number of slugs can increase during a period of set-aside (Clarke 1993, 1994, Hancock *et al.*, 1992). Several of the features of the management of set-aside land have the potential to increase slug numbers, such as not ploughing in the surface brash after harvest, leaving cuttings *in situ* and producing a 'green cover', all of which potentially improve the environment for slugs (Glen *et al.*, 1988).

Two set-aside experiments have been established at the Leeds University Farms, to examine the effects of rotational and guaranteed set-aside on the abundance of slugs and their natural enemies. This paper describes some interim results of research into the effects of the rotational set-aside on the number and dispersion of slugs.

METHODS

The rotational set-aside has been established in an arable field at the Leeds University Farms, Bramham, North Yorkshire, (SE445415), following a spring barley crop, on Wothersome series soil, with a south-west aspect.

Set-aside plots

Four MAFF-approved methods of managing set-aside were investigated; (i) sowing perennial ryegrass (*Lolium perenne* cv. Parcour), (ii) sowing mustard (*Sinapis arvensis* cv. Tilney), (iii) natural regeneration following a minimum post-harvest cultivation (by spring-tine harrowing) (cultivated) and (iv) natural regeneration with no post-harvest cultivation (uncultivated). Seeds were sown at 2 g m^{-2} for mustard, and 4 g m^{-2} for ryegrass. Four replicates of each treatment were set up in sixteen 10 m x 3 m plots in a 4 x 4 grid in a randomised block design. Regularly rotovated 2-m-wide strips between the plots acted as barriers to migration of slugs between the plots. The plots were set up in October 1995 and were sampled through the following growing season. The vegetation was considered to be established by the end of November.

Monitoring of slug populations

Both absolute and relative methods of monitoring slug populations were used (Southwood, 1971). During establishment of the plots in autumn of 1995, relative measures of the number of slugs were determined by refuge traps, and from spring until the end of July absolute measures of slug numbers were determined by surface searching of the plots (South, 1993). Slugs were identified using Eversham (1988), Cameron *et al.* (1983), and Kerney & Cameron (1979).

Refuge trapping

The refuge traps consisted of a piece of roofing felt (20 cm x 30 cm) weighed down by a house brick. Four unbaited traps were set in each plot, initially along the centre line of each plot, at 2.5 m, 4.5 m, 6.5 m and 8.5 m from the base of the plot, and were left *in situ* for ten days. The slugs were collected on both the tenth and eleventh days. The traps were then all moved to either the left or right of the centre line and the procedure repeated. Sampling began 12 October 1995 and there were three sampling periods, on 21 and 22 October, 1 and 2 November, and 10 and 11 November. The slugs were stored in a freezer before weighing. The data were analysed by nested analysis of variance with replicate nested within treatment (Minitab for Windows).

Mapping of slug dispersion

Surface mapping and counting of slugs were used to determine both the abundance and dispersion of the slugs because methods such as soil-core washing or flooding (South, 1964, Hunter, 1968) which determine the numbers present in the soil profile, are destructive and time consuming.

A pilot mapping study was carried out on 14-15 January 1996 and regular measurements commenced on 14 April. These were suspended after 5 June as the vegetation became too tall to search through, and resumed after the compulsory cut on 17 July. This involved the detailed searching of the soil surface and vegetation of entire plots overnight, when the slugs were active. Mapping began approximately half an hour after sunset and four plots, one from each treatment, were mapped in random order on each date. Any slug found was identified, and its position recorded. This technique gives an absolute measure of population density (South, 1993), although it is biased towards those species that are surface-active. The location of each slug was determined trigonometrically, initially by using two tape measures attached to canes, placed a known distance apart, at each end of the base of the plot. So that no damage was done to the plots during the searching, a system of a mobile plank and tressles was used to support the observer above the vegetation. Later the tape measures were replaced by an instrument (SONIN Combi Pro C.S.T., Forestry Supplies, U.S.A), which uses i.r. and sonar to measure distances. The data were analysed using nearest-neighbour-analysis software written by B. Huntley.

RESULTS

Refuge trapping

A total of 69 slugs was collected from the refuge traps (Table 1). *Deroceras reticulatum* was the only limacid collected and was the most common slug (n = 37). It was more common in the uncultivated plots (n = 25) than the other treatments (n = 8, 3, and 1 for ryegrass, mustard, and cultivated plots respectively). There were no significant differences in the wet masses of *D. reticulatum* between treatments (one-way ANOVA, F = 2.02, df = 3, $P = 0.13$) with a mean mass of 0.36 ± 0.03 g per slug (mean $\pm$ 1 s.e). Of the 20 arionids collected, *Arion subfuscus*

Table 1. The number of slugs caught under refuge traps during the establishment of set-aside.

| Treatment | Sampling period | | | | |
	21-22 Oct	1-2 Nov	10-11 Nov	Total	Mean $\pm$ 1 S.E.
Ryegrass	5	7	1	13	4.3 ± 2.2
Mustard	2	4	0	6	2.0 ± 1.4
Cultivated	1	6	3	10	3.3 ± 1.8
Uncultivated	17	21	2	40	13.3 ± 7.1

was the most common (n = 10) with a mean mass of 0.43 ± 0.07 g per slug. These were most common in the uncultivated treatment, followed by the ryegrass, mustard and cultivated plots (n = 4, 3, 2 and 1 respectively). *Milax gagates* was the most common milacid (n = 11), with only 1 individual of *Tandonia budapestensis* present. More milacids were collected in the uncultivated treatment (n = 8) than the other treatments (n = 2, 1 and 1 for ryegrass, mustard, and cultivated plots respectively). There were significant differences in mass per slug between the genera (one-way ANOVA, F = 3.32, df = 2, P = 0.04). The milacids were significantly heavier than either the limacids or arionids, with a mean mass of 0.52 ± 0.10 g per slug (LSD, P < 0.05).

There were 25, 38 and then 6 slugs collected in the consecutive trapping sessions. Apart from the last session, the uncultivated treatment always had the most slugs, and the cultivated and mustard treatment generally had the least (Table 1). There was a significant difference in the total number between the treatments (nested ANOVA, df = 3, F = 6.57 P = 0.007), but not between the replicates (nested ANOVA, df = 12, F = 1.85, P = 0.066).

Spatial analysis

The locations of 381 slugs were mapped. Numbers varied widely between each date, with virtually no slugs active on some nights (Table 2). On most occasions the ryegrass plots contained the most surface-active slugs, with densities ranging from 1.17 to 0.13 slugs m^{-2} (0.61 ± 0.13 slugs m^{-2}). The cultivated plots generally had the lowest number of surface-active slugs (0.21 ± 0.05 slugs m^{-2}). *D. reticulatum* was the most common slug, accounting for 75% of those observed. It was the only slug present in 19 of the 32 mapped plots, and was observed more frequently in the ryegrass plots (n = 121) than the other treatments (n = 39, 35, 66, for mustard, cultivated and uncultivated respectively). Forty five arionids were observed, more frequently in the ryegrass plots (n = 20), with *A. subfuscus* the most common species. Thirty nine milacids, all *M. gagates*, were observed, most frequently in the mustard plots (n = 22).

Table 2. The number of slugs recorded during spatial mapping.

Treatment	Date of mapping								
	14/1	11/4	24/4	9/5	22/5	5/6	17/7	31/7	Total
Ryegrass	35	27	4	17	24	6	21	12	146
Mustard	28	11	0	27	8	0	9	4	87
Cultivated	6	11	0	6	7	1	8	12	51
Uncultivated	31	10	0	18	18	4	10	6	97

Table 3. Nearest neighbour statistic (R)[#] for all slug species.

Treatment	Date of mapping							
	14/1	11/4	24/4	9/5	22/5	5/6	17/7	31/7
Ryegrass	1.26*	1.13	1.68*	0.99	0.81	1.17	0.76	0.94
Mustard	0.96	1.67*	-	0.77	1.35	-	1.02	0.89
Cultivated	1.38	1.17	-	0.75	1.58*	-	0.67	1.22
Uncultivated	1.10	0.66	-	1.31*	1.23	1.98*	0.72	0.70

[#] Degree of dispersion: aggregated R = 0; random R = 1; regular R = 2.15
* $P \le 0.05$

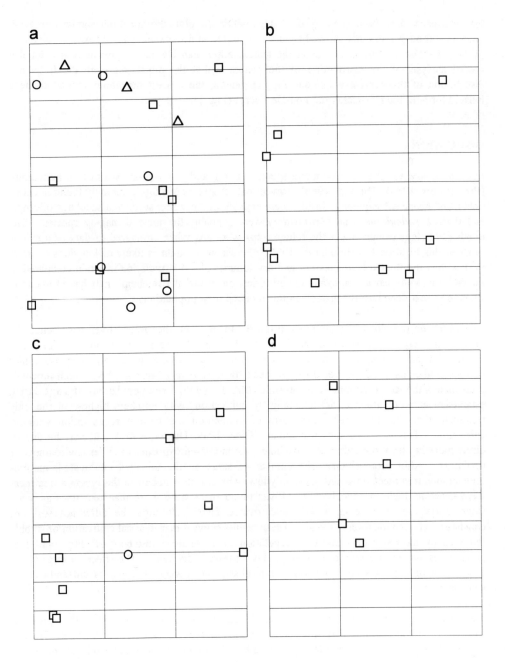

Figure 1. Examples of the dispersion of individual slugs in four set-aside plots; each plot is 10 m long x 3 m wide (not to scale). Treatments: a) ryegrass, plot 11, 9 May; b) mustard, plot 16, 17 July; c) cultivated, plot 15, 11 April; d) uncultivated, plot 9, 31 July.
Symbols: □ Limacidae, O Arionidae, △ Milacidae.

On most occasions, the dispersion of all slugs within the plots did not differ significantly from random, although plots that showed significantly different dispersions tended towards regularity (Table 3). The low numbers among the genera other than Deroceras precludes any detailed spatial analysis of the distribution of the genera within the plots, but from the frequency distribution of the nearest neighbour for the genera, the nearest neighbour tended to be a limacid for both the Deroceras and non-Deroceras (Fig. 1).

DISCUSSION

D. reticulatum was the most common species of slug and is a well-known pest in agriculture (Port & Port, 1986). Duthoit (1964) showed that all species of slugs commonly found in cereal fields, and hence subsequent set-aside, were capable of damaging seeds and seedlings of crops, and that D. reticulatum (and Arion ater) were probably the most damaging species. The numbers of slugs caught under the refuge traps were low when compared to data of previous years at this location (Griffiths et al., 1996). Persistent drought is likely to kill slugs (Port & Port, 1986) and the exceptionally hot and dry summer of 1995 (with 563 mm rain compared to the 641.2 mm 44-year average, and the 8th driest on record at Bramham), may have forced the slugs to retreat deep into the ground and/or reduced their populations.

Vegetation on the plots differed between the treatments. Of the 'green cover' treatments, the mustard proved not to be winter hardy at this location this year, with almost all the plants killed, whereas the ryegrass formed a good cover. Mustard was not resown because, under current guidelines, a new cover does not need to be resown in spring when establishment is poor, unless the land is to remain in set-aside (MAFF 1995b). The populations of slugs during establishment were affected detrimentally by cultivation, with numbers highest in the only treatment that did not receive any form of cultivation i.e. natural regeneration with no cultivation. This is a well-known response (Glen, 1989, Hunter, 1967), and may be due to direct mortality from the cultivating machinery, or increased exposure to either environment or predators. After establishment, the number of slugs on the minimally cultivated, natural-regeneration treatment remained constantly low, whereas the numbers in the ryegrass treatment were generally high. Hancock et al. (1992) found there were more than ten times as many slugs in ryegrass plots compared to uncultivated fallow, although the difference was not consistent between locations or years. These results suggest that minimal cultivation of stubble followed by natural regeneration of vegetation gives reduced numbers of slugs, and that populations may not be able to recover quickly, at least in dry years. The effect on slugs of the destruction of the set-aside vegetation, and subsequent sowing of crops, is currently under investigation.

The dispersion of slugs on the surface was, in general, random. Contrary to this, previous work has generally found that slugs have an aggregated dispersion (South, 1965, Hunter, 1966). However, the evidence for aggregation by D. reticulatum on arable land has not always been conclusive (South, 1993) and no evidence of aggregation was found in woodland for this and seven other species of slug (Jennings & Barkham, 1975). A positive correlation between dispersion of D. reticulatum and grass tussocks has been demonstrated (South, 1965) and it was thought that the aggregation found in this grassland may have been due to higher survivorship of slugs in the tussocks. Similar very local differences may have been responsible for those set-aside plots which tended towards a regular dispersion, but if present, such effects are clearly weak. Intra- and inter-specific antagonistic behaviour has been found amongst slugs (Rollo & Wellington, 1979) which, in cage experiments, has led to a regular dispersion (Rollo & Wellington, 1983). This antagonistic behaviour has been related to acquisition of high quality shelter, and the homing ability of slugs (Rollo & Wellington, 1983). The dispersion of slugs during the day may therefore be different to that during the night. Even if the distribution of all slugs was random or regular, negative interactions between genera would become apparent if individuals of one genus were most often near to individuals of the same genus. This was not apparent in our results suggesting that, where the distribution was regular, it was a consequence of interactions between individuals irrespective of their taxon.

ACKNOWLEDGEMENTS

We thank D Hardy and D T Corry of the Department of Biology's Field Research Unit for support and B Huntley, University of Durham, for allowing us to use his Nearest Neighbour software. This research was funded by a MAFF Research studentship to LDI and SGC.

REFERENCES

Cameron, R A D; Eversham, B C; Jackson, N C S (1983) A field key to the slugs of the British Isles. *Field Studies*. **5**, 807-824.

Clarke, J H (1992) *Set-Aside*. BCPC Monograph 50, BCPC: Farnham.

Clarke, J H (1993) *Management of set-aside land: Research progress*. Leaflet AR13. MAFF: London.

Clarke, J H (1994) *Management of set-aside land: Research progress 1994/1995*. Leaflet AR13. MAFF: London.

Clarke, J H (1995) *Management of set-aside land: Research progress 1995 update*. Leaflet AR13. MAFF: London.

Duthoit, C M G (1964) Slugs and food preferences. *Plant Pathology*. **13**, 73-78.

Eversham, B C (1988) *Identifying British Slugs*. Unpublished Booklet. 19 pp.

Floyd, W D (1992) Political aspects of set-aside as a policy instrument in the European community. In: *Set-Aside* J H Clarke (ed.). BCPC Monograph No. 50, pp. 13-20.

Glen, D M; Wiltshire, C W; Milsom, N F (1988) Effects of straw disposal on slug problems in cereals. *Aspects of Applied Biology 17: Environmental Aspects of Applied Biology*. **2**, 173-179.

Glen, D M; Milsom, N F; Wilthire, C W (1989) Effect of seed bed conditions on slug numbers and damage to winter wheat in a clay soil. *Annals of Applied Biology*. **115**, 177-190.

Griffiths, J; Phillips, D S; Compton, S G; Wright, C; Incoll L D (1996) Slug number and slug damage in a silvoarable agroforestry landscape. *Journal of Animal Ecology.* (in press).

Hancock, M; Ellis, S; Green, D B; Oakley, J N (1992) The effects of short- and long-term set-aside on cereal pests. In: *Set-Aside* J H Clarke (ed.). BCPC Monograph No. 50, pp. 195-200.

Hunter, P J (1966) The distribution and abundance of slugs on an arable plot in Northumberland. *Journal of Animal Ecology.* **35**, 543-557.

Hunter, P J (1967) The effects of cultivations on slugs of arable land. *Plant Pathology.* **16**, 153-156.

Hunter, P J (1968) Studies on the slugs of arable grounds I. Sampling methods. *Malacologia.* **6**, 369-377.

Jennings, T J; Barkham, J P (1975) Slug populations in mixed decidous woodland. *Oecologia.* **20**, 279-286.

Kerney, M P; Cameron R A D (1979) *A Field Guide to the Land Snails of Britain and North West-Europe.* Collins: London.

MAFF (1992) *Arable Area Payments: Explanatory Booklet.* Booklet AR2. MAFF: London.

MAFF (1995a) *Arable Area Payments: Explanatory Guide Part 1.* Booklet AR27. MAFF: London.

MAFF (1995b) *Arable Area Payments: Explanatory Guide Part 2.* Booklet AR27. MAFF: London.

Port, C M; Port, G R (1986) The biology and behaviour of slugs in relation to crop damage and control. *Agricultural Zoology Reviews.* **1**, 255-299.

Rollo, C D; Wellington, W G (1979) Intra and interspecific agonistic behaviour among terrestrial slugs (Pulmonata: Stylomatophora). *Canadian Journal of Zoology.* **57**, 845-855.

Rollo, C D; Wellington, W G (1983) Consequences of competition on the reproduction and mortality of three species of terrrestrial slugs. *Reseach in Population Ecology.* **25**, 20-43.

South, A (1964) Estimation of slug populations. *Annals of Applied Biology.* **53**, 251-258.

South, A (1965) Biology and ecology of *Agriolomax reticulatus* (Müll.) and other slugs: spatial distribution. *Journal of Animal Ecology.* **34**, 403-417.

South, A (1993) *Terrestrial Slugs: Biology, Ecology, Control.* Chapman and Hall: London.

Southwood, T R E. (1971) *Ecological Methods with Particular Reference to the Study of Insect Populations.* Chapman and Hall: London.

POPULATION DYNAMICS OF THE MEDITERRANEAN SNAIL, *CERNUELLA VIRGATA*, IN A PASTURE-CEREAL ROTATION IN SOUTH AUSTRALIA.

G H BAKER

Division of Entomology, CSIRO, PMB 2, Glen Osmond, South Australia 5064

ABSTRACT

The mediterranean helicid snail, *Cernuella virgata*, is an introduced pest of grain crops and pastures in south-eastern Australia. The abundance of *C. virgata* was monitored in four fields on Yorke Peninsula, South Australia from 1984 to 1995. All four fields were managed as pasture-cereal rotations. Snails were sampled in autumn (adult populations) and spring (juvenile populations). Snail numbers were highest in spring, especially in pastures, and in 1992 when autumn rainfall was above average. The numbers of juvenile snails produced per adult were inversely related to the density of the adults. Adult snails were smaller in size in dense populations in pastures, but not in crops. Snail contamination of barley delivered to a silo on Yorke Peninsula in early summer was also greatest when rainfall in the preceding autumn was high.

INTRODUCTION

The mediterranean snail, *Cernuella virgata* (Helicidae), along with other helicid snails (*Theba pisana* and *Cochlicella* spp.), is an introduced pest of grain crops and pastures in south-eastern Australia (Baker, 1986, 1989a). These snails aestivate on the heads, pods and stalks of cereals and grain legumes, thereby clogging machinery and contaminating grain at harvest. Contaminated grain may be rejected or down-graded upon delivery to silo. Snail contamination poses a serious threat to the marketing of Australian grain in international markets. Snails also feed upon young crops and pastures, and foul herbage with their slime.

Various aspects of the life history, ecology and activity of *C. virgata* in pastures and cereal crops have been reported by Baker (1988a,b, 1989a,b, 1991, 1992) and Bull *et al.* (1992). Adult *C. virgata* mate and oviposit in autumn and winter. Largest clutches are produced early in the breeding season. Spring populations are mostly juveniles. Life cycles vary between habitats and may be annual or biennial.

Current methods used by farmers to control snails are either expensive (e.g. molluscicide) or environmentally unsuitable (e.g. burning fields prior to sowing crops). Burning fields reduces nutrient return to the soil from decomposing vegetation and encourages soil erosion. Biological control methods have been sought through surveys for natural enemies within the snail's native distribution. Parasites (e.g. sarcophagid flies) are now being tested in quarantine for their host-specificity against endemic Australian snails (prior to seeking permission for their release) (Hopkins & Baker, 1993, Coupland, 1994, 1995, Coupland & Baker, 1995).

In order to further understand the factors influencing the abundance of *C. virgata* on farms in Australia and, in particular, to develop a database of natural variation in numbers prior to the release of biological control agents (to assist evaluation of the agents' effectiveness), the population densities of *C. virgata* have been monitored in four fields, all managed as pasture-cereal rotations, on one farm in South Australia for 12 years. The results of these population censuses are reported here.

MATERIALS AND METHODS

Snails were sampled in four fields on a farm near Weetulta, Yorke Peninsula, South Australia. *C. virgata* was the only snail species found there. The four fields were paired, with fields A & B and C & D adjacent to each other. The fields were generally used in alternate years for pasture (sheep grazing) and barley (*Hordeum vulgare*) or wheat (*Triticum aestivum*) production (Fig. 1). In one year, peas (*Pisum arvense*) were grown in field B, rather than a pasture. Occasionally in field D, crops were grown in two successive years. Common pasture plants included medics (*Medicago* spp.) and various grassy and broad leaf weeds.

Each year, in autumn (May) and late spring (November), snails were collected by hand from the soil surface and vegetation within 25 quadrats (each 0.25 m^2), placed within a 50 x 50 m plot, 10 m from the edge of each field. The plots for the paired fields were adjacent to each other and separated by a barbed wire fence. The two pairs of plots were approximately 500 m apart. Sample coordinates within each plot were selected on a stratified random basis. Fields A & B were sampled each year from 1985 to 1995. Fields C & D were sampled from 1984 to 1995. Fields were usually burnt and harrowed in March-April prior to sowing crops. The autumn samples of snails were always taken after the harrowing. Crops were usually harvested in November-December. Most spring samples of snails were taken before harvest.

Average annual rainfall at Maitland, the nearest reliable weather station (25 km from the farm) is 504 mm. Maitland has about 100 mm more rainfall per year than the farm, but annual and seasonal patterns are similar. Summers are hot and dry, winters are cool and wet. Average daily maximum temperatures at Maitland in summer (January) and winter (July) are 32.1°C and 16.3°C respectively.

Harvested barley is currently down-graded from malting to animal feed quality (or rejected) if > 6 snails are present in a 200 g sample when the grain is delivered by farmers to South Australian silos in early summer. The Australian Barley Board has kept records of the tonnages of barley downgraded because of snail contamination at the silo at Port Giles on Yorke Peninsula from 1990 to 1995. Data are also available for 1983, which was regarded by the grains industry as a particularly bad year for snails. Annual variations in snail contamination at the Port Giles silo are compared here with the field counts of snails at Weetulta.

Fig. 1. (opposite) Numbers of *Cernuella virgata* (m^{-2}) in four fields near Weetulta, South Australia in autumn and spring, 1984 - 1995. Annual rainfall (mm), recorded 25 km away at Maitland, is included. P = pasture, C = cereal crop, C* = pea crop, N/A = data not available.

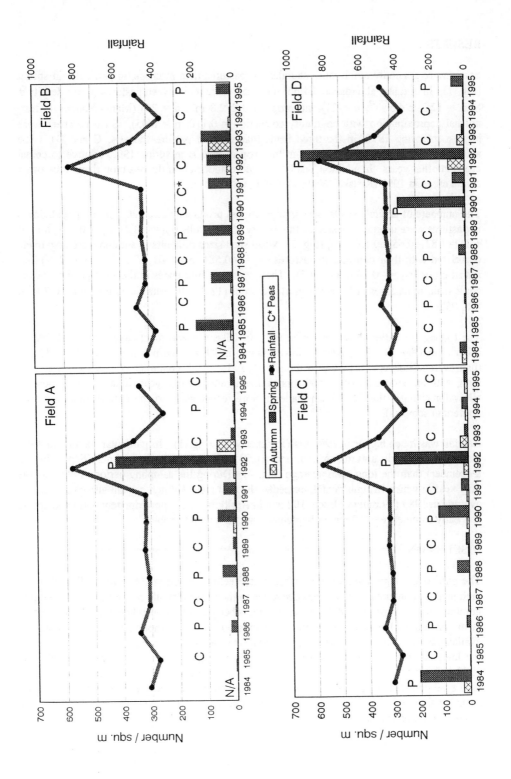

RESULTS

Snails were generally much more abundant in pastures than in crops, in particular in spring (Fig. 1). In autumn, abundance did not differ significantly between pastures (mean density ± S.E. = 12.3 ± 4.5 m^{-2}, n = 21) and crops (9.0 ± 2.8 m^{-2}, n = 25) (Mann-Whitney U test, normal approximation with continuity correction = 0.110, p > 0.05). Snail numbers were highest in spring 1992, in the three fields that were in pasture then (A, C and D). The abundance of snails was also higher in the crop in field B in spring 1992, relative to cereal crops in other years. In addition, snails were more abundant in the pea crop than in the other cereal crops in 1991 (Kruskal Wallis H = 20.0, p < 0.001).

The numbers of juvenile snails (spring populations) produced per adult (autumn populations) in pastures were inversely related to the density of the adults (r_{21} = - 0.710, p < 0.001; log (Y + 1) = 1.82 - 0.65 log (X + 1)) (Fig. 2). Mean shell sizes of adults in autumn were negatively correlated with their densities in pastures (r_{21} = - 0.581, p < 0.01; Y = 13.1 - 0.007 X), but not in crops (r_{25} = - 0.234, p > 0.05). There were no differences in the average sizes of the adult snails in pastures (12.6 ± 0.3 mm) and crops (12.8 ± 0.3 mm) in autumn (U = 0.706, p > 0. 05).

The tonnage of barley down-graded at the Port Giles silo was much greater in 1992 than in other years (Fig. 3). This peak in down-grading in 1992 could not be attributed to a higher than average delivery of barley to the silo (Fig. 3). Snail contamination in 1983 was also high, with an equivalent of 12,400 tonnes downgraded in that year (data corrected here to allow for changes that have been made in more recent years to the critical limits for snail content in samples).

Rainfall was much higher in 1992 than in other years (Fig. 1). In particular, rainfall in 1992 was higher in early autumn (March and April) and spring - early summer (September to December) than the long-term average (Fig. 4). During the sampling period, 1992 was the only year in which autumn rainfall exceeded the average. Autumn rainfall also exceeded the average in 1983 (69 mm in March, 102 mm in April), but was approximately normal in spring and early summer (e.g. 62 mm in September and 32 mm in October).

DISCUSSION

The particularly high numbers of C. virgata at Weetulta in spring 1992 were reflected in the high level of contamination of grain deliveries to the Port Giles silo in the same year. Higher numbers of C. virgata in 1992 compared with other years have also been recorded on other farms on Yorke Peninsula (G. Baker, unpublished data). Although no on-farm data are available for 1983, snail contamination at the Port Giles silo was comparable in that year with the level recorded in 1992. Autumn rainfalls were higher than average in both 1992 and 1983. It seems probable that this wetter than average weather in autumn is the cause of the high snail densities in spring. Wet weather in autumn would allow earlier breeding by adult snails. This may in turn lead to greater oviposition and / or enhance the survival of the young snails that are produced.

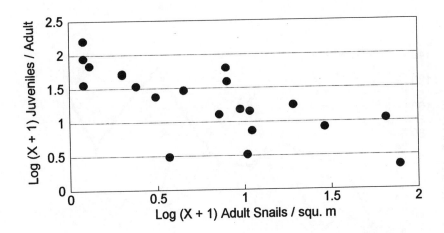

Fig. 2. Ratio of juvenile *Cernuella virgata* produced per adult as a function of adult density (snails m^{-2}) in the 21 "pastures" at Weetulta between 1984 and 1995. Data have been transformed to log (x + 1).

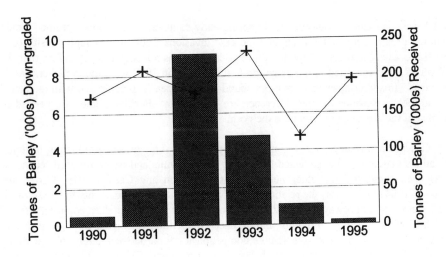

Fig. 3. Tonnages of barley delivered to Port Giles silo, South Australia (line) and down-graded because of snail contamination (bars) during 1990 - 1996.

121

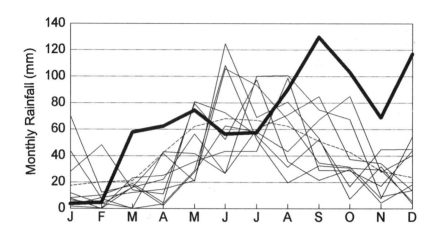

Fig. 4. Monthly rainfalls (mm) recorded at Maitland, South Australia during 1984 - 1995. Dashed line = long term average, bold line = 1992, thin lines = other years.

Rainfall was also higher than average in spring-early summer, 1992. Such weather would encourage greater activity by snails and increase invasions from sites of high density (e.g. pastures) to adjacent areas of low density (e.g. crops). It is not surprising therefore that the numbers of snails recorded in the cereal crop at Weetulta in 1992 were much higher than in other years. In addition, snails were more abundant in the pea crop compared with the cereal crops in 1991. Reasons for this difference are not known. Further studies are needed to determine if the observed difference is the norm.

Adult *C. virgata* produced fewer young when crowded. This poorer production in dense populations may in part be explained by smaller size of adults and hence reduced fecundity, as has been suggested for other helicid snails (Cameron and Carter, 1979, Carter and Ashdown, 1984). However, the relationship between adult size and fecundity in *C. virgata* is at best weak, and not always demonstrable, across the range of adult sizes commonly found in the field (Baker, 1991). Possibly, interference competition (mediated through inhibitory signals in slime trails or faeces), scramble competion for food between adults or the young they produce (Baker, 1991), or cannibalism amongst young (Baur, 1988) can also help explain the poorer production in dense populations.

Although snails were much more abundant in pastures than in crops in spring, differences in density were not detectable in autumn. Large numbers of snails are killed when pasture residues are burnt and harrowed prior to sowing crops (Baker, 1989a,b). These deaths, plus movements of snails between fields in spring and autumn (Baker, 1989a, 1992), obscure previous differences in abundance between fields. In addition, there was no difference in the sizes of adult snails in pastures and crops in autumn. Their gonads are mature at this time in both habitats (Baker, 1989b). The reason for the greater production of young in pastures compared with crops is therefore puzzling. Similar differences in production of young have been observed for *T. pisana* and *C. acuta* (Baker and Hawke, 1990, Baker *et al.*, 1991). Possibly, food is in short supply for the newly hatched young in fields used for crops or the herbicides used to control weeds in such fields are lethal to young snails.

ACKNOWLEDGEMENTS

I wish to thank Bonnie Vogelzang, Bruce Hawke, Vicki Barrett and Penny Carter for their many hours helping me collect the snails. The research was funded in part by the Australian Grains Research and Development Corporation and the International Wool Secretariat.

REFERENCES

Baker, G H (1986) The biology and control of white snails (Mollusca : Helicidae), introduced pests in Australia. *CSIRO Division of Entomology Technical Paper* No. 25, 31 pp.

Baker, G H (1988a) The life history, population dynamics and polymorphism of *Cernuella virgata* (Mollusca : Helicidae). *Australian Journal of Zoology*. **36**, 497-512.

Baker, G H (1988b) The dispersal of *Cernuella virgata* (Mollusca : Helicidae). *Australian Journal of Zoology*. **36**, 513-520.

Baker, G H (1989a) Damage, population dynamics, movement and control of pest helicid snails in southern Australia. In : *Slugs and Snails in World Agriculture*. I F Henderson (ed.). BCPC Monograph No. 41, pp. 175-185.

Baker, G H (1989b) Population dynamics of the white snail, *Cernuella virgata* (Mollusca : Helicidae), in a pasture-cereal rotation in South Australia. *Proceedings of the V Australasian Conference on Grassland Invertebrate Ecology*. D & D Printing, Melbourne, pp. 177-183.

Baker, G H (1991) Production of eggs and young snails by adult *Theba pisana* (Muller) and *Cernuella virgata* (da Costa) (Mollusca : Helicidae) in laboratory cultures and field populations. *Australian Journal of Zoology*. **39**, 673-679.

Baker, G H (1992) Movement of introduced white snails between pasture and cereal crops in South Australia. In : *Pests of Pastures : Weed, Invertebrate and Disease Pests of Australian Sheep Pastures*. E Del Fosse (ed.). CSIRO Information Services, Melbourne, pp. 115-120.

Baker, G H; Hawke, B G (1990) Life history and population dynamics of *Theba pisana* (Mollusca : Helicidae) in a cereal-pasture rotation. *Journal of Applied Ecology* **26**, 16-29.

Baker, G H; Hawke, B G; Vogelzang, B K (1991) Life history and population dynamics of *Cochlicella acuta* (Muller) (Mollusca : Helicidae) in a pasture-cereal rotation. *Journal of Molluscan Studies* **57**, 259-266.

Baur, B (1988) Population regulation in the land snail *Arianta arbustorum* : density effects on adult size, clutch size and incidence of egg cannibalism. *Oecologia* **77**, 390-394.

Bull, C M; Baker, G H; Lawson, L M; Steed, M A (1992) Investigations of the role of mucus and faeces on interspecific interactions of two land snails. *Journal of Molluscan Studies*. **58**, 433-441.

Cameron, R A D; Carter, M A (1979) Intra- and interspecific effects of population density on growth and activity in some helicid land snails (Gastropoda : Pulmonata). *Journal of Animal Ecology* **48**, 237-246.

Carter, M A; Ashdown, M (1984) Experimental studies on the effects of density, size and shell colour and banding phenotypes on the fecundity of *Cepaea nemoralis*. *Malacologia* **25**, 291-302.

Coupland, J B (1994) Diptera associated with snails collected in south-western and west-mediterranean Europe. *Vertigo*. **3**, 19-26.

Coupland, J B (1995) Susceptibility of helicid snails to isolates of the nematode *Phasmarhabditis hermaphrodita* from southern France. *Journal of Invertebrate Pathology*. **66**, 207-208.

Coupland, J B; Baker, G H (1995) The potential of several species of terrestrial Sciomyzidae as biological control agents of pest helicid snails in Australia. *Crop Protection*. **14**, 573-576.

Hopkins, D C; Baker, G H (1993) Biological control of white and conical snails. In : *Pest Control and Sustainable Agriculture*. S A Corey, D J Dall & W M Milne (eds). CSIRO Information Services, Melbourne, pp. 246-249.

ECOLOGY OF TERRESTRIAL MOLLUSCS IN RELATION TO FARM LAND-USE PRACTICE

J WARD BOOTH, G B J DUSSART & A PAGLIA

Christ Church College, Canterbury, North Holmes Rd., Canterbury, Kent CT1 1 QU

ABSTRACT

A multifactorial investigation of macromollusc distributions was undertaken near Saffron Walden in South East UK. The categories under investigation were (1) three soil types - sand, chalk and clay; (2) three types of land use - pasture comprising rough and grazed grass, crops comprising mostly wheat, and woodland comprising oak, maple and hazel; (3) two site locations - ditch and flat, where a ditch was a topographical depression and a flat was an area of flat ground. There were higher numbers of molluscan species in the ditch category than the flat, though this topographical difference was less marked when live individual molluscs were counted. Dead molluscs made up the greater proportion (68%) of the samples. Cropping, especially on sandy soil seemed to be an important factor affecting both numbers of species and individuals of macromolluscs.

INTRODUCTION

It is a truism of molluscan ecology that rich molluscan communities can be found in particular types of geographical areas such as woodland. Boycott (1934) offered a wide-ranging account of factors affecting the distribution of molluscs and Ellis (1926) notes that snails with calcareous shells are found chiefly on more alkaline soils, limestone districts being the most favourable. It can also be assumed that organisms are adapted to particular niches - for example, the land snail *Ena montana* appears to be adapted to ancient woodland on calcareous soil (Boycott, 1939). More recently, Kerney & Cameron (1979) note that a rich woodland, with rocky outcrops and lime-rich soil will yield forty or more species. They also raise the issue of the impact of human activity on molluscan communities, noting that forestry practice can reduce molluscan numbers in woodland, and grazing, trampling and ploughing can reduce molluscan numbers in more open conditions. Nevertheless, agricultural and horticultural practice is plagued by the more anthropochoric species such as the field slug *Deroceras reticulatus* (Glen, Jones & Fieldsend, 1990) and there are more specific studies which allow the precise prediction of aspects of molluscan ecology such as slug activity (Young, Port, Emmett & Green, 1991).

There is thus a considerable knowledge about particular species such as *Deroceras*, and a large anecdotal literature. Malacologists 'know' that there are more molluscs in areas of certain geology and land-use but can such differences be seen in an experimental context ? The aim of the present work was therefore to test whether the mainly anecdotal observations of authors such as Boycott could be quantified, particularly in relation to agricultural land-use.

MATERIALS AND METHODS

Eighteen areas of farmland in the region of Saffron Walden in Essex were classified on the basis of a 3 x 3 x 2 experimental design *i.e.:*- (1) three soil types - sand, chalk and clay; (2) three types of land use - pasture comprising rough and grazed grass, crops comprising mostly wheat and woodland comprising mainly oak, maple and hazel; (3) two site locations - ditch and flat. Ditches and flats were located at the edges of the areas denoted by the above factors; a ditch was a narrow, usually dry, channel bordering a field margin and a flat was an area of flat ground. Each of the eighteen areas was sampled by 3 replicate x 15 minutes searches for macromolluscs in May-June 1996 and soil samples were taken in order to measure micromollusc density; local environmental factors such as soil temperature, humidity and pH were also measured. The molluscs were identified to species where possible and only the distribution of macromolluscs will be discussed here. All samples were taken in similar weather conditions.

RESULTS

A total of twenty-one species were found in varying degrees of abundance, slugs representing less than one per cent of the molluscs found. On average, live snails represented 22% (SD 15) of the molluscs encountered. The live species included *Arianta arbustorum, Arion circumscriptus, Aegopinella pura, Azeca goodalli, Candidula gigaxii, Candidula intersecta, Cepaea hortensis, Cepaea nemoralis, Cernuella virgata, Clausilia* sp., *Cochlicella acuta, Cochlicopa lubrica, Cochlodina laminata, Columella edentula, Discus rotundatus, Ena obscura, Monarcha cantiana, Oxychilus helveticus, Trichia hispida, Trichia striolata,* and *Vitrina pellucida.* In most cases, sample standard deviations were smaller than means, *e.g. Monarcha cantiana* in certain of the sites (Table 1) .

Table 1. Mean numbers of *Monarcha cantiana* found in three 15 minute searches
at a selection of sites.

Location			Live snails		Dead snails	
			Mean	SD	Mean	SD
Clay	Pasture	Flat	4.0	1.0	4.3	3.0
Clay	Pasture	Ditch	3.5	2.1	14.6	5.5
Chalk	Pasture	Flat	0	0	1	0
Chalk	Pasture	Ditch	9.7	7.7	19.0	7.0
Sand	Pasture	Flat	1.7	0.6	25.5	10.6
Sand	Pasture	Flat	19.3	9.0	5.0	0

In a preliminary analysis to identify general patterns, the replicates were pooled to give data on the total number of species and individuals encountered (Fig. 1) .

Fig. 1 (next page) Distributions of numbers of molluscs found in the target areas. For the purpose of this exercise, data from three replicate samples have been pooled. Fig. 1a-1d represents 'all molluscs'. Fig. 1e-1h represents live molluscs only.

Fig. 1a

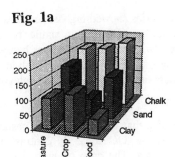

No. of individuals - Ditch

Fig. 1b

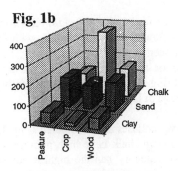

No. of individuals - Flat

Fig. 1c

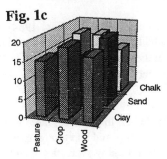

No. of species - Ditch

Fig. 1d

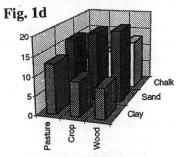

No. of species - Flat

Fig. 1e

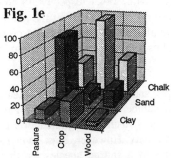

No. of individuals - Ditch

Fig. 1f

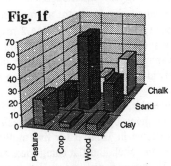

No. of individuals - Flat

Fig. 1g

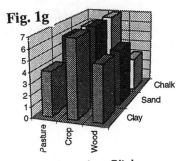

No. of species - Ditch

Fig. 1h

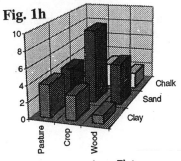

No. of species - Flat

Pooled data for living and dead molluscs in the ditch categories showed highest densities on chalk and lowest densities on clay, with little effect of land-use (Fig. 1a); a similar picture could be seen in the flat category (Fig. 1b). There were few significant differences between soil types or land-use for the living and dead species in the ditch area (Fig. 1c), though in the flat area (Fig. 1d), there were consistently higher numbers of species in the sand category, irrespective of land-use.

When live molluscs are considered (Fig. 1e), it can be seen that the categories of pasture-sand and crop-chalk have high numbers of individuals in the ditch area. By contrast, the crop-sand category has the highest numbers in the flat area. The crop category has the highest numbers of living species in the ditch area, irrespective of soil type (Fig. 1g) whereas the crop category on sandy soil has the highest number of species in the flat area.

An rct three-dimensional contingency analysis of mutual independence (Zar, 1984) was applied to the data for 'all molluscs' and significant variability in the numbers of individuals was revealed ($\chi^2 = 317.7$; df = 12; P<0.001) but not in the number of species ($\chi^2 = 10.1$; df = 12; P>0.5). A similar result was obtained for the live individual molluscs ($\chi^2 = 184.8$; df = 12; P<0.001) but not for the number of species ($\chi^2 = 9.9$; df = 12; P>0.5).

The data were also analysed for partial independence (Table 2). In all cases, there was full independence for the species data, suggesting that factors of crop, soil and topography were acting independently in relation to total number of niches. However, the results for individuals suggest interdependence between these factors, which operate in the order soil /crop /topography for both 'all molluscs' ($\chi^2 = 276.2$, 264.6, 184.1 respectively) and live molluscs ($\chi^2 = 182.0$, 162.0, 144.2 respectively). These results suggest that the number of individual molluscs on a particular soil type is strongly affected by the factors of crop and topography. By contrast, the number of individual molluscs on a particular ditch/flat is less strongly affected by the factors of crop and soil.

The diversity index of Menhinick (1964) is a simple and direct method of measuring diversity, given by (number of species)/(square root of number of individuals) and gave the results shown in Table 3 for live molluscs. There was no statistically significant difference between the ditch and flat categories with respect to diversity. However, Figs. 1c *cf* 1d and 1g *cf* 1h indicate, as expected, that the flat areas support a shorter species list than the ditches, though the distribution patterns are different.

A similar survey approach was used contemporaneously at two additional locations approximately 100 km to the South, on farms near Edenbridge in the Weald and Elham in Kent. Table 4 compares the numbers of species and individuals found on chalk at these locations. It is immediately clear that more molluscs were found in the northern site at Saffron Walden, Essex ($\chi^2 = 97.8$; df = 4; P<0.001), though species numbers did not vary across locations.

Table 2. Results of χ^2 tests for partial independence

		Factor	df	χ^2	P
'All molluscs'	Individuals	Crop	10	264.6	P<0.001
		Soil	10	276.2	P<0.001
		Topography	8	184.1	P<0.001
	Species	Crop	10	4.8	NS
		Soil	10	9.9	NS
		Topography	8	8.5	NS
Live molluscs	Individuals	Crop	10	162.0	P<0.001
		Soil	10	182.0	P<0.001
		Topography	8	144.2	P<0.001
	Species	Crop	10	8.7	NS
		Soil	10	9.0	NS
		Topography	8	3.9	NS

Table 3. Diversity indices for live molluscs located in the various experimental categories.

		Flat	Ditch
Clay	Crop	2.1	1.7
Clay	Pasture	1.7	1.5
Clay	Wood	1.1	2.3
Chalk	Crop	0.5	1.2
Chalk	Pasture	0.8	1.2
Chalk	Wood	1.2	0.8
Sand	Crop	1.6	1.7
Sand	Pasture	1.3	1.1
Sand	Wood	1.0	1.6

Table 4. Total numbers of individuals and species of 'all molluscs' found at three locations in the South East of England.

Individuals	Ditch			Flat		
	Essex	Weald	Elham	Essex	Weald	Elham
Pasture	203	37	8	115	28	39
Crop	209	7	19	368	0	33
Wood	235	8	78	163	0	147

Species	Ditch			Flat		
	Essex	Weald	Elham	Essex	Weald	Elham
Pasture	17	5	3	9	1	7
Crop	17	4	9	10	0	7
Wood	13	3	10	13	0	11

DISCUSSION

The species list is relatively short which is probably a function of the focus on macromolluscs. Another artefact of the sampling method is the relatively low variation between replicates which is probably due to the searcher covering a substantial area of about ten square metres for one sample. Any tendency for the molluscs to have clumped distribution patterns would therefore be obscured.

The patterns for 'all molluscs' differ from those for 'live molluscs only'. For example, in both ditch and flat categories, 'all molluscs' show a trend to higher molluscan numbers on chalk with no apparent influence of land-use. It is possible that the high alkaline soils buffer against corrosion of the shells which therefore accumulate in these conditions. There appears to be no significant pattern in the number of living and dead species in the ditch category (Fig. 1c) but the pattern of higher species numbers on sand in the flat areas (Fig. 1d) is a little surprising. It may relate to the observations of Glen et al., (1993) and Glen, Milsom & Wiltshire (1989) that soil tilth can be an important factor in crop damage by molluscs. The tilth developed on a sandy soil might promote mollusc activity, mobility and niche partitioning. However, Ferguson, Barratt & Jones (1988) note that mollusc density can be controlled by stock management and this factor might explain why numbers of both individuals and species tended to be lower on the pasture category. In many cases, the crop category appeared to lead to either higher numbers of individual molluscs, or more species. For example, in the flat-crop-sandy category, the numbers of individuals reached sixty five; also, crop-sandy was the highest scoring category for number of living species. It is possible that crop management promotes the development of communities of anthropochoric molluscs by providing a wider range of niches within fine soil aggregates. Walden (1981) notes that when conditions improve, species with similar ecology tend to accumulate rather than replace each other, indicating that competition is of a subordinate importance. This factor may outweigh other factors such as soil type. Conclusion drawn from Fig. 1 are to some extent confirmed by the partial independence analysis, the latter showing that the presence of a ditch is an important factor which affects the abundance of molluscs in this survey and interacts with other factors of crop and soil.

The results for different geographic locations showed differences between regions for total numbers of individual molluscs but not for numbers of species. This suggests that latitudinal factors might be important in relation to the carrying capacity of the habitats.

In conclusion, it appears that despite the coarse categorisation of the factors of soil type, land-use and topography, the density and species composition of these macromolluscs is partially explainable, particularly with regard to local topography and cropping.

REFERENCES

Boycott, A E (1934) The habitats of land Mollusca in Britain. *Journal of Ecology.* **12**, 1-38.

Boycott, A E (1939) Distribution and habitats *of Ena montana* in England. *Journal of Conchology.* **21**, 153-159.

Ellis, A E (1926) *British Snails* Oxford University Press, London. 1-275.

Ferguson, C; Barratt, B & Jones P (1988) Control of the grey field *slug Deroceras reticulatus* (Müller) by stock management prior to direct drilled pasture establishment. *Journal of Agricultural Science.* **111** , 443-449.

Glen, D; Jones, H & Fieldsend, J (1990) Damage to oilseed rape seedlings by the field slug *Deroceras reticulatus* in relation to glucosinolate concentration. *Annals of Applied Biology.* **117**, 197-207.

Glen, D; Milsom, N & Wiltshire, C (1989) Effect of seed bed conditions on slug numbers and damage to winter wheat in a clay soil. *Annals of Applied Biology.* **115,** 177-190.

Glen, D; Spaull, A; Mowat, D; Green, D & Jackson, D (1993) Crop monitoring to assess the   risk of slug damage to winter wheat in the UK. *Annals of Applied Biology.* **122,**   161-172.

Kerney, M & Cameron, R A D (1979) *Land snails of Britain and North West Europe.* Harper Collins, London. 1-288.

Menhinick, E (1964) A comparison of some species individual diversity indices applied to samples of field insects. *Ecology.* **45,** 859-861.

Walden, H (1981) Communities and diversity of land molluscs in Scandinavian woodlands. 1 High diversity communities in taluses and boulder slopes in SW Sweden. *Journal of Conchology.* **30,** 351-372.

Young, A; Port, G; Emmett, B & Green, D (1991) Development of a forecast of slug activity: models to relate slug activity to meteorological conditions. *Crop Protection* **10,** 413-415.

Zar, J (1984) *Biostatistical analysis* 2nd. Ed. Prentice Hall, New Jersey 1-697.

THE USE OF REFUGE TRAPS IN ASSESSING RISK OF SLUG DAMAGE: A COMPARISON OF TRAP MATERIAL AND BAIT

A G YOUNG, G R PORT, A D CRAIG

Department of Agricultural and Environmental Science, University of Newcastle, Newcastle upon Tyne, NE1 7RU, UK

D A JAMES, T GREEN

Rhône-Poulenc Agriculture, Fyfield Road, Ongar, Essex, CM5 0HW, UK

ABSTRACT

The relative efficiency of a number of trap and non-toxic bait material combinations for retaining slugs were investigated in cereal fields in north-east England. A number of materials and baits were found to be more efficient than commonly used combinations, and hardboard squares baited with a poultry food, layers mash, was the best combination for retaining slugs.

INTRODUCTION

Slugs are major pests of cereal and potato crops in the UK. Field work associated with the control of pest slug populations often uses trapping as a means of obtaining an indication of species composition, a relative assessment of slug activity or indirectly assessing population size. Traps have a vital role to play in the on site assessment of slug activity density and have long been used by farmers as the basis for decisions on the need for chemical control.

A number of trap types have been used, the commonest being some form of shelter laid on the soil surface which may or may not cover bait. Glen and Wiltshire (1986) used inverted plastic plant pot saucers (18 cm diameter) covering molluscicide bait in their work on the assessment of populations from bait trap catches and showed they provided a reasonable estimate of the number of large slugs in the top 10 cm of soil, but also that the traps gave no indication of the number of small slugs present. Byers and Bierlein (1984) used roof shingles (30.5 cm square) lain over areas of alfalfa to assess residual populations of slugs following molluscicide applications.

Young (1990) compared two commonly used trap types in arable land and found that inverted plant pot saucers consistently retained more slugs than ceramic tiles when baited with bran. Plant pot saucers and similar forms of trap have been widely used to forecast slug damage in crops (Oakley, 1984) and under these circumstances they are assumed to provide a reasonably accurate short-term assessment of the level of slug activity at the soil surface.

Until recently trapping was an unreliable way of estimating the size and composition of slug populations, as the proportion of the population sampled was unknown. New techniques using defined area traps (Ferguson, Barratt and Jones, 1989) allow this assessment to be carried out without resorting to more laborious and destructive methods such as soil washing (Pinder, 1974) or soil flooding (South, 1964). Defined area traps simply enclose a population, which is sampled (removed) as slugs move to the surface. Refuge traps such as those described in this study are only 'traps' if they are baited with molluscicide, i.e., slugs can enter, but not leave. Recent studies (Howling and Port, 1989 and Wareing and Bailey, 1989) have shown that slugs may move quite some distance after consuming molluscicide baits and care must be taken in interpreting data collected in this way. Baiting with bran reduces their status to that of a refuge, as slugs can both enter and leave, however in keeping with common usage the term 'trap' will in this report refer also to refuge 'traps'.

Traps are only useful as a decision tool under a restricted range of circumstances. Firstly they must only be used when environmental conditions are suitable for slug activity. Secondly traps of unsuitable material or with unpalatable bait will fail to retain slugs and subsequent assessments will be unreliable. This paper describes the results of a number of successive field trials to compare the numbers of slugs retained under refuge traps of a wide range of material and non-toxic bait with currently popular trap/toxic bait combinations. A preliminary screen of the materials was followed by two more detailed studies with a restricted range of materials at a number of sites in the north east of England.

EXPERIMENTAL WORK

In the experiments all traps were arranged in a randomised block design with three replicates of each treatment and baited with bran. Baits were replaced three times each week, unless otherwise stated. Slug counts were analysed (following square-root transformation) using analysis of variance followed by multiple range tests where appropriate.

Experiment 1: Initial screen of a wide range of trap material and non toxic baits

Initial screen of the materials in Table 1 were done in a oilseed rape stubble at the University's field station at Close House, Heddon on the Wall, Northumberland (Grid Ref. NZ127 659) between 9/11 and 25/11/94. The initial screen of non-toxic baits (Table 2) was carried out using 15 cm square ceramic tiles at the same site and time. Traps were visited daily at around 9 am, slugs counted, identified to species and returned to the plot.

Experiment 1: Results

Significant differences were found between trap type for all species present. Interpretation of 27 treatments is difficult as the number of combinations is large. However it is possible to rank each treatment in order of decreasing effectiveness (i.e., 1=best, 27=worst), for each species. Adding ranks gives a total score, and an indication of the most effective material for the species at this site. This method adds equal weight to the ability of the trap to retain each species, despite the dominance of *Deroceras reticulatum* in the sample. It would be unwise to choose traps on the basis of its ability to retain a single species as the species composition of remote sites is often not known. Where possible, attempts were made to ensure that the majority of traps were the same dimensions as the Bayer mat traps. However, twenty of the trap types differed in size from this standard and therefore a correction was made, by dividing the perimeter length of each trap into that of the standard. The resulting quotient was used to multiply up the trap catches from all the non-standard size traps and the analysis of ranks was repeated. The results are shown in Table 3.

The majority of slugs present at this site were *Deroceras reticulatum*. This species was retained in traps baited with carrot, apple, cabbage and layers mash significantly more than with other baits. The two *Arion* species showed slightly different preferences in that *Arion distinctus* preferred bran, blank molluscicide pellets, apple and beer, whilst *Arion fasciatus* preferred all baits in preference to carrot. Total slug numbers tended to reflect the dominance

Table 1. Trap materials.

Trap material	Trap size (d=diameter)	Trap material	Trap size (cm) (d=diameter)	Trap material	Trap size (cm) (d=diameter)
Ceramic tile	15x15	Black foam rubber carpet underlay	35 x 52	Plastic seed tray	38 x 24
Plastic plant pot saucer	16.5 d	Cardboard	50 x 50	Plastic tube	15.5 x 2d
Carpet	44 x 46	Plywood	30 x 50	Egg tray, cardboard	29 x 29
Black plastic sheet	50 x 50	Terracotta plant pot saucer	18.5 d	Lonza slug trap	24 x 24
Felt carpet underlay	41 x 30	Twinwall plastic sheet (painted black)	18.4 x 18.4	Plastic seed tray, inverted	38 x 24
Bubble pack (small, covered with black plastic)	50 x 50	Bayer mat trap (capillary matting between polythene)	44 x 46	Bubble pack (large, covered with black plastic)	50 x 50
Hessian sack	54 x 44	Plastic gutter	30 x 11	Plastic conduit	17.6 x 2.3
Hardboard, shiny side up	40 x 40	Green foam rubber underlay	50 x 50	Plastic dustbin lid	49 d
Brick	24 x 11.5	White plastic sheet	50 x 50	Linoleum	41 x 28

Table 2. Non-toxic bait materials.

Bait material	Comments	Bait material	Comments
Bran	wheat bran	Potato	sliced, raw
Blank molluscicide	commercial product without a.i.	Cabbage	whole leaves
Layers mash	commercial poultry food	Beer	in a shallow dish
Wheat	whole grains	Cat food	tinned, meat based
Apple	sliced	Carrot	sliced

Table 3. Most effective trap materials for all species and overall ranks, following compensation for trap size.

Trap material	Rank	Trap material	Rank
Linoleum	1	Lonza slug trap	6
Hardboard, shiny side up	2	Terracotta plant pot saucer	7
Black foam rubber carpet underlay	3	Bubble pack (large, covered with black plastic)	8
Plywood	4	Cardboard	9
Plastic dustbin lid	5	Ceramic tile	10

of *Deroceras reticulatum* in the population and as a result only wheat seed and cat food were significantly poorer in retaining slugs.

Experiment 2: Assessment of promising materials and baits

Having screened the above materials and baits, two further trials were carried out. One trial was done in rotational set-aside, at Harlow Hill, Northumberland (Grid Ref. NZ084 680) between 31.3 and 11.4.95. The second trial was done in a field of winter rape, at Heddon on the Wall, Northumberland, (Grid Ref. NZ138 657), between 13.4 and 4.5.95. Traps were visited daily, slugs counted, identified to species and returned to the plot. Six different trap types (bubble pack with black polythene cover, Bayer mat trap, black foam rubber carpet underlay, cardboard egg tray, ceramic tile and hardboard) were used and three baits (carrot, apple and layers mash). Baits were replaced every second day. Trap types, sizes and baits are described in Tables 1 and 2, all test traps were baited with bran, all test baits were held under tile traps. Four replicates of each treatment were used.

Experiment 2: Results

Harlow Hill. Of the 19,007 slugs recorded, all were *Deroceras reticulatum* except 9 *Arion ater* which were excluded from the analysis. The results are shown in Figures 1 and 2. Figure 1 shows the influence of trap type on the numbers of *D. reticulatum* caught. The Bayer mat trap, hardboard and underlay all performed equally well at this site. The ceramic tile provided smallest numbers of slugs, but this is partly an effect of size. Of the baits, layers mash was significantly better than either carrot or apple.

Heddon on the Wall. 10,430 slugs were recorded at this site. *Deroceras reticulatum* was predominant (10,140), but there were also numbers of *Arion fasciatus* (191), *Arion ater* (86) and *Arion circumscriptus* (13). The results are shown in Figure 1. The Bayer mat trap retained the largest numbers of *D. reticulatum*, but was not significantly different from either the egg carton or the bubble pack. In turn these two were not significantly different from hardboard or underlay. As at Harlow Hill the tile trap provided the smallest catch. Catch of *Arion fasciatus* and *Arion ater* showed no differences between trap types, possibly because of low numbers of slugs. In contrast to the results from Harlow Hill, there were no significant differences between baits at the Heddon site for *Deroceras reticulatum*. However, *Arion fasciatus* and *Arion ater* both showed a preference for carrot. Whether these differences are due to the slug species involved, population differences between the sites or due to site differences (the Heddon site was in oilseed rape) is not clear. There were significant day to day variations in how the traps and baits performed. For example the results for egg carton were more variable than some of the other trap types. This was probably due to the susceptibility of this trap to drying of the underside, reducing the catch. At Heddon, in oilseed rape, this trap performed far better presumably because it was less likely to dry out.

Of the trap types there was little difference between the Bayer mat, the hardboard the underlay and the bubblepack in terms of the total catch. Few of the traps are without their own limitations in the field however. The Bayer mat trap is expensive, and must be kept moist, the underlay and the bubble pack is unlikely to be robust enough to withstand field conditions. Of the two remaining materials, hardboard is less costly, more readily available and more robust. For those reasons, in addition to its ability to retain slugs, this material was selected for use in the final experiment.

Figure 1. Effect of trap material on slugs retained using bran bait (bars with same letter (within site) are not significantly different at p=0.05)

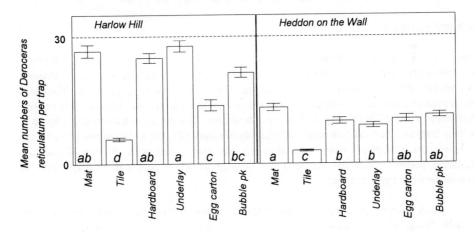

Figure 2. Effect of bait material on slugs retained in tile traps (bars with same letter (within site) are not significantly different at p=0.05)

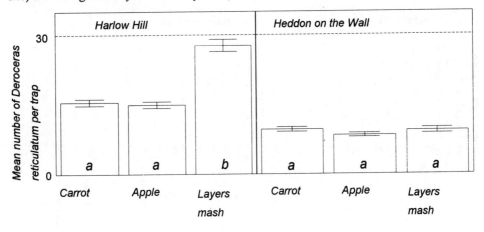

Experiment 3. Final assessment of trap/bait combinations

This experiment aimed to evaluate the performance of the favoured trap design together with a number of baits at several sites. One trial was done in August (1995) at a time when slug populations need to be assessed in order to evaluate the risk of damage in subsequent crops. The second trial was done in November (1995) when slug activity is monitored in order to determine the need for molluscicide applications. Both trials were done in the same five fields at Heddon on the Wall, Stocksfield and at Harlow Hill, Northumberland (grid refs, NZ 137 657, 085 663, 081 667, 071 684 and 038 631). Traps were visited daily, slugs counted, identified to species and returned to the plot. Two trap types were used (yellow painted and unpainted hardboard squares 40 x 40 cm) and three baits (bran, Genesis - 4%w/w Thiodicarb molluscicide and layers mash). In addition 15 cm ceramic tile traps baited with Thiodicarb molluscicide were included as a control. Baits were replaced every other day. Five replicates of each treatment (ten replicates of the 15 cm ceramic tile traps) were used.

Experiment 3: Results

August 1995. Trapping was done on 13 nights giving a total of 2,600 trap nights. However, only 2,689 slugs were recorded probably due to the extremely warm, dry weather conditions. All of these slugs were *Deroceras reticulatum*. Whilst there were significant differences between numbers recorded from day to day and site to site this did not influence the relative performance of the different trap and bait combinations. The results are shown in Figure 3. Only two trap and bait combinations differed significantly (**P**<0.05), unpainted traps baited with layers mash trapped more slugs than 15 cm tiles baited with 4%w/w Thiodicarb molluscicide pellets. Over this period about 66% of all trap-nights had no slugs with no difference between the trap types.

Figure 3. mean numbers of slugs retained in different combinations of trap and bait (Y=yellow painted hardboard trap, U=unpainted hardboard trap, bars with same letter (within date) are not significantly different at p=0.05)

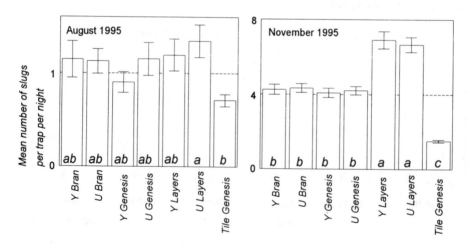

November 1995. Trapping was done on 11 nights at all sites and continued for a further 3 days at a single site where initial captures were low, giving a total of 2,320 trap nights. A total of 10,593 slugs were recorded, all were *Deroceras reticulatum* with one exception (*Arion* sp.) Again there were significant differences from day to day, but this did not influence the relative performance of the different trap and bait combinations. There was some evidence of relative trap performance differing from site to site, but inspection of the data showed this was not important. The results are shown in Figure 3. All the hardboard traps performed better than the 15 cm tiles. Whether the trap was painted or not had no effect on the numbers of slugs. Traps baited with layers mash collected significantly (P<0.05) more slugs than those baited with bran or 4%w/w Thiodicarb molluscicide . Over this period about 13% of hardboard trap-nights had no slugs whereas for tile traps 33% of trap-nights were zeroes.

CONCLUSIONS

With farmers and advisors under increasing pressure to reduce pesticide use to a minimum and to use more ecologically friendly methods of pest control the timing and appropriateness of molluscicide pellet applications are under particular scrutiny. The need for adequate test baiting has been apparent for many years in order to ensure slug activity in fields is large enough to justify pellet applications. The means by which this has been done has not been rigorously studied and there are certainly no proven economic thresholds of slug activity. In selecting materials we have attempted to choose materials which will be suitable resting places for slugs and in the case of non-toxic baits, suitable as foods. We have concerned ourselves with non-toxic baits in particular since the use of large quantities of toxic bait in a single trap constitutes an unacceptable risk to wildlife and domestic animals and, in any case, contravenes guidelines on the use of pesticides.

Of the trap and bait combinations tested many were shown to be suitable but impractical in the field. Above all, traps should be cheap, easy to put out in a field, easy to find the following morning and consistent. Ceramic tile traps have been shown to be poor at retaining slugs (Young 1990) and this was reiterated here. A number of materials were shown to be as useful as the Bayer mat traps, but much cheaper and simpler to use. Hardboard squares appear to fit the criteria well, all of the hardboard traps, painted or unpainted, performed with similar efficiency. Layers mash, a poultry product readily available to farmers, cheap and resistant to decay, proved to be superior to other conventionally used baits.

We believe that the use of the hardboard/layers mash trap will provide a significant advance in the control of slugs, yet they must be used in the context of an holistic approach to slug control which takes into account an acknowledged threshold of slug activity in combination with prevailing weather conditions in order to ensure that pellets are applied at the right time in the field.

ACKNOWLEDGEMENTS

We wish to acknowledge the co-operation of Mr Arthur Watson of Heddon Banks farm, Heddon on the Wall, Mr Lockey of Harlow Hill farm, Harlow Hill and Mr Moffitt of Peepy farm, Stocksfield for allowing us to use their land for this work. We acknowledge the support of Rhône-Poulenc Ltd for financial support.

REFERENCES

Byers, R A and Bierlein, D L (1984) Continuous alfalfa: Invertebrate pests during establishment. *Journal of Economic Entomology.* 77, 1500-1503.

Ferguson, C M; Barratt, B I P; Jones, P A (1989) A new technique for estimating density of the field slug (*Deroceras reticulatum* (Müller)). In: *Slugs and Snails in World Agriculture*, IF Henderson (ed.). BCPC monograph No. 41, pp. 331-336.

Glen, D M and Wiltshire, C W (1986) Estimating slug populations from bait trap catches. Proceedings *1986 British Crop Protection Conference* - Pests and Diseases, pp. 1151-1158.

Howling, G G and Port, G R (1989) Time lapse video assessment of molluscicide baits. In: *Slugs and Snails in World Agriculture*, IF Henderson (ed.). BCPC Monograph No. 41, pp. 161-166.

Oakley, J N (1984) *Slugs and Snails*. Leaflet No. 115, MAFF. HMSO, London, 12p.

Pinder, L C V (1974) The ecology of slugs in potato crops with special reference to the different susceptibility of potato cultivars to slug damage. *Journal of Applied Ecology*. **11**, 439-451.

South, A (1964) Estimation of slug populations. *Annals of Applied Biology*. **53**, 251-258.

Wareing, D R and Bailey, S E R (1989) Factors affecting slug damage and its control in potato crops. In: *Slugs and Snails in World Agriculture*, IF Henderson (ed.). BCPC Monograph No. 41, pp. 113-120.

Young, A G (1990) Assessment of slug activity using bran baited traps. *Crop Protection*. **9**, 355-358.

ORGANIC MATTER PARTITIONING IN *ARION ATER*: ALLOMETRIC GROWTH OF SOMATIC AND REPRODUCTIVE TISSUES THROUGHOUT ITS LIFESPAN

J M TXURRUKA, O ALTZUA

University of the Basque Country, Faculty of Sciences, Department of Plant Biology and Ecology, Apdo 644, 48080 Bilbao, Spain

M M ORTEGA.

University of the Basque Country, Faculty of Sciences, Department of Animal Biology and Genetics, Apdo 644, 48080 Bilbao, Spain

ABSTRACT

Growth of seven body organs (dorsal body wall, foot, digestive gland, empty gut, ovotestis, albumen gland and accessory sex organs), as well as the total body growth, of *Arion ater* (L.) have been modelled. Sigmoidal curves have been fitted to every organ with the exception of the ovotestis where a bell shaped curve has been chosen. Intrinsic growth rate is lower for the somatic fractions (0.030-0.058 mg mg^{-1} day^{-1}) which determine the trend exhibited by total body weight (0.028 mg mg^{-1} day^{-1}); values for reproductive organs increase by a factor of 2-3. The study of the inflexion points and maxima suggest that growth in the different sections is tightly coordinated, the shift from somatic increase to germinal development occurring when ovotestis growth is triggered.

INTRODUCTION

Arion ater is a semelparous species with determinate growth (*sensu* Perrin *et al.*, 1993) and a "bang-bang" strategist, at first devoting all excess resources to growth, and then switching the allocation towards reproduction. Thus, the life cycle is better understood when divided into two broad phases. In the "immature phase" the reproductive organs remain undifferentiated and/or small, relative to the somatic structures, whereas in later developmental stages, the "mature phase", once the switch has occurred, the reproductive organs experience rapid growth implying the massive and explosive production of gametes, followed by copulation, egg-laying, and, after a short post-reproductive period, death.

Distinct periods in the life cycle of *Arion* have been described by several authors (references in South, 1992) who divide the immature phase of slugs into two developmental stages, using for this purpose a similar terminology. The first, or "infantile", stage corresponds to the stage of slow growth, usually occurring during winter, and the second or "juvenile" stage showing the highest growth rate, takes place during spring, with the general rise in temperature and day lengthening. Maturation of the reproductive tract of *Arion ater* was studied by Smith (1966) who described three phases. The first one corresponds to the "juvenile" stage and the remaining two may be assigned to the broadly defined "mature phase", with male and female stages.

Animal growth can often be described by sigmoidal models. In *Arion*, in particular, an initial phase, characterized by slow growth, is followed by an accelerating (exponential) period in which the animal reaches its maximum rate of weight increase. In this work an attempt to model growth of *Arion ater* has been made, taking into consideration body partitioning into functional subsystems. So, in the somatic components, digestive structures and locomotive as well as reserve organs, have been individually studied and the same procedure has been followed when analyzing the reproductive organs distinguishing among ovotestis, accessory sex organs (ASO) and albumen gland.

MATERIALS AND METHODS

Random sampling of a population of *Arion ater* was undertaken over 19 consecutive months (3 weeks periodicity except in winter) in a fixed area ($\approx$ 1250 m^2) of mixed woodland and pasture in Bizkaia (43°23' N, 2°40' W, Basque Country). Collection started in early August 1994 and finished in late February 1996. This period of time allowed us to record two complete growth cycles of the reproductive tract. 830 slugs were collected, kept one day in a laboratory fed on lettuce, weighed, killed by immersion in liquid nitrogen, packed individually and conserved at -18 °C in hermetic flasks until dissection. 675 have been dissected into four somatic parts (dorsal body wall (DBW, it is the part of the body wall above the well defined and coloured foot fringe), foot, digestive gland and empty gut), and three parts of the reproductive system (ovotestis, albumen gland and ASO (all the remaining parts of the reproductive tract)). They were freeze-dried (15 °C, 18 h) and ashed (450 °C, 14 h) for organic determination. Wet, dry and ash weights were recorded. The "entire body weight" equals the sum of the weights of the seven parts.

Modelling growth for the different body parts has been performed by fitting experimental data to either a sigmoidal curve or to a bell shaped curve, by means of the iterative algorithm (Levenberg-Marquardt method; Deltagraph Pro3. DeltaPoint, Inc., Monterey, CA). To make the curve fitting of the genital organs possible, it has been necessary to exclude data corresponding to the first 250 days for albumen gland and ASO (when weight $\approx$ 0). The following equations have been used:

Sigmoidal: $W = k/1 + e^{(a-bt)}$

where W= weight (mg) of organic matter at age t, k= asymptote of the maximal weight, e= 2.71828 (base of natural logarithms), a= a parameter of integration defining the position of the curve relative to the origin, b= intrinsic rate of natural increase (growth in mg mg^{-1} day^{-1}), and t= age in days.

Bell shaped curve: $W = a*e^{[(-1/2)\ [(t-b)/(c-b)]^2]}$

where W= weight (mg) of organic matter at age t, a= maximal weight, e= 2.71828 (base of natural logarithms), t= age in days, b= age at maximal weight, and c= age at ascending inflexion point of the curve.

Exact age is difficult to ascertain from field studies. In this population of *Arion ater*, hatching occurs probably continuously from mid autumn to mid winter: On October 24 we found 7 infant slugs with a mean live weight of 1027 mg (range, 388-2099 mg) and 51 females, and three weeks earlier we had found 55 females and no newly hatched infants. So we supposed that the infantile slugs found in October 24 hatched on some day between October 4 and 24. We therefore arbitrarily fixed the generation's first day of life at October 9. Thus, we have not estimated the age of individuals, but the mean age of the generation. In this sense, modelled growth curves should describe the growth of the different body organs of the year class.

RESULTS

Parameters which define the various growth curves are summarized in Tables 1 and 2. The equations explain between 50-80% of the variation, and the intrinsic growth rates of the reproductive organs greatly exceed those of the somatic organs. Curves based on data obtained for the somatic tissues of the 1995 generation and the reproductive organs of both the 1994 and 1995 generations are shown in Figures 1 and 2. Incomplete sampling of the somatic growth phase during 1994 prevented its modelling, but the maximal weights attained are similar for both years (i.e. Anova results not significantly different).

Comparison between trends for genitalia, showed no differences between years: ANOVA results for maximal weights of albumen gland (n=26, Scheffe F-test = 0.00193) and of the ASO (n=75, Scheffe F-test = 3.307) and t test of the parameters of the ovotestis curve (n=507, t value = 1.2253) were inconclusive.

Table 1. Parameters of the sigmoidal growth curves (Organic matter content of organs vs estimated generation age).

Variable	k	a $\pm$ CL$_{95\%}$	b $\pm$ CL$_{95\%}$ (x10^{-3})	n	r^2
Entire body	1959.79	6.789 $\pm$ 0.110	0.0287 $\pm$ 2.051	423	0.608
Somatic sections					
Foot	410.36	10.436 $\pm$ 0.108	0.0479 $\pm$ 2.010	423	0.542
Dorsal body wall	626.79	12.306 $\pm$ 0.119	0.0577 $\pm$ 2.204	423	0.472
Digestive gland	371.50	6.966 $\pm$ 0.134	0.0313 $\pm$ 2.488	423	0.493
Empty gut	102.38	6.905 $\pm$ 0.102	0.0303 $\pm$ 1.902	423	0.613
Reproductive organs					
A S O (1994)	270.53	6.794 $\pm$ 0.195	0.0968 $\pm$ 6.884	130	0.698
A S O (1995)	234.59	9.743 $\pm$ 0.141	0.1245 $\pm$ 5.536	194	0.805
Albumen gland (1994)	361.45	13.915 $\pm$ 0.224	0.1359 $\pm$ 8.111	135	0.508
Albumen gland (1995)	323.81	20.188 $\pm$ 0.210	0.1898 $\pm$ 10.364	168	0.640

Except for the ovotestis, which is best described by a bell shaped curve, the growth of the remaining organs as well as the total body fitted sigmoidal growth curves. Body organs showed different rates of approach to their maxima, indicating allometric relationships with body weight. A more detailed picture may be obtained from the inflexion points for each curve (Table 3). The inflexion point represents the age at which each organ reaches half of its maximum organic weight. Moreover, the slope of the tangent to the curve at this point gives the value of the highest growth rate (Table 4). The somatic tissues all reach their inflexion points within about two weeks of each other, just when the rapid development of ovotestis starts (Figure 2, Table 3). Thereafter, withdrawal of organic matter from these tissues takes place at a high rate and extensive weight loss occurs from day 380 (Table 5).

Table 2. Parameters of the bell shaped growth curves of the ovotestis .

Variable	a	b	c	n	r^2
Ovotestis (1994)	103.89	332.0	305.3	163	0.547
Ovotestis (1995)	96.35	335.6	308.3	345	0.725

Table 3. Timing of the major events of the different body tissues of *Arion ater* L calculated according to the parameters in Tables 1 and 2.

Estimated age (days)	Event(s)
213	Inflexion point for dorsal body wall
218	Inflexion point for the foot
223	Inflexion point for the digestive gland
225	Beginning of the rapid development of the ovotestis
229	Inflexion point for the gut
236	Inflexion point for total body weight
300	Beginning of the rapid development of the ASO
308	Inflexion point for the ovotestis
328	Inflexion point for the ASO
335	Maximal weight of the ovotestis; appearance of rudimentary albumen glands
356	Inflexion point for the albumen gland
362	Inflexion point in the degrowth of the ovotestis
380	Maximal weights of the foot, DBW, digestive gland, gut and ASO
390	Maximal weight of the albumen gland and of the total body

From the equation parameters in Tables 1, 2 and 5, it is possible to calculate the amount of matter incorporated in, or transferred to, a particular tissue in a given period. So from age 200 d (when non-reproductive tissues are growing at near maximal rates and the ovotestis is still very small) to age 300 d (when the rapid development of the ASO begins), organic matter in the slug has increased by 1178 mg, but only 50 mg have gone to reproductive organs. On the other hand, from age 300 d to age 400 d (when females are laying eggs), despite an increase of 500 mg in the organic matter of the reproductive organs, total body weight has only increased by 250 mg, indicating that the somatic tissues have lost a similar amount. In quantitative terms, the dorsal body wall is the main storage tissue, losing 19% of its organic content between day 300 and 400. Significantly, at day 300, the ASO and albumen gland together represent 0.4% of total body weight, and this percentage rises to 27.6% by day 400.

The explosive growth of the albumen gland appears as the major contributor to this increase: growth rates for this organ greatly exceed those of the remaining tissues, even when taking the slug as a whole (Table 4).

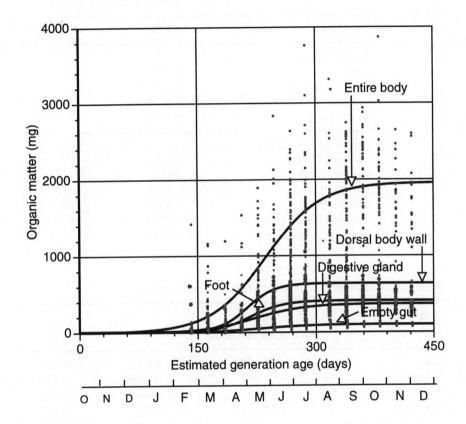

Figure 1. Growth curves of the somatic tissues and entire body.

DISCUSSION

The timing differences in growth of the various body tissues found in this work (Figures 1 and 2), confirm the separation of body size and maturation of the reproductive tract.. This same conclusion had been reached by Smith (1966), but is in contrast to Lüsis (1961), and indicates energy gained from the environment is being channelled, in sequence first to somatic tissue building and, later, to reproduction. Our modelling procedures allow some generalizations about the life cycle of *Arion ater* in our area of study. Four stages can be recognized, reproductive activity occurring exclusively in the final two:

Table 4. Maximal growth rates.

Body tissue	Growth rate (mg organic matter day^{-1})
Total body	14.0741
Somatic tissues	
Foot	4.9156
Dorsal body wall	9.0349
Digestive gland	2.9049
Empty gut	0.7756
Reproductive tissues	
Ovotestis (1994)	2.3604
Ovotestis (1995)	2.1382
A S O (1994)	6.5434
A S O (1995)	7.3007
Albumen gland (1994)	12.2741
Albumen gland (1995)	15.3582

Table 5. Rate of weight lost in several tissues (time > 380 d)

Section	Lost rate (mg organic matter day^{-1})	p	r^2
Foot	1.279	0.0703	0.050
Dorsal body wall	3.308	0.0049	0.117
Digestive gland	1.401	0.0412	0.063
Empty gut	0.333	0.0054	0.117
Accessory sex organs	1.591	0.0001	0.207

• Infantile: From hatching until late February, involving about 150 days of generation age. Mean weight of the population increases steadily and the ovotestis/entire body weight ratio remains very low (less than 0.001). These results are coincident with studies undertaken in France (Abeloos, 1944).

• Juvenile: From late February to late June, up to a generation age of 225 days. Weight increase proceeds at an accelerating rate, showing the largest growth rates on lifespan basis: 14.07 mg of organic matter per day which imply total body weight increases by a factor of 10 (Figure 1). At the end of this stage, growth of the ovotestis begins and the shift from somatic growth (which, in fact, reaches the inflexion point) to germinal growth occurs.

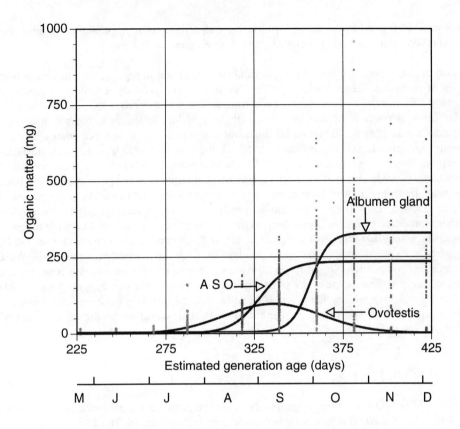

Figure 2. Growth curves of the reproductive tissues.

• Male (corresponding to the mid-spermatozoa stage of the ovotestis described in Smith (1966)): From late June to the middle of September, up to a generation age of 350 days. Growth of somatic structures has nearly ceased, asymptotic weights having been attained (Figure 1). At the end of this stage, the ovotestis reaches its maximum weight -largely due to sperm accumulation (Smith, 1966)- and exponential growth of the albumen gland is beginning to occur (Figure 2). ASO, presumably male, appear well developed reaching 90% of their final weight; copulation and differentiation of female ASO take place (Smith, 1966), and during this period the ovotestis/entire body weight ratio (1/20) matches values reported by Abeloos (1944)).

• Female stage (corresponding to the "oocyte" stage described in Smith (1966)): Concludes with death in late November or December. The ovotestis/entire body weight ratio drops to a value smaller than 1/300 (1/200 in wet weight in Abeloos (1944)). In contrast, albumen gland/entire body weight ratio reaches its highest value: 1/6 (1/7 in wet weight in Abeloos (1944)). No degrowth occurs in the albumen gland, in agreement with Abeloos' findings (1944). Female ASO differentiation is complete, and fertilisation and oviposition are taking place. From October on, two generations overlap (laying females and newly hatched slugs).

Abeloos (1944) and Laviolette (1954) observed oviposition at the beginning of autumn, and Smith (1966) found few newly hatched slugs in November.

In conclusion, a constancy in both timing and final inversion in reproductive tissues in both years of study has been found. Globally, investments in reproductive structures represent roughly 30% of total body weight. Considering that as a semelparous species, reproductive effort in *Arion* ought to be maximized in a single reproductive season in order to guarantee population persistence, and that no simultaneous competition for assimilated energy between somatic and reproductive tissues has been found, the striking parallelism between our ratios of reproductive tissue to body weight and those of Abeloos' (1944) in France deserve some comments. Considerations of metabolic welfare would require a fixed proportion of the body resources (both in terms of biosynthetic and reserves capabilities) to accomplish enhanced egg production (Laviolette, 1954; Smith, 1966). Although, obviously, larger size implies higher reproductive potential, switching to the reproductive stage occurs at a wide range of sizes, suggesting that this process is subject to strict neuroendocrinal control (Joose, 1988), triggered by season and, probably, dependent on previous nutritional conditions. As a consequence, oviposition would be constrained by reserve accumulation, since reproduction occurs largely at the expense of the somatic tissues, when digestive capabilities (South, 1992) and nutritional availability has been noticeably reduced in this population. Additionally, future reproductive success, phylogenetically adjusted, could be predicted from actual allometric relationships.

REFERENCES

Abeloos, M (1944) Recherches expérimentales sur la croissance. La croisance des mollusques Arionidés. *Bulletin Biologique de la France et de la Belgique*. **78**, 215-256.

Joose, J (1988) The hormones of molluscs. In: *Endocrinology of Selected Invertebrate Types*. H Lanfer & G H Downer (eds). New York: A. R. Liss Inc. Invertebrate Endocrinology Vol. 2, pp. 89-140.

Laviolette, P (1954) Rôle de la gonade dans le déterminisme humoral de la maturité glandulaire du tractus génital chez quelques gastéropodes *Arionidae* et *Limacidae*. *Bulletin Biologique*. **88**, 310-332.

Perrin, N; Sibly R M; Nichols N K (1993) Optimal growth strategies when mortality and production rates are size-dependent. *Evolutionary Ecology*. **7**, 576-592.

Smith, B J (1966) Maturation of the reproductive tract of *Arion ater* (Pulmonata: Arionidae). *Malacologia*. **4**, 325-349

South, A (1992) *Terrestrial slugs. Biology, ecology and control*. London: Chapman & Hall. 428 pp.

Session 5
Chemical Control

Session Organiser
and Chairman   T J Martin

THE EFFICACY OF METALDEHYDE FORMULATIONS AGAINST HELICID SNAILS: THE EFFECT OF CONCENTRATION, FORMULATION AND SPECIES.

J B COUPLAND

CSIRO Division of Entomology, European Laboratory, Campus de Baillarguet, 34982 Montferrier sur Lez, France

ABSTRACT

Large differences in efficacy of 8 metaldehyde formulations were seen when tested against the helicid pest snails, *Theba pisana*, *Cernuella virgata* and *Cochlicella acuta*. Of all three species *C. acuta* was very resistant to all formulations tested while *T. pisana* and *C. virgata* were susceptible to two different formulations. This variation is important in assessing suitable formulations for large scale use when there is more than one target snail species.

INTRODUCTION

Metaldehyde was first used as a molluscicide in 1934 in South Africa (Gimingham, 1940) though was not used in England until 1936. There is much uncertainty as to how the molluscicidal properties were discovered with some stories verging on the unbelievable. What is known for certain is that metaldehyde is an effective molluscicide which has found broad-use throughout the world.

In general it has been found that snails are more susceptible than slugs to metaldehyde though there are marked interspecific differences (Godan, 1983). These differences in susceptibility should therefore be taken into consideration when developing a suitable formulation for the control of more than one species of snail or slug.

In Australia, four species of snail are pests of both crops and pastures and have caused serious economic damage (Baker, 1989). The species *Theba pisana*, *Cernuella virgata* and *Cochlicella acuta* are serious pests of both crops and pastures while *Cochlicella barbara* is a serious pest of pastures. Control of these snails using both chemical and biological methods seems to be the most appropriate solution to this problem.
To determine an appropriate formulation of metaldehyde to use in Australia, *T. pisana*, *C. virgata* and *C. acuta* were tested against 8 different formulations of metaldehyde.

MATERIALS AND METHODS

Test species

Eight products were tested against adult *Theba pisana* (> 15 mm), *Cernuella virgata* (> 15 mm) and *Cochlicella acuta* (> 10 mm) under controlled laboratory conditions. Snails were collected on Oct. 5, 1995 from two areas, *C. acuta* and *T. pisana* were collected from near Grabels, France while *C. virgata* were collected from Baillarguet, France.

Formulation characteristics

The formulations were chosen so that both the effect of metaldehyde concentration and formulation type could be tested. Thus, the formulations F1-F4 were similar in physical characteristics and composition, though, varied in their concentration of metaldehyde. The remaining formulations V1, V2 and G1, G2 varied in both concentration, physical characteristics and composition of inerts (Table 1).

Table 1. Characteristics of the formulations tested against *T. pisana*, *C. virgata* and *C. acuta*.

Formulation	% metaldehyde (wt/wt)	Type	Particles/ gram ± SE
F 1	0	Pellet	46.8 ± 1.1
V 1	1 %	Pellet	15.2 ± 1.5
F 2	2 %	Pellet	51.0 ± 2.1
F 3	3 %	Pellet	48.6 ± 2.3
V 2	5 %	Pellet	50.0 ± 0.7
F 4	6 %	Pellet	46.2 ± 2.3
G 1	1.5 %	Granule	116.2 ± 5.3
G 2	7 %	Granule	557.8 ± 25.6

Testing procedure

For each of 5 replicates per species, 14 individuals of either *T. pisana* or *C.virgata* or 30 of *C. acuta* were placed in a covered ventilated plastic box (330x220x30mm) on a thin layer (4mm) of moistened vegetable based sponge material (Spontex, 92022 Nanterre, France). In each replicate either 7 g (for *T. pisana* and *C. virgata*) or 4 g (for *C. acuta*) of finely diced carrot were introduced along with 1.25 g (at the beginning of the assay) of metaldehyde product. Each day the number of dead snails were recorded and removed.

All remaining uneaten carrot was removed daily and replaced with fresh carrot. The remaining uneaten carrot was then dried for 24 h at 100° C and then weighed. To control for variation in carrot dry weight, 8 replicate samples were wet weighed and then dried for 24 h at 100°C and then re-weighed for each daily batch of carrot used. The mean conversion factor was then calculated from these replicates and used for conversion of the bioassay samples. All experiments were carried out at 15°C in a 12h:12h light:dark regime in a controlled environment chamber and ran for 10 d.

Statistical analysis

Mortality curves were estimated from data grouped from each of the five replicates.. Snail mortality rates for each product were compared by a two-sided t-test on the slope of the logistic equation after initial analysis in GLIM using binomial errors. Damage is given as the mean percentage damage. Due to the inherent variation in the technique any damage less than 10% can be viewed as no damage.

RESULTS

Mortality

The products varied widely in their efficacy in killing the different snail species. The control, F1, was eaten readily (all product eaten by day 6 for *T. pisana* and *C. virgata*) by all species of snail tested. The other products were eaten in varying amounts with lower concentration formulations generally being more attractive.

Surprisingly, when tested against *C. virgata* , F1 (control) was not significantly different to V1 (1.1 %) and G1 (1.5%)(Table 2). which indicates that low doses of metaldehyde are inefficient in killing *C. virgata*. Furthermore, was no difference between any products when tested against *C. acuta* (Table 3) indicating that *C. acuta* is extremely resistant to all formulations in lab tests. In contrast, the most efficient product against *T. pisana* was V2 (4.8%) (Table 4), while F4 (5.9%) was the most effective against *C. virgata*.

Table 2. Results of mortality logistic regressions against *C. virgata* and final percent mortality at day 10 (M10).

Product	Alpha	SE	Beta	SE	*	M10
F1	-3.91	0.36	0.24	0.05	a	17 %
V1	-2.38	0.19	0.29	0.03	a	56 %
F2	-4.09	0.30	0.49	0.04	c	61 %
F3	-2.44	0..30	0.42	0.03	bc	77 %
V2	-2.45	0.19	0.51	0.03	c	84 %
F4	-3.06	0.23	0.67	0.04	d	94 %
G1	-2.88	0.22	0.35	0.03	ab	60 %
G2	-2.79	0.21	0.45	0.03	c	76 %

*rows with the same letter are not significantly different, double sided t-test.

Table 3. Results of mortality logistic regressions against *C. acuta* and final percent mortality at day 10 (M10).

Product	Alpha	SE	Beta	SE	*	M10
F1	-5.86	0.51	0.36	0.06	a	8 %
V1	-3..13	0.17	0.31	0.02	a	40 %
F2	-2.97	0.16	0.30	0.02	a	43 %
F3	-2.89	0..16	0.27	0.02	a	39 %
V2	-3.31	0.17	0.37	0.02	a	51 %
F4	-3.42	0.18	0.35	0.02	a	47 %
G1	-3.55	0.19	0.32	0.02	a	35 %
G2	-3.20	0.16	0.34	0.02	a	48 %

*rows with the same letter are not significantly different, double sided t-test.

Table 4. Results of mortality logistic regressions against *T. pisana* and final percent mortality at day 10 (M10).

Product	Alpha	SE	Beta	SE	*	M10
F1	-4.08	0.46	0.12	0.07	a	6 %
V1	-2.76	0.22	0.27	0.03	b	41 %
F2	-2.54	0.20	0.29	0.03	b	54 %
F3	-1.85	0.17	0.32	0.02	b	74 %
V2	-1.82	0.17	0.46	0.03	c	90 %
F4	-1.66	0.16	0.35	0.03	b	80 %
G1	-2.24	0.18	0.29	0.03	b	60 %
G2	-1.87	0.17	0.35	0.03	b	76 %

*rows with the same letter are not significantly different

Feeding

All but F1 effectively stopped feeding on carrot by all the snail species tested. Only F1 had snail feeding damage after day 1 of the experiments. Observations of the testing arenas indicated that ingestion or absorption of metaldehyde inhibited feeding in the pellet formulations while the granular formulations inhibited both movement and feeding within the arenas.

Table 5. Carrot feeding of helicid snails. Cumulative carrot eaten in g during the 10 day trial.

Product	C. virgata	C. acuta	T. pisana
F1	11	2	6
V1	4	2	5
F2	7	1	4
F3	4	2	3
V2	3	2	4
F4	4	2	3
G1	4	2	4
G2	3	1	3

DISCUSSION

The results indicate that there is large variation in the killing effectiveness of different formulations between snail species, though all formulations except F1 effectively stopped feeding damage. The variation in killing efficiency between species may be due to several factors. The feeding behaviour and food resources of the different species is quite different. Indeed, *T. pisana* is primarily a herbivore while *C. virgata* and *C. acuta* are primarily detrivores (Baker, 1986). Therefore, different formulations may vary in attractiveness and ultimate effectiveness. Size is also known to be a factor in the killing effectiveness in snails and slugs. In general, small slugs are more resistant to metaldehyde though the reasons for this are a matter of debate (Godan, 1983). Indeed Godan (1983) found that the large snail, *Helix pomatia* (L.) was much more resistant than the smaller snails, *Cepea nemoralis* L.and *Helix arbustorum* L. In this study, the small snail *C. acuta* was very resistant to all the metaldehyde formulations tested. This is a similar result to Richardson & Roth (1965) who found that the related conical snail, *C. barbara* was more resistant to methyl bromide than the larger *T. pisana*. Thus it seems that species-specific differences in susceptibility in snails may be independent of body size.

The metaldehyde content of the different formulations affected their overall effectiveness. Low concentration formulations (0 - 3%) were in general less effective in killing snails. However, the highest concentration G2 (7 %) was not as effective as V2 a 5% formulation against *T. pisana* and F4 a 6% against *C. virgata*. This indicates that while low concentration formulations may not give a killing dose, high concentration formulations may be repellent and that there is an optimum metaldehyde concentration, which may vary with snail species. The good efficacy of V2 also indicates that bait components other than the concentration of metaldehyde influence both the attractiveness and effectiveness of a formulation.

The granules based products were effective in immobilising the snails (perhaps by contact poisoning) but produced lower mortality rates. There use is therefore somewhat limited in broadacre crops but may have importance in strip mollusciciding where movement of snails into the crop is impeded by a line of molluscicide. The small particle size and concomitant large number of particles per gram of formulation make them particularly appropriate for this application method. The pellet formulations generally had around the same number of pellets per gram (46.2 - 51 pellets/g) though the ineffective formulation V1 had only 15.2 pellets per gram. Therefore overall effectiveness of pellets in the field will be due more to a combination of their attractiveness and the number of pellets per square metre based on application rates.

While laboratory tests are not always applicable to field situations they give an idea of the potential of different formulations. This study indicates that different species are susceptible to different formulations of metaldehyde. Development of molluscicide formulations should take into account these differences with more focus being on formulations that have good efficacy against several species especially in areas where there is a more than one pest snail targeted.

ACKNOWLEDGEMENTS

The financial contribution of LONZA in carrying out this study is gratefully acknowledged.

REFERENCES

Baker, G H (1986) The biology and control of white snails (Mollusca: Helicidae), introduced pests in Australia. *C.S.I.R.O. Technical Paper* No. 25, 31 pp

Baker, G. H (1989) Damage, population dynamics, movement and control of pest helicid snails in southern Australia. In: *Slugs and Snails in World Agriculture*. I F Henderson (ed.). BCPC Monograph No. 41, pp.175-185.

Gimingham, C T (1940) Some recent contributions by English workers to the development of methods of insect control: slugs. *Annals of Applied Biology.* **27**, 167-168.

Godan, D (1983) *Pest slugs and snails: biology and control*. Springer-Verlag, Berlin.
Richardson, H H; Roth, H (1965) Methyl bromide, sulfuryl fluoride and other *Economic Entomology.* **58**, 690-693.

TESTING BAIT TREATMENTS FOR SLUG CONTROL

R H MEREDITH

Bayer plc, Crop Protection Business Group, Elm Farm Development Station, Great Green, Bury St Edmunds, Suffolk IP31 3SJ

ABSTRACT

Data are presented on the performance of new slug bait formulations based on methiocarb, in comparison with a range of commercially available treatments. Results are discussed within the context of the various trial designs used, the merits of which are also considered. Treatments were tested for the potential to attract and kill slugs in baited traps, then further evaluated for crop protection in arable situations, using large replicated plots. All treatments demonstrated the ability to kill slugs and protect crops of wheat, barley and oilseed rape. One trial in oilseed rape stubble, using a special design of small concentric plots, showed consistent evidence that the efficacy of methiocarb was improved by the addition of a phagostimulant or an attractant. In this trial the precision to differentiate between treatments was helped considerably by the large number of slugs, which were evenly distributed and very active.

INTRODUCTION

Chemically treated baits are a long established means of protecting agricultural and horticultural crops from slug attack. The first of these was in widespread commercial use by 1940 (Gimingham, 1940) and contained metaldehyde, a polymer of acetaldehyde, which poisons and immobilises slugs so that they die through dehydration (Cragg and Vincent, 1952). Since then carbamate chemistry has provided two additional active ingredients, methiocarb (Martin et al., 1969) and thiodicarb (Yang & Thurman, 1981); these compounds are stomach poisons and in common with other carbamates they inhibit cholinesterases (Young and Wilkins, 1989). In all baits, the chemical is present at a low concentration (2-6%) in pellets typically made from a cereal base. Although frequently used with great success, the benefits from such baits are inherently linked to slug behaviour and its complex relationship with a whole range of factors including species, physiology, agronomy and weather; this subject has been reviewed by Martin and Kelly (1986). A critical role is of course played by the properties of the baits themselves e.g. attractiveness, palatability and toxicity. Throughout the decades of commercial use, the research and development of molluscicidal baits have continued and improvements to formulations have been made. Described and discussed in this paper are some field evaluations of new candidate treatments from Bayer plc, made in comparison with commercial products. Trial design and the particular difficulty of obtaining consistent results in molluscicide trials are also considered.

MATERIALS AND METHODS

Slug traps

Trials were located on closely mown swards or in crops abutting good slug habitat. Traps consisted of a 5 g heap of pellets protected by a 15 cm x 15 cm tile supported at about 2 cm above the ground by three pegs. Traps were arranged in a line with 1 m between centres and a guard trap containing 4% methiocarb pellets at either end. Dead or moribund slugs were counted in the morning for at least four days after trial initiation, each species being recorded separately. The trial design was randomised blocks with four replicates.

Field control of slug damage (large plots in arable crops)

Trials were located in winter crops of cereals or oilseed rape, where soil conditions and/or previous cropping favoured slug activity e.g. heavy, wet, coarse seedbeds following oilseed rape. Plots were 8 m x 8 m arranged as randomised blocks with four replicates. Treatments were applied using a hand held mechanical spreader or a 'pepper-pot' applicator, usually before crop emergence. Protection of the crop by treatments was typically assesssed by counting the number of slug damaged and undamaged plants in five 1 m lengths of row per plot. Grain yield was also recorded in one wheat trial.

Small concentric plots used in a high pest pressure situation

Plots were located along 3 m wide strips of volunteer oilseed rape. Each plot was 3 m x 3 m with a central area of 1 m x 1 m hand cultivated and sown with oilseed rape, cv. Lictor. There were four replicates, each on a different strip of volunteer rape, in a randomised block design. The trial was surrounded by a 1 m wide band of 4% methiocarb pellets, to prevent further slug migration into the area. Treatments were applied using a hand held applicator, before emergence of the sown crop but when volunteers were at 4-8 leaves and approximately 25% ground cover. Crop establishment was assessed by counting the number of plants in each central sown area at 12 and 26 days after application. Slug control was assessed by removing and counting dead or moribund slugs from $0.5 \ m^2$ fixed quadrats (two per plot from the volunteer area and one per plot from the sown area) at 1, 2, 5, 7, 12, and 14 days after application; all plots were searched thoroughly, including the untreated control.

Slug bait treatments

Methiocarb 2% RB (w) - wheat base, 7.5 kg/ha ('Decoy')
Methiocarb 2% RB (wx+ph) - wheat base + phagostimulant, 5.5 kg/ha (experimental)
Methiocarb 2% RB (wx) - wheat base, 5.5 kg/ha (experimental)
Methiocarb 4% RB (w+ph) - wheat base + phagostimulant, 5.5 kg/ha ('Exit')
Methiocarb 4% RB (w+at) - wheat base + attractant, 5.5 kg/ha (experimental)
Methiocarb 4% RB (w) - wheat base, 5.5 kg/ha ('Draza')
Methiocarb 4% RB (b) - barley base, 5.5 kg/ha (experimental)

Metaldehyde 6% RB - 8 kg/ha
Metaldehyde 4% RB - 15 kg/ha
Thiodicarb 4% RB - 5 kg/ha

The above rates are the amounts of formulation for topical application; metaldehyde and thiodicarb treatments were commercially available formulations applied at recommended rates. Under tile traps all treatments were applied as 5 g piles.

RESULTS

Slug trap data

All treatments attracted and killed slugs of both *Arion* and *Deroceras* species at a range of locations and situations (Table 1). There was a general tendency for methiocarb baits to have the greatest effect but there were rarely significant differences between the various formulations of this active ingredient.

Table 1. Numbers of dead/moribund slugs per baited tile trap

Bait treatment	*Arion* species		*D. reticulatum*	
	range	mean	range	mean
Methiocarb 2% (w)	1-26	11.4	4-14	8.3
Methiocarb 4% (w)	1-25	12.4	1-12	8.6
Methiocarb 4% (b)	1-28	11.6	4-16	9.9
Metaldehyde 6%	1-16	9.5	0-13	5.6
Thiodicarb 4%	1-17	6.0	1-13	6.0

Results are from 5 trials (*Arion* spp) or 6 trials *(Deroceras reticulatum)*
All treatments were applied in autumn 1993 at a rate of 5g per trap

Field control of slug damage

At four sites where slugs were very active after application, treatments prevented considerable crop damage in terms of grazing and plant loss, with benefits being particularly dramatic at a site in Devon, where the crop of oilseed rape was saved from almost complete destruction (Table 2). In a winter wheat trial near Hereford, slug control during crop establishment later resulted in significant increases in yield relative to the untreated crop (Table 3). In these trials all formulations based on methiocarb performed well, generally being more effective than baits made from other active ingredients. The principal species at all locations was *Deroceras reticulatum*.

Table 2. Crop protection following topical application of slug bait treatments

Assessment	Crop establishment				Crop damage		
Location	Kent	Hereford	Kent	Devon	Hereford	Kent	Devon
Crop	wheat	wheat	wheat	OSR	wheat	wheat	OSR
Application date	11 Nov 1993	15 Nov 1993	23 Sep 1994	13 Sep 1994	15 Nov 1993	23 Sep 1994	13 Sep 1994
Days after application	116	B65	20	16	B65	20	16
Untreated units	plants/m^2				percent grazing		
Treatment units	percent relative				percent reduction		
Bait treatment							
Untreated	(87)	(153)	(148)	(2)	(44)	(73)	(79)
Methiocarb 2% (w)	137	326^a	---	---	100^a	---	---
Methiocarb 4% (w+ph)	---	---	136	1700^a	---	78^{ab}	26^a
Methiocarb 4% (w)	127	315^a	118	1647^a	100^a	80^{ab}	33^a
Methiocarb 4% (b)	115	342^a	---	---	100^a	---	---
Metaldehyde 6%	91	261^a	133	1626^a	75^b	85^a	10
Metaldehyde 4%	---	---	126	1505^a	---	90^a	0
Thiodicarb 4%	125	297^a	126	1368^b	85^b	64^b	22^a

Suffixes indicate significantly different from the untreated; values from the same location sharing a common suffix are not significantly different (LSD, P<0.05)
Days after application: "B" = 2nd application
Calculations of % relative and % reduction were made with reference to the untreated

Table 3. Establishment and yield in winter wheat

Assessment	Establishment	Grain yield
Treatment units	% relative	% relative
Bait treatment		
Untreated	153 plants/m^2	6.01 t/ha
Methiocarb 2% (w)	326^a	142^{ac}
Methiocarb 4% (w)	315^a	146^{ac}
Methiocarb 4% (b)	342^a	145^{ac}
Metaldehyde 6%	261^a	134^{ab}
Thiodicarb 4%	297^a	134^{ab}

Results from 1 trial initiated near Hereford on 15 Nov 93
^{a} = group mean significantly different from the untreated (P<0.001)
^{b,c} = group means significantly different from each other (P<0.01)
(compared by orthogonal contrasts)
Calculations of % relatives were made with reference to the untreated

Crop protection and effects against slugs in small concentric plots

The trial area was heavily populated by *Deroceras reticulatum* although *Arion* species, principally *A. hortensis*, were also present. Considerable numbers of dead or moribund slugs were found in treated plots, with only isolated individuals in untreated plots. All treatments prevented almost complete crop destruction in the central sown area. Given the severe pest pressure, all methiocarb baits performed well, being generally more effective than those based on other active ingredients. Of particular interest was how all aspects of the data showed the efficacy of methiocarb baits to be enhanced by the addition of a phagostimulant or an attractant (Table 4, Figs.1-2). There was also a tendency for benefits from the phagostimulant to be greater in the more sheltered volunteer area.

Table 4. Establishment of oilseed rape

Location	Suffolk
Application date	30 Sep 92
Days after application	26
Bait treatment	plants/m^2
Untreated	0.8^c
Methiocarb 2% (w) 7.5 kg/ha	41.0^{ab}
Methiocarb 2% (wx+ph) 5.5 kg/ha	51.0^{ab}
Methiocarb 2% (wx) 5.5 kg/ha	28.5^{ab}
Methiocarb 4% (w+ph) 5.5 kg/ha	59.3^{ab}
Methiocarb 4% (w+at) 5.5 kg/ha	65.7^a
Methiocarb 4% (w) 5.5 kg/ha	56.7^{ab}
Metaldehyde 4% 15 kg/ha	61.5^{ab}
Thiodicarb 4% 5 kg/ha	25.8^{bc}

Values sharing a common suffix are not significantly different from each other (LSD, P<0.05)
N B pest pressure was severe in this trial

Fig 1: 14 day cumulative kill of *Deroceras reticulatum* in oilseed rape, Suffolk 1992.

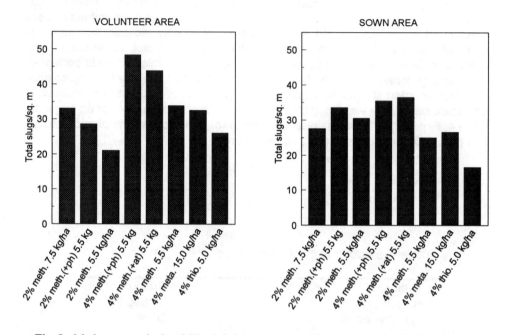

Fig 2: 14 day cumulative kill of *Arion* spp. in oilseed rape, Suffolk 1992.

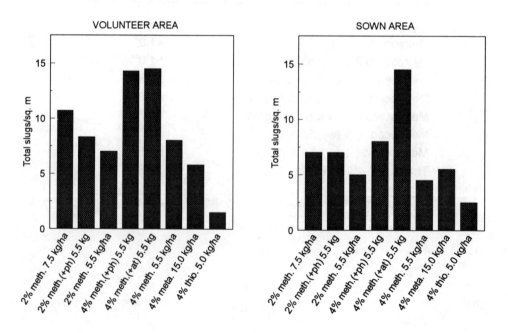

DISCUSSION

The baited slug trap is an artificial situation, not closely resembling practical usage, but nonetheless it serves as a good starting point in the field evaluation of new treatments. As a technique it provides data on the potential to attract and kill slugs, two essential criteria for a successful treatment. Care must of course be taken when interpreting such information as there are limitations to the design. For instance, under normal circumstances the pest will have the choice between crop and pellet, which might influence its decision to feed on the pellet. It would also not be presented with a pile of bait but individual pellets, increasing the probability of sub-lethal poisoning. The results of Howling & Port (1989) suggest that slugs are less likely to die *in situ* after ingesting methiocarb bait compared with those ingesting metaldehyde; thus the results with methiocarb might also underestimate its relative efficacy if more slugs migrated from traps to die. Despite this potential drawback, methiocarb pellets tended to show better efficacy relative to other treatments in tile trap evaluations.

Although numbers of dead slugs found in traps or on the soil surface are a good indication of an effective treatment, the most important measure of success is the ability to protect a crop, as shown by the large plot trials in arable crops. Despite the selection of high risk locations, sufficient slug activity for differentiating between treatments can be elusive, reflecting the complexity of factors that influence slug behaviour. However, in most seasons there are a few trials where the pest pressure is uniform and strong enough to allow such comparisons. This is particularly important when investigating improvements to the formulation of methiocarb, which already exists as an effective, commercially available bait. In the trials presented, protection of the crop was clearly required and all treatments demonstrated good molluscicidal activity. As might be expected, the greatest benefits were obtained in milder, wetter areas in the west of the country and in oilseed rape, which is particularly susceptible to attack.

One of the most successful trials for differentiating between treatments, especially between methiocarb formulations, was that in oilseed rape when small concentric plots were used. The total plot size was the minimum according to EPPO guidelines (Anon. 1986) and the central sown area was therefore less, but this trial provided much useful information due to the large number of slugs which were evenly distributed and highly active. However, for the sown area in particular, larger plots would have been desirable, as the variability of crop establishment data allowed only the most and least effective treatments to be separated with statistical significance. Some variability in plant emergence did come from preparing the seedbed by hand cultivation, although the avoidance of machinery probably improved results with other parameters by minimising disturbance of the slugs and their environment.

To conclude, several techniques were employed to evaluate new slug bait treatments, each method having its own strengths and weaknesses. Ultimately, across all trials, obtaining clear and consistent results depended upon maximising the factors that influence slug activity. This was not always attainable, particularly with the weather being such a major influence. In situations where a molluscicide was undoubtedly required, effective slug control was obtained with a range of methiocarb formulations. In a trial where pest pressure was particularly strong and even, there was consistent evidence that the inclusion of a phagostimulant or an attractant improved the efficacy of methiocarb baits. Laboratory studies by Bowen *et al.*, (1996) also confirm the benefits of using additives with methiocarb.

ACKNOWLEDGEMENTS

I would like to express thanks to all colleagues within Bayer plc for their contributions towards this paper but particularly to Trevor Martin for his many years of enthusiasm and inspiration in matters molluscicidal.

REFERENCES

Anon. (1986) Guideline for the biological evaluation of molluscicides - Slugs in field crops. In: *European and Mediterranean Plant Protection Organisation (EPPO) Bulletin.* **16**, 189-196.

Bowen, I D; Antoine S; Martin T J (1996) Reformulation studies with methiocarb (these proceedings).

Cragg, J B; Vincent M H (1952) The action of metaldehyde on the slug *Agriolimax reticulatus* (Muller). *Annals of Applied Biology.* **39**, 392-406.

Gimingham, C T (1940) Pests of vegetable crops. *Annals of Applied Biology.* **27**, 167-168.

Howling, G G; Port G R (1989) Time-lapse video assessment of molluscicide baits. In: *Slugs and Snails in World Agriculture.* I F Henderson (Ed.) BCPC Monograph No. 41, pp 161-166.

Martin, T J; Davis M E; Morris D B (1969) Development work with methiocarb in Great Britain. *Proceedings 5th British Insecticide and Fungicide Conference (1969).* **2**, 434-441

Martin, T J ; Kelly J R (1986) The effect of changing agriculture on slugs as pests of cereals. *Proceedings 1986 British Crop Protection Conference - Pests and Diseases.* **2**, 411-424.

Yang, H S ; Thurman D E (1981) Thiodicarb - a new insecticide for integrated pest management. *Proceedings 1981 British Crop Protection Conference - Pests and Diseases.* **3**, 687-697.

Young, A G; Wilkins R M (1989) The response of invertebrate acetylcholinesterase to molluscicides. In: *Slugs and Snails in World Agriculture.* I F Henderson (ed.) BCPC Monograph No. 41, pp. 121-127.

CONTROL OF FIELD SLUGS *(DEROCERAS RETICULATUM)* IN WINTER WHEAT USING SEEDS TREATED WITH PESTICIDES

A ESTER, A DARWINKEL
Research Station for Arable Farming and Field Production of Vegetables (PAGV), PO Box 430, 8200 AK Lelystad, The Netherlands

H W G FLOOT
Experimental Farm 'Ebelsheerd', Nieuw-Beerta, The Netherlands

J H NIJËNSTEIN
Cebeco Zaden B V, PO Box 10000, 5250 GA Vlijmen, The Netherlands

ABSTRACT

From 1993 to 1996, six field and several laboratory experiments were carried out in winter wheat *(Triticum aestivum* L.) to assess the controlling effect of seed treatment on slug damage under prevailing growing conditions in The Netherlands. Fields with a heavy marine clay soil have been selected, where high populations of the field slugs commonly caused considerable damage to germination and seedling growth.

The phytotoxicity and efficacy of formulations methiocarb, metaldehyde, thiocyclam hydrogen oxalate, bromoxynil and ioxynil, at two and three dosages of seed film-coating, were tested and compared with the conventional broadcast-application of molluscicide pellets.

Seeds treated with methiocarb and metaldehyde consistently decreased slug damage and were at least as effective as molluscicide pellets. The other pesticides tested were less effective due to an insufficient control of slugs or to phytotoxic effects.

INTRODUCTION

Slugs damage winter wheat by eating the germ causing hollowing of the newly-sown seed, by consuming the shoots below ground level and by shredding the leaves. Seed hollowing is the most important form of damage as it prevents germination (Moens, 1983).

Damage to cereals by slugs (such as *Deroceras reticulatum* Müller) can be a real problem for growers, particularly in the Northern regions of The Netherlands, where the soil is a very heavy marine clay. Damage to slugs can vary from year to year, depending on seasonal weather, the kind of soil and natural enemies (Moens, 1980; Glen *et al.*, 1989; Glen, 1989). Severe damage usually occurs in years with a high level of precipitation in summer and autumn. In the past, 10-25% of wheat fields were attacked. Damage varied from light to so severe that redrilling was necessary (Ester and Darwinkel, 1993).

Until now farmers can reduce crop damage by spreading pelleted baits containing a molluscicide. Application can take place either by mixing the baits at sowing with the cereal seeds (Green *et al.*, 1992), or by spreading the baits after slugs or slug-damage is observed. None of these methods appears to be fully effective (Ester, 1989).

Using molluscicides as seed treatment could be a more effective way of controlling slugs and is less injurious to other soil organisms than a pelleted toxic bait.

Laboratory tests of seed treatments against slugs have been reported previously (Scott *et al.*, 1984; Scott, Griffiths and Stephenson, 1977). This paper deals with further laboratory tests and field trials of some compounds and evaluation of the most effective materials.

MATERIALS AND METHODS

Laboratory experiments

Experiments were carried out in a growth chamber at a constant temperature. Two different experiments were carried out in earthenware pots measuring 20 cm diameter at the top and 18 cm deep. Pots were filled with peat compost up to 7 cm and then covered with 2 cm silver sand. The surface area of the sand was 177 cm². The pots were covered with a polyethylene gauze of 1.35 x 1.35 mm to prevent the slugs from escaping.

In order to maintain the moisture content of the silver sand at its field capacity, the pots were placed in plastic boxes measuring 30 x 45 x 13 cm containing 2 cm tapwater. These boxes containing two pots each were covered with a transparent lid. This resulted in a relative humidity of 85% inside the boxes. Each pot contained 40 wheat seeds of a single treatment, placed embryo-up on the sand surface, in five concentric circles of eight seeds each.

Deroceras reticulatum weighing 400-600 mg, which were collected from traps in a clover field at the Experimental Station at Lelystad were chosen. Slugs which appeared unhealthy were rejected. Following collection the slugs were kept at 15°C in a dark/light cycle of 10/14 h without food for two days which allowed the gut to be completely cleaned, faeces produced in experiments were only white due to feeding on wheat seed.

The assessments were made after two, three and/or seven, eight days, by counting the number of white slug faeces and the percentage of seeds attacked.

Seeds and seed treatment

All seeds were provided by Cebeco Zaden B.V. For the laboratory experiments, 500 g seeds were treated in a plastic bag, to which the pesticide and 2-5% water were added and the bag shaken vigorously until the distribution was even. After treatment the seeds were dried for two days in a laboratory fumehood, after which they were stored at 10°C and 40% r.h.

Experiment A

This experiment was carried out with winter wheat seeds cv. Ritmo. The moisture content of the seeds was approximately 14%, the thousand kernel weight was 48 g.

In this experiment the effective products, which were published by Ester and Nijenstein (1995) were retested. Apart from these products, several other promising active ingredients became available. Among these methiocarb is known for its molluscicide action and bromoxynil and ioxynil (herbicides) both repel slugs (Port *et al.*, 1986). Stephenson (1967, cited in Scott *et al.*, 1977) reported that ioxynil, apart from being a slug repellent, acts as well as a contact and stomach poison to slugs. Thiocyclam hydrogen oxalate (further mentioned as thiocyclam) is known as an insecticide with a slug repellent action (Port *et al.*, 1986). These seeds were not coated.

The experiment was done with four replications at a temperature of 16°C in April 1993. Three slugs were added to each pot. Assessments were made after two and seven days.

Experiment B

This experiment was carried out with seeds of spring wheat cv. Baldus. Thousand kernel weight was 38 g. Together with the molluscicides, a fungicide (Beret at 4 g of formulated product) was added. In this experiment metaldehyde in eight dosages was tested and compared to methiocarb and thiocyclam. Nacret as a coating agent was added as an extra factor. It is a seed coating agent which helps to keep the pesticide load on the seeds during handling and drilling. In addition, filmcoating may influence phytotoxicity. Nacret was used as a sticker (coating) at a rate of 2 g/kg seed. Pesticide and Sacrust (when applied) were added as a mixture.
The experiment was carried out at a temperature of 14°C in March 1995, with four replicates. Each pot contained five slugs. The assessments were made after three and eight days.

Field trials

Seeds and seed treatment

All seeds were provided by Cebeco Zaden B.V. Seed treatments were applied to 10 kg lots of seed. Seed treatment was carried out by rotating a drum for three minutes.
The moisture content of the seeds was approximately 14%, the thousand kernel weight was 52 gram. Before applying other pesticides to the seeds, the seeds were first treated with the fungicide quazantine at a rate of 0.7 g a.i./kg seed; the 'untreated' only received fungicide.

Trials in 1994/1995 and 1995/1996

After an orientating experiment in 1993, during the two seasons of 1994/95 and 1995/96 field experiments were done in the Northern region of The Netherlands. The soil is a heavy marine clay with 63% silt, where usually high densities of slugs are present.
The experiments were randomised in three replicates (1994) and four replicates (1995), with plots of 108 m² each. The winter wheat cv. Ritmo was sown at the end of September 1994 and mid October 1995, the row distance was 12.5 cm between rows and 2 - 3 cm deep. In this field trial the effective products of the laboratory experiments A and B were retested. In addition thiocyclam and methiocarb were tested in half doses as well.
In the 1995/1996 field trial only the molluscicide metaldehyde was used at rates of 3, 4.5 and 6 g a.i./kg seed in comparison to pelleted baits.
The baits pelleted with methiocarb were applied three times 200 g a.i./ha each (1994/1995) and metaldehyde two times 448 g a.i./ha (1995/1996) as recommended for Dutch growers, to provide a benchmark. Controls without pesticide were filmcoated seeds with Neo-Voronit only. All the experiments were laid out in randomised block design and sown with a pneumatic drill.
Field emergence was assessed in November and December 1994, by counting the number of plants at 4 places of 0.25 m²/plot. The crop damage in each plot was assessed and the percentage of attacked plants measured. The plant development has been assessed by estimating crop stand by scoring the density from 0-10. The yield data were determined by harvesting the

grains in August 1995. In November 1995 and January and April 1996 the plants were counted and the development of plants were also determined.

RESULTS

Efficacy of pesticides

Experiment A

In the laboratory experiment A only methiocarb significantly reduced the number of seeds attacked at both rates and on both dates (Table 1). Bromoxynil, ioxynil and thiocyclam also significantly reduced the attack by slugs at all rates tested after two days. However, after seven days the lowest rate of ioxynil and thiocyclam were no longer effective. The number of faeces produced after two days correlated very well with the percentage of seeds attacked. The number of faeces clearly increased at higher levels of seed damage.

Table 1.   Attacked seeds (%), number of faeces produced, non-germinated seeds and intact plants after two and seven days with different pesticide seed treatments in 1993. (Experiment A).

Pesticide	g a.i./kg seed	2 days		7 days		Intact plants (%)
		Seeds attacked (%)	Faeces (no.)	Seeds attacked (%)	Non-germinated seeds (%)	
Untreated	-	38	18	95	4	1
Bromoxynil	0.25	26	12	60	40	0
	0.50	11	5	63	38	0
Ioxynil	0.20	15	10	77	19	4
	0.40	3	1	54	44	2
Methiocarb	1,00	7	4	29	11	60
	2.00	5	2	32	4	64
Thiocyclam	1.0	19	5	74	13	13
LSD (α=0.05)		21	8.2	27	18	21

Experiment B

The fungicide and coating agent tested had no effect on slug damage (Table 2). At the rate thiocyclam was tested, it was not able to protect the seeds and seedlings conclusively. Refering to untreated, methiocarb clearly reduced the number of attacked plants, but this protection was still insufficient. Metaldehyde at rates over 3.2 g a.i./kg seed appeared to be very effective in protecting seeds from slug damage. The protection was not always consistent and increased at higher rates of metaldehyde applied.

Table 2. Attacked seeds (%), number of faeces produced, non-germinated seeds and intact plants after three and eight days at different pesticide seed treatments in 1995. (Experiment B).

Pesticide		g a.i./kg seed	3 days		8 days		
			Seeds attacked (%)	Faeces (no.)	Seeds attacked (%)	Non-germina-ted seeds (%)	Intact plants (%)
Untreated	no fungicide no filmcoating	0	37	17	100	0	0
Untreated	plus fungicide no filmcoating	0	45	16	100	0	0
Untreated	plus filmcoating	0	47	18	100	0	0
Metaldehyde		0.8	6	0	71	8	22
		1.6	3	1	56	8	37
		2.4	4	0	48	4	48
		3.2	2	0	31	9	61
		4.0	1	1	41	2	58
		4.8	1	0	21	5	74
		6.4	1	0	25	6	69
		8.0	1	0	13	2	86
Methiocarb		1.0	10	4	64	8	28
Thiocyclam		1.5	0	0	58	33	10
LSD (α=0.05)			17.5	7.7	21.8	8.6	21.6

Field trial 1994/95

Crop development

After seed treatment with methiocarb 1 g a.i. and metaldehyde 8 g a.i./kg seed, a significantly higher number of plants were found in November and December (Table 3). Also treatment with methiocarb at a rate of 0.5 g a.i. gave more plants than the untreated plots on 20 December. Thiocyclam, ioxynil and bromoxynil had a similar number of plants as the untreated plots. Methiocarb pellets applied twice, methiocarb seeddressing 1 g a.i./kg seed and metalde-hyde 8 g a.i./kg seed gave a significant lower percentage of attacked plants than the untreated plots. On December 20, seed treated with methiocarb at 0.5 and 1 g a.i., thiocyclam at 1 g a.i., bromoxynil at 0.25 g a.i. and metaldehyde at 1.6 and 8 g a.i./kg seed and plots where methio-carb pellets were applied, showed significantly better plant development compared with the untreated plots.

Grain yield

The yield of the plots, where seeds were treated with methiocarb at 0.5 and 1 g a.i., bro-moxynil at 0.25 g a.i. and metaldehyde at 1.6 and 8 g a.i./kg seed, were not significantly dif-ferent from the plots where methiocarb pellets were applied in the drillfurrow. These yields

Table 3. Number of winter wheat plants/m², plants attacked by slugs (%), crop stand (score 0-10) and the yield (kg/100 m²) with different pesticide treatments in 1994/95.

Pesticide	g a.i./kg seed	3 Nov. No. of plants	20 Dec. No. of plants	Plants attacked (%)	Crop stand	Yield
Untreated	0	194	193	49	5.9	93.9
Untreated[3]	0	172	173	42	5.5	90.5
Methiocarb 4% gran.[1][2]	200	199	207	18	7.0	99.4
	200	227	239	37	7.3	102.3
Bromoxynil	0.25	184	183	40	7.0	100.6
	0.50	145	151	45	6.2	91.8
Ioxynil	0.20	204	212	39	6.1	94.1
	0.40	178	190	42	6.5	98.5
Metaldehyde	1.60	215	192	46	7.0	100.4
	8.00	279	279	23	8.4	105.5
Methiocarb	0.50	218	245	35	8.3	104.2
	1.00	264	275	20	9.0	105.8
Thiocyclam[3][3]	0.50	192	185	53	6.5	98.7
	1.00	174	205	30	7.0	97.7
LSD (α=0.05)		62	72	23.7	1.1	7.4

[1] broadcast application; [2] furrow drilling; [3] filmcoated seeds with Panoctine 35

Table 4. Plant number (at different dates) and crop stand (at April 4; score 0-10) at increasing metaldehyde seed treatment and pellets in 1995/96.

Molluscicide	g a.i./kg seed	14 Nov.	27 Nov.	15 Jan.	4 April	Crop stand 4 april
Untreated	0	16	53	79	87	4.5
Metaldehyde	3	32	103	134	143	6.3
	4.5	33	128	151	146	6.9
	6	38	114	156	166	7.5
Metaldehyde 2x	448 g/ha	43	107	136	136	6.3
LSD (α=0.05)		15.7	47.8	62.4	31.6	1.0

were all significantly higher than that from untreated plots. Seeds treated with thiocyclam, ioxynil and bromoxynil 0.5 g a.i./kg seed as well as broadcast applied methiocarb pellets, had no significant effect on yield, if related to the untreated plots.

Field trial 1995/96

In November 1995 and April 1996 metaldehyde seed treatment and metaldehyde pellets twice applied, gave a significantly higher number of plants compared to the plots where untreated seeds were drilled (Table 4). Due to a high field variation, the plant numbers were not significantly different in January 1996. Metaldehyde at a rate of 6 g a.i./kg seed resulted in the best crop development. But also metaldehyde pellets twice applied and the metaldehyde 3 g a.i./kg seed (Table 4) were clearly better than the untreated plots.

DISCUSSION

The active ingredients tested here have been used by other researchers as well. Among the active ingredients, metaldehyde used as seed treatment was very effective for wheat. This compound is toxic to slugs by ingestion and by absorption by the foot of the mollusc. The chemical causes an increase in the secretion of slime causing immobilisation and eventual death by loss of water (Cremlyn, 1991). In our laboratory-experiments with metaldehyde, slugs could crawl over treated seeds and take up metaldehyde through the foot. As a consequence, a lot of slime was produced and the slugs were immobilised. A further indication of such destruction is provided by the reduction in food intake (Table 2), such a finding after the ingestion of metaldehyde by *Deroceras reticulatum* was made by Bieri and Schweizer (1990) and Bourne *et al.* (1988), who also observed a reduced production of faeces as found in Experiment B also. Gould (1962) used metaldehyde at a rate of approximately 10 g/kg seed and this proved to be very effective against slugs up to ten days in boxes and in the field. Scott *et al.* (1977, 1984) tested different compounds at 2 g a.i./kg seed. Ioxynil and bromoxynil proved to be very effective, methiocarb exerted some control, but metaldehyde was not effective at all in their experiments. However, our experiments proved that metaldehyde was effective, but at somewhat higher rates. This may have been the reason for the poor performance of metaldehyde in Scott's experiments. Metaldehyde on wheat was effective at 3.2 g a.i./kg seed. Charlton (1978) found methiocarb to be phytotoxic in the sense that it retarded emergence slightly. In our experiments and that of Gould (1962) no phytotoxicity was found, even if 10 g a.i./kg seed was used. Scott *et al.* (1977) mentioned bromoxynil and ioxynil to be very phytotoxic; in our experiments ioxynil at 0.4 g a.i./kg seed was found to be moderately phytotoxic. Thiocyclam appeared to be only slightly phytotoxic at a rate of 2 g/kg seed in their experiments, whereas we found phytotoxicity at 1.5 g a.i./kg seed. Efficacy and phytotoxicity are usually narrowly related: higher rates give higher levels of protection, but often also an increase in phytotoxicity. Often a compromise has to be chosen, an equilibrium to be found. However, it appears that slug protection can be achieved without problems of phytotoxicity. Based on the laboratory experiments and field trials, only metaldehyde looks promising because of its excellent efficacy and no phytotoxicity. This compound is toxic to slugs by ingestion and by absorption by the foot of the mollusc. The chemical causes an increase in the secretion of slime causing immobilisation and eventual death by loss of water (Cremlyn, 1991; Triebskorn and Ebert, 1989). Treated seeds with metaldehyde are hardly toxic to birds even when birds eat large amounts of treated grains. Methiocarb showed clear effects on slug damage. However, methiocarb is a broad pesticide and has negative effects on soil organisms and birds. Seed treatments provide a more convenient and cheaper method of control than baits or sprays. Toxic chemicals are likely to be effective at smaller rates in this form and should therefore be less of an environmental hazard. As the seeds of wheat are most susceptible to slugs, seed treatments provide an effective control measure in this crop (Port and Port, 1986). Seed treatment places the chemical where it is needed to control slugs and so should be more effective and less injurious to other soil organisms than a pelleted toxic bait. Further requirements that must be met are relative safety to mammals, so that it presents little hazard during treatment, handling and sowing and to the germinating seed.

REFERENCES

Bieri, M and Schweizer, H (1990) The lethal effect of metaldehyde on the grey field slug (*Deroceras reticulatum* Müller) on water saturated contact surfaces in relation to the quantity of active ingredient and ambient temperature. ANNP-Second international Conference on pests in agriculture, Versailles, pp. 151-158.

Bourne, N B, Jones, G W and Bowen, I D (1988) Slug feeding behaviour in relation to control with molluscicidal baits. *The Journal of Molluscan Studies* **54**, pp. 327-338.

Charlton, J F L (1978) Slugs as a possible cause of establishment failure in pasture legumes oversown in boxes. *New Zealand Journal of Experimental Agriculture* **6**, pp. 313-317.

Cremlyn, R J (1991) Molluscicides. Agrochemicals-preparation and mode of action, Chichester, Wiley, pp. 309-313.

Ester, A (1989) Slakken bestrijden eist een geïntegreerde aanpak! *Groenten en Fruit* **49**, · pp. 70-71.

Ester, A and Darwinkel, A (1993) Damage by slugs in horticulture and agriculture in The Netherlands. *IMMP Newsletter* **1**, pp. 4-5.

Ester, A and Nijënstein, J H (1995) Control of the field slug *Deroceras reticulatum* (Müller) (Pulmonata: *Limacidae*) by pesticides applied to winter wheat seed. *Crop Protection* **5**, pp. 409-413.

Glen, D M (1989) Understanding and predicting slug problems in cereals. In: *Slugs and Snails in World Agriculture*. I.F. Henderson (ed.) BCPC Monograph, No. 41, pp. 253-262.

Glen, D M, Milson, N F and Wiltshire, C W (1989) Effects of seed-bed conditions on slug numbers and damage to winter wheat in a clay soil. *Annals of Applied Biology* **115**, pp. 177-190.

Gould, H J (1962) Tests with seed dressings to control grain hollowing of winter wheat by slugs. *Plant pathology* **11**, pp. 147-152.

Green, D B, Corbett, S J, Jackson, A W and Nowak, K J (1992) Surface versus admixed applications of slug pellets to winter wheat. In: *Brighton Crop Protection Conference: Pests and Diseases 1992*, **2**, pp. 587-596.

Moens, R. (1980) Het slakkenprobleem in de plantenbescherming. *Landbouwtijdschrift*, **1**, pp. 113-128.

Moens, R (1983) Proeven betreffende de bescherming van tarwezaden tegen slakken. *Landbouwtijdschrift*, **4**, pp. 1279-1292.

Port, C M and Port, G R (1986) The biology and behaviour of slugs in relation to crop damage and control. *Agricultural Zoology Reviews*, **1**, pp. 255-299.

Scott, G C, Griffiths, D C and Stephenson, J W (1977) A laboratory method for testing seed treatments for the control of slugs in cereals. In: *Brighton Crop Protection Conference: Pests and Diseases 1977*, **1**, pp. 129-134.

Triebskorn, R and Ebert, D (1989) The importance of mucus production in slugs; reaction to molluscicides and the impact of molluscicides on the mucus producing system. In: *Slugs and Snails in World Agriculture*. I.F. Henderson (ed.). BCPC Monograph No. 41, pp. 373-378.

ACTIVE DURATION OF MOLLUSCICIDES

A CHABERT

Association de Coordination Technique Agricole,
4, place Gensoul 69287 Lyon Cedex 02, FR

ABSTRACT

The aim of this study was to measure the duration of molluscicidal activity of compounds used against *Deroceras reticulatum*. This study was conducted under semi - field conditions, using twenty five, 1 m² netting cages. Pellets of the compound under test were placed in the cage on the first day (D) and were left throughout the 18 day experiment. At the same time 20 slugs were added. From day D+1 to D+6, all dead and living slugs were counted. After the counts of D+6 all dead and living slugs were removed and 20 new slugs were added. The mortality of the new slugs were assessed for a further six days and then a third set started. Three experiments were conducted using this method: comparison of compounds, influence of rainfall, and influence of the pellet number. For each product the highest effectiveness was observed during the first set, and then decrease after that. In these conditions the best product retained good activity for 12 days. Active duration of the compounds decreased with increased rainfall.

INTRODUCTION

In this study, duration of molluscicidal activity was measured for some compounds, in order to clarify their directions for use. In semi-field experiments, it has been established that mortality mainly occurs during the first 5 days after treatment. Afterwards, there is a very slight evolution in mortality rates (Maurin *et al.*,1989; Chabert et *al.*,1994). In the field, mortality mainly occurs during the first 4 days after treatment (Glen *et al.*,1986).

However, compound pellets remain visible on the ground long after treatment. The aim of this study, therefore was to determine whether these compounds remain molluscicidal after this 6 day period and if so, to assess the period during which the compounds remain active. Moreover, in most cases, the cumulative mortality rate never reaches 100%. It was therefore necessary to determine whether this limited effectiveness is due to a loss of pellet activity or to other factors.

MATERIALS AND METHODS

Experimental design

The study was conducted using twenty five 1m² netting cages. These were placed outdoors in a grassy stand. The netting base of each cage was covered with a 5 cm layer of horticultural sand (50 l/cage). At the centre of each cage was a tile to provide a shelter for the slugs.

Treatment conditions

Table 1 shows the characteristics of the non-coded substances used in this study. To each rate there corresponded a certain mean number of pellets. Compounds were applied by hand. Control cages received no treatment (non-treated control).

Table 1. Characteristics of compounds and instructions for use.

Product name	Active substances	Quantity (kg/ha)	Pellets/m²
Mesurol RF Bayer France S.A.	methiocarb 4%	3	20
Métarex RG Desangosse S.A.	metaldehyde 5%	7	36

Three tests were made, corresponding to the different treatments tested, as follows:

Test A : Comparison of compounds (6 replicates)
1 - Non-treated control
2 - methiocarb 4% (3 kg/ha)
3 - metaldehyde 5% (7 kg/ha)
4 - Compound A

Test B : influence of rainfall (4 replicates)
1 - Non-treated control, drought
2 - metaldehyde 5% (7 kg/ha), drought
3 - methiocarb 4% (3 kg/ha), drought
4 - Non-treated control, moist
5 - metaldehyde 5% (7 kg/ha), moist
6 - methiocarb 4% (3 kg/ha), moist

Test C : influence number of pellets (5 replicates)
1 - metaldehyde 5% (7 kg/ha)
2 - metaldehyde 5% (14 kg/ha)
4 - Compound B quantity N
5 - Compound B quantity 2N

Test procedure

Test chronology is presented in Figure 1. Compound pellets were introduced on day T, with 20 slugs *(Deroceras reticulatum)* at the same time. Further slugs were introduced twice at 6 day intervals. At each renewal, dead and living slugs were removed. For the three slug introductions, counts of dead and living slugs were made at D+1 and D=6.
At each introduction of slugs, four lettuce leaves were places in the corners of each cage, and renewed at the same time as the slugs.

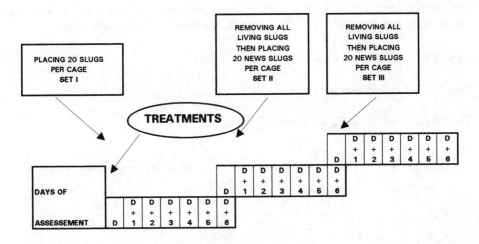

Fig. 1 Timing of the tests

For test B (influence of rainfall), the cages were covered, in order to avoid natural rainfall. Some of them were not watered, the others were sprayed at a rate of 5 mm/day during the first 6 days of treatment: i.e., 30 mm in all. To simulate rainfall, a spray-nozzle was used. Flow was checked before each watering.

During daily counting, all slugs with a moribund aspect were placed on a box-lid until the next day to verify death, in which case they were then placed at the bottom of a box. In this way, the numbers of living slugs, of those dead during the day, and the total number of dead slugs were counted and written down daily. During the last phase of test B, counts were made only on D+3 and D+6, as the mortality rate showed very little change.

Climatic conditions

The weather station was located about 100 metres from experimental site.

Table 2. Climatic conditions.

	Test A	Test B	Test C
Region of	Bordeaux	Bordeaux	Lyon
Date	Oct-Nov 1993	October 1994	April 1996
Mean temperature	10.65°C	13.9°C	12.7°C
Minimum temperature	0.4°C	6.8°C	2.6°C
Maximum temperature	22°C	25.5°C	23.8°C
Total precipitation	9 mm	78.5 mm*	36 mm**
Relative humidity	92.9%	92.5%	67.6%

*Cages were sheltered from rainfall during this trial **Rainfall was mainly at the end of the trial

The climatic conditions of tests A and B were characterised by a high air humidity level during all the experimental period. During test C, the air humidity level was lower.

RESULTS

Test A : Comparison of compounds

The results are shown in Figure 2 and in Table 3.

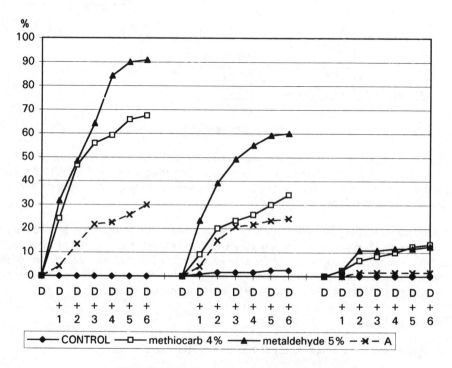

Fig. 2 Test A : Comparison of compounds
cumulative percentage mortality

Table 3. Test A, percentage mortality at D+6 and results of statistical analysis

Product	Set I	(1)	Set II	(1)	Set III	(1)
Control	0	D	2.50	C	0	B
Methiocarb 4%	67.50	B	34.17	B	13.33	A
Metaldehyde 5%	90.83	A	60.00	A	12.50	A
Product A	30.00	C	24.17	B	1.67	B
R S D	9.87		9.14		6.89	
C V	21.5%		30.5%		64.5%	

(1) Grading using the Newman-Keuls test on the arcsin of the percentage cumulative mortality. C V, coefficient of variation. R S D, residual standard deviation.

During the first test set, the highest effectiveness was found with metaldehyde 5% (90.3%). For methiocarb 4%, the cumulative mortality rate was 67.7%, and product A proved of low effectiveness. During the second test set, metaldehyde 5% was the only product to reach a mean effectiveness of 60%. Methiocarb 4% proved weak , at 34.17%.

For the third test set, all products showed very low activity. Thus, overall metaldehyde 5% retained a good level of activity for 12 days, the persistence of activity of methiocarb 4% being less.

Test B : Influence of rainfall

Results are shown in Figure 3 and Table 4.

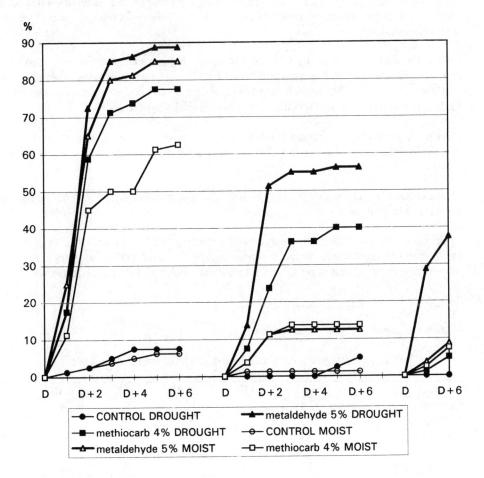

Fig. 3 Test B: Influence of rainfall
cumulative percentage mortality

Table 4. Test B, percentage mortality at D+6 and results of statistical analysis

Products	Set I	(1)	Set II	(1)	Set III	(1)
Control D[a]	7.5	B	5.00	BC	0	B
Metaldehyde 5% D	88.75	A	56.25	A	37.5	A
Methiocarb 4% D	77.50	A	40	A	5	A
Control M[b]	6.25	B	1.25	C	0	B
Metaldehyde 5% M	85.00	A	12.50	B	8.75	A
Methiocarb	62.50	A	13.75	B	7.5	A
R S D	9.87		7.27		9.31	
C V	21.5%		32.3%		75.9%	

[a], drought conditions. [b], moist conditions. (1) Grading using the Newman-Keuls test on the arcsin of the percentage cumulative mortality. C V, coefficient of variation. R S D, residual standard deviation.

During the first test set in drought conditions, the highest effectiveness was found with metaldehyde 5% (cumulative mortality rate 88.7%). For methiocarb 4%, the cumulative mortality rate was 77.5%. In moist conditions, the rates were 85% for metaldehyde 5% and 62.5% for methiocarb 4%. This difference was statistically significant.

During the second test in drought conditions, metaldehyde 5% showed an effectiveness level of only 56.25% and methiocarb 4% of 40%. In moist conditions, the two products achieved only just 10% mortality.

Metaldehyde 5% was the only product to remain somewhat effective (37.5%) during the third test set (12 to 18 days after treatment).

In drought conditions, the active duration of metaldehyde 5% was about 12 days, and for methiocarb 4% slightly less. In moist conditions, the two products did not remain active for more than one week. It is important to note that the relative air humidity in test B was high.

Test C : Influence of number of pellets

Results are shown in Figure 4 and Table 5.

Table 5. Test C, percentage mortality at D+6 and results of statistical analysis

Products	Set I	(1)	Set II	(1)	Set III	(1)
Control	3	C	0	C	4	B
Metaldehyde 5% N	90	A	80	A	56	A
Metaldehyde 5% 2N	83	A	79	A	61	A
Product B N	25	B	36	B	42	A
Product B 2N	34	B	35	B	41	A
R S D	9.41		5.15		10.04	
C V	22.2%		13%		27.1%	

(1), Grading using the Newman-Keuls test on the arcsin of the percentage cumulative mortality. C V, coefficient of variation. R S D, residual standard deviation.

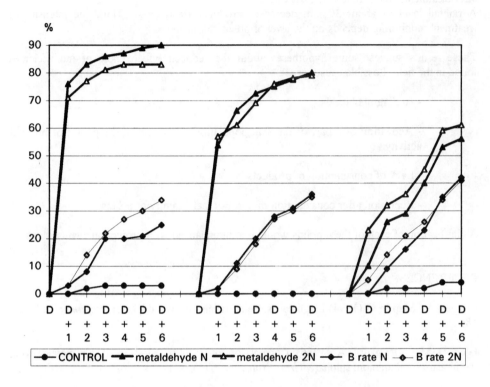

Fig. 4 Test C: Influence of the pellets number, cumulative percentage mortality.

Climatic conditions in this trial were drier than in the others. In this situation, metaldehyde 5% remained active longer than in test A. The levels of effectiveness were very high during the first two test sets. Only the third test set showed a decrease in effectiveness. It seems that there is no correlation between the number of pellets and the level and persistence of effectiveness. The only relevant factor seems to be the behaviour of the product; however, the pellets need to be able to be eaten by the slugs.

DISCUSSION

For the products showing the highest level of activity, the duration of effectiveness was about ten days in moist conditions. In drought conditions, this duration was longer. This result confirms the importance of using molluscicides at the time of sowing. In this condition, the product remains active between sowing and germination. In very dry conditions, this period can be lengthened, by delayed germination. Even so, such a climatic condition can deter the activity of slugs.

On the other hand, in moist conditions, it is necessary to check slug activity as of one week after treatment, and to renew the treatment.
A rainfall level of about 30 mm decreases product effectiveness. Thus, the intensity of treatment monitoring depends on the level of precipitation.

These results suggest some hypotheses about the reduced effectiveness of molluscicides found in the field. Possible causes are:

- High rainfall levels.

- Activation of a part of the slug population at the end of the period of product activity.

- Lack of consumption of products.

- Repulsion after consumption of a sub-lethal quantity of pellets.

A clarification of each of these points would permit enhanced crop protection against slugs.

REFERENCES

Chabert A ; Maurin G (1994) Les limaces, ne nous laissons pas dépasser. *Phytoma*. **467**, 35-38.

Glen D M ; Orsman I A (1986) Comparison of molluscicides based on metaldehyde, methiocarb or aluminium sulphate. *Crop Protection* **5** (6), 371-375.

Maurin G ; Lavanceau P ; Fougeroux A. (1989) A method for the study of molluscicides. In *Slugs and Snails in World Agriculture*. I.F. Henderson(ed.). BCPC mono. N° **41,** 167-171.

THE EFFECTS OF METHIOCARB SLUG PELLETS ON THE EARTHWORM *LUMBRICUS TERRESTRIS* IN A LABORATORY TEST

P WELLMANN

RWTH Aachen University of Technology, Department of Biology (Ecology, Ecotoxicology, Ecochemistry), 52056 Aachen, Germany

F HEIMBACH

Bayer AG, Crop Protection, Environmental Biology, 51368 Leverkusen, Germany

ABSTRACT

The effects of methiocarb slug pellets on *Lumbricus terrestris* were studied in three laboratory funnel tests. The funnels, each with one randomly selected earthworm, were filled with standard artificial soil (upper diameter 18 cm). The normal agricultural rate (40 pellets/m²) and a 10-fold higher rate were used (10 replicates per treatment) to ascertain the effects under laboratory test conditions. The measured parameters were mortality, behavioural symptoms, weight gain, and activity (food consumption, moved or withdrawn pellets, movement of small wooden toothpicks on the soil surface). Powdered cow manure was offered as food (no food in the third study). The duration of each test was 21 days.

The results show no relevant differences between the control and the normal dose for each of the measured parameters. In contrast, measured results clearly indicate some symptoms (weight gain, behaviour, activity) for the 10-fold normal application rate. Also, earthworm mortality lies in the normal range of natural mortality. According to the results of these studies no harmful effects of methiocarb slug pellets on earthworm populations are to be expected under field conditions if the normal rate is used.

INTRODUCTION

In agriculture, the crop, cultivation methods, fertilizers and pesticides influence the soil ecosystem and the composition of the soil flora and fauna (Martin & Forrest, 1969; Martin, 1993). Several laboratory methods have been described to determine effects of chemicals on earthworms (Heimbach, 1984; Reinecke, 1992), e.g. the Contact Filter Paper Test (OECD, 1984), the Artificial Soil Test (OECD-standard test, 1984), the Artisol Test, and the Daniel-funnel Test (Bieri *et al.*, 1989). While the Artificial Soil Test has been accepted worldwide as a standard test method, other test methods may be applicable to special problems associated with hazard assessments of some pesticides to earthworms. The test substance is mixed into the soil of the Artificial Soil Test and *Eisenia fetida* is used as the test species. The result of these tests correlate quite well with those of standardised field tests (Heimbach, 1992). Some pesticides e.g. methiocarb slug pellets, on the other hand, are applied on the soil surface in agriculture. The exposure of *L. terrestris* to this scenario cannot be reliably simulated in the Artificial Soil Test, as this earthworm species feeds on organic material on the soil surface.

The Bieri-funnel test, wherein the slug pellets are surface applied and *L. terrestris* is exposed acording to its natural way of life, simulates field application under standardised laboratory conditions in a better way. Nevertheless, the results published by Bieri *et al.* (1989) on methiocarb slug pellets cannot be used to predict effects under natural agricultural conditions as the application rate was far too high in comparison to the field use of this pesticide.

MATERIAL AND METHODS

Test organisms

For the test, adults of *L. terrrestris* were bought locally from a fishing-tackle shop. They were conditioned under the test situation in a plastic bucket seven days before they were placed into the test funnels. Only adult organisms (> 1.5 g) were individually placed in a test funnel after weighing. After a further week of adaption in the test funnels under test conditions (without slug pellets), the study started. Earthworms without any activity (food consumption, movement of wooden toothpicks) during this period were excluded from the study.

Soil

The funnels were filled immediately before the introduction of test organisms with an artificial soil (OECD 1984). This soil consists of 69% fine quartz sand, 20% kaolinite clay, 10% finely grounded sphagnum peat, and about 1% $CaCO_3$ to adjust the pH. These components were mixed together with water to prepare the test soil. The actual pH during the studies was 6.0 ± 0.1, its moisture content $34.9 \pm 1.3\%$ and its water holding capacity $53.9 \pm 3.8\%$.

Funnels

To simulate the field application rate of methiocarb slug pellets, the funnels used in our study were significantly larger than those of Bieri *et al.* (1989) and the proposed test guideline. Plastic funnels with an upper diameter of 25 cm (upper soil diameter 18 cm) were used for the test. A transparent plastic disk (Macrolon®) prevented the earthworm from escaping and the soil from drying, a small hole in the centre allowed air to penetrate. A vertical hole in the centre of the soil from the surface to the lower end of the funnel was bored to introduce the worm. The lower opening of the funnel led into a silicon tube (length 30 cm, inner diameter 1 cm) with a rubber plug at the end. A removable dark tube and black adhesive tape at the end protected the base of the funnel and the plastic tube from light.

Test substance and treatments

For the test, methiocarb slug pellets (Fl. No. 00313/1654) containing 4% a.i. were used. Three treatments per test were selected, a normal application rate according to agriculture use, the 10-fold of this rate and an untreated control. Ten worms were used per treatment rate, one worm per funnel. Treatment rates were randomly distributed to test funnels. For the 10-fold rate in the first study only 8 test funnels were available. The methiocarb pellets were placed onto the soil surface. Corresponding to the soil surface area of a funnel (254 cm²) and the application dose, the pellets were applied in a defined pattern (Fig. 1), one pellet per funnel for the standard dose rate and 10 pellets for the 10-fold rate. The normal agricultural use rate is 5.6 kg/ha (UK), equivalent to 40 pellets/m². The actual rate for the study (254 cm² soil surface) results to exactly one pellet per funnel (39.4 pellets/m²).

Chloroacetamide was used as a reference. In the study, chloroacetamide was mixed into the soil according to the OECD-Guideline No. 207 (1981) and adult *L. terrestris* were introduced (3 worms per test vessel with 500 g d.w. artificial soil, 3 replicates per dose). After a period of 14 days, the surviving worms were counted. The result of this study was an LC_{50} of 36 mg/kg d.w. soil. It falls within the normal range of about 15-50 mg/kg for *Eisenia fetida* as evaluated by international ring tests.

Food

Cow manure (taken from a stable nearby, air-dried and finely powdered) was used as a food supply. 0.7 g of it was laid at a defined position on the soil surface of each funnel (Fig. 1). The food was moistened with a few ml of tap water. In a third study, no food was offered.

Study performance

For the study, the methiocarb slug pellets were laid out on the soil surface of the test funnels one week after introduction of earthworms into the funnels. 1 pellet per funnel was used for the normal rate and 10 pellets were used for the 10-fold rate at defined positions. 2-3 times per week the test systems were evaluated, i.e. the activity of earthworms was recorded by movement of the wooden toothpicks (length 7 cm, diameter 2 mm), the number of moved or withdrawn pellets was recorded and the amount of food consumed was roughly estimated. If no food was left, the same amount of powdered cow manure as at the start of the test was offered again. Also, earthworms on the soil surface were noted as well as any abnormal behaviour or signs of inactivity at the soil surface.

To record earthworm activity toothpicks, pellets and food were laid out in a regular design on the soil surface as indicated in Figure 1 (moved or withdrawn pellets were not replaced during the study).

Test funnels were exposed under constant climatic conditions (15 ± 1°C, 80% relative humidity and a light-dark cycle of 12:12 hours). After a duration of 3 weeks the funnels were emptied, the earthworms were investigated for symptoms or abnormal behaviour and then their weight was determined.

For statistical calculations of the test results, the Fisher-test (p=0.05) for the mortality and behaviour symptoms and the t-test (p=0.05) for weight gain, food consumption, and movement of wooden toothpicks were used.

Fig. 1. Position of exposed slug pellets, toothpicks and food on the soil surface of each test funnel for the standard rate (left) and the 10-fold rate (right)

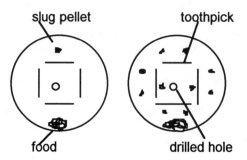

slug pellet   toothpick

food   drilled hole

RESULTS

Three studies have been performed according to the above mentioned methods. The results of these studies on mortality, observed activities and weight gain of the test organisms are summarised in Tables 1 to 3.

Table 1. **Mortality** [%] of test organisms after 3 weeks exposure, means and standard deviations

	control	normal rate	10-fold rate
first study	0	20	25
second study	0	10	0
mean	0	15 ± 7	13 ± 18
study without food	0	10	10

According to the exact binomial test after Fisher (p=0.05) the results do not show any significant differences between control and normal rate or 10-fold rate.

Table 2. Number of surviving test worms with observed **behavioural symptoms** [%] after three weeks exposure, means and standard deviations

	control	normal rate	10-fold rate
first study	0	20	38
second study	0	0	10
mean	0	10 ± 14	24 ± 19
study without food	0	0	10

According to the exact binomial test after Fisher (p=0.05) the results do not show any significant differences between control and normal rate or 10-fold rate.

Table 3. **Weight gain** [%] of surviving test worms after 3 weeks exposure, %-increase (+) or decrease (-), means and standard deviations

	control	normal rate	10-fold rate
first study	+15.4 ± 10.9	+15.7 ± 14.9	+ 2.3 ± 11.7*
second study	+12.1 ± 9.0	+ 7.2 ± 10.9	-12.1 ± 7.4*
mean	+13.8 ± 2.3	+11.5 ± 6.4	- 5.0 ± 9.9*
study without food	+ 4.8 ± 6.8	- 8.0 ± 13.1*	-18.0 ± 4.2*

* differs significantly to the control (t-test, p=0,05)

These results clearly indicate some symptoms (weight gain, behaviour) of earthworms at the 10-fold rate in all studies. In contrast, there were no significant differences between the weight gain and behavioural symptoms at the normal rate and the findings of the control except of a slight weight reduction in the third study which is not considered as a biological relevant effect. Some mortality was observed at the normal (10 and 20%, respectively) and the high rate (0 to 25%); 10% mortality is considered as a natural mortality rate in test-guidelines for earthworms (e.g. OECD No. 207, 1984). As two out of three studies do not show higher mortality at the normal rate, this mortality cannot be discussed as a treatment related effect. This statement is especially confirmed by a similar mortality rate at the 10-fold rate without an indication of a dose-response-relationship.

Activity

The activity of the earthworms was recorded 8 times throughout the first study and 6 times throughout the second and third one. The results of these observations are given in Tables 4 to 6.

Table 4. **Food consumption** [g] of earthworms during 3 weeks of exposure, means per funnel and standard deviations

		control	normal rate	10-fold rate
a) surviving worms without symptoms	first study	3.3 ± 0.9	3.3 ± 1.1	2.3 ± 0.4
	second study	2.2 ± 0.2	2.3 ± 0.3	2.0 ± 0.2
	mean	2.7 ± 0.7	2.7 ± 0.8	2.1 ± 0.3*
b) surviving worms with symptoms	first study	-	3.9 ± 0.5	2.1 ± 0.7
	second study	-	-	1.4 ± 0.0
	mean	-	3.9 ± 0.5	1.9 ± 0.7
c) mean of earthworms of a) and b)		2.7 ± 0.9	2.8 ± 0.9	2.1 ± 0.4*
d) dead earthworms	first study	-	2.5 ± 0.5	1.4 ± 0.0
	second study	-	2.1 ± 0.0	-
	mean	-	2.3 ± 0.4	1.4 ± 0.0

- = no earthworm in this category / * differs significantly to the control (t-test, p=0,05)

Table 5. Number of **moved or withdrawn pellets** [%], means per funnel and standard deviations

	normal rate unmoved/moved/withdrawn	10-fold rate unmoved/moved/withdrawn
a) surviving worms without symptoms		
first study	0.0 / 0.3 / 0.7	0.3 / 5.0 / 4.7
second study	0.0 / 0.1 / 0.9	0.4 / 4.4 / 5.1
study without food	0.0 / 0.3 / 0.7	0.0 / 3.8 / 6.2
b) surviving worms with symptoms		
first study	0.0 / 0.5 / 0.5	0.7 / 4.3 / 5.0
second study	- / - / -	0.0 / 7.0 / 3.0
study without food	- / - / -	0.0 / 6.0 / 4.0
c) mean of earthworms of a) and b)	0.0 / 0.3 ± 0.2 / 0.7 ± 0.2	0.2 ± 0.3 / 5.1 ± 1.2 / 4.7 ± 1.1
d) dead earthworms		
first study	0.0 / 0.0 / 1.0	0.0 / 6.5 / 3.5
second study	0.0 / 1.0 / 0.0	- / - / -
study without food	0.0 / 1.0 / 0.0	0.0 / 6.0 / 4.0

- = no earthworm in this category

Table 6. Activity of earthworms on the soil surface measured by **movement of wooden toothpicks** (sum of all recorded activity units [0 = no activity, 1 = low activity, 2 = high activity] within three weeks of exposure, means per funnel and standard deviations

		control	normal rate	10-fold rate
a) surviving worms without symptoms	first study	11.6 ± 3.4	11.0 ± 2.3	7.7 ± 3.2
	second study	10.3 ± 1.8	10.2 ± 1.9	7.1 ± 2.3
	mean	10.9 ± 2.8	10.5 ± 2.0	6.8 ± 2.7*
	study without food	11.3 ± 1.1	10.6 ± 1.2	8.1 ± 2.5*
b) surviving worms with symptoms	first study	-	11.5 ± 3.5	7.7 ± 4.9
	second study	-		8.0 ± 0.0
	mean	-	11.5 ± 3.5	7.8 ± 4.0
	study without food	-		9.0 ± 0.0
c) mean of earthworms of and a) and b)		11.0 ± 2.3	10.5 ± 1.8	7.9 ± 2.4*
d) dead earthworms	first study	-	7.5 ± 4.9	5.0 ± 1.4
	second study	-	6.0 ± 0.0	-
	mean	-	7.0 ± 1,1	5.0 ± 1.4
	study without food	-	12.0 ± 0.0	2.0 ± 0.0

- = no earthworm in this category / * differs significantly to the control (t-test, p=0,05)

Earthworm activity was high at all treatment levels in all studies with or without food supply: the movement of the wooden toothpicks exposed on the soil surface and the number of moved or withdrawn pellets was high. The movement of toothpicks was not influenced at the 1-fold rate compared to the control (t-test, p=0.05). In contrast, the 10-fold rate reduced this earthworm activity significantly. The number of moved or withdrawn pellets demonstrated that the earthworms had been exposed to the slug pellets at both application rates: 70% of the pellets had been withdrawn from the soil surface into the soil at the 1-fold rate and 51% at the 10-fold rate, respectively. The same trend was observed for the surviving earthworms with or without symptoms as well as those that died.

The food consumption was high at all treatment levels. Only the test organisms of the 10-fold rate consumed significantly less food than the control organisms. The results do not indicate any differences between earthworms with or without symptoms. The amount of food consumed was reduced only for those organisms which died during the study period.

Test organisms had been exposed for three further weeks after the study period under the experimental conditions. The final evaluation (mortality, weight gain, symptoms) did not show significant differences between treatments and control, except for a slightly higher weight reduction at the 10-fold dose. These findings confirm the presented results of the studies and the conclusions drawn from it.

DISCUSSION

In this study, the effects of methiocarb slug pellets on the earthworm *Lumbricus terrestris* were studied. To simulate the natural environment of these earthworm species, the "funnel test" was used. Simulating the normal use of these slug pellets in agriculture, one pellet at the normal rate and 10 pellets at the 10-fold higher rate were distributed on the soil surface of each test funnel (equivalent to 40 pellets/m²).

After three weeks of exposure, the number of dead earthworms and those with behavioural symptoms, as well as the weight gain of organisms and their activities were not influenced by methiocarb slug pellets at the normal rate as compared to the results of the control. On the other hand, the 10-fold rate clearly influenced earthworm activity, food consumption, weight gain, and behavioural symptoms. Some mortality lower than the natural mortality rate was observed in all studies without any indication of a dose-response relationship. On the other hand, the results indicate clearly that the test organisms got in contact with the slug pellets throughout the studies.

These findings are in good agreement with those of Bieri *et al.* (1989), where effects of methiocarb slug pellets had been observed on the earthworm *L. terrestris* at very high exaggerated doses, wherein the lowest application rate tested was equivalent to 400 pellets/m², i.e. the 10-fold normal field rate. According to the results presented in this paper, ecological relevant effects on adult *L. terrestris* have not to be expected under field conditions if methiocarb slug pellets are used at the recommended rate, as observed in some other studies under field conditions (Martin and Forrest, 1969; Heimbach, 1990; Martin, 1993).

ACKNOWLEDGEMENTS

We would like to thank Mr Ronald Müller and Ms Gabriele Pesch for capable technical assistance.

REFERENCES

Bieri, M; Schweizer, H; Christensen, K; Daniel, O (1989) The effects of methaldehyd and methiocarb slug pellets on *Lumbricus terrestris* LINNÉ. In: *Slugs and Snails in World Agriculture*. I F Henderson (ed.). BCPC Monograph No. 41. pp. 237-244.

OECD (Organization for Economic Cooperation and Development) (1984) Earthworm, acute toxicity test. OECD Guideline for testing of chemicals, No. 207. April 4, 1984.

Heimbach, F (1984) Correlations between three methods for determining the toxicity of chemicals to earthworms. *Pest. Sci.* **15**, 605-611.

Heimbach, F (1990) The effect of Mesurol® slug pellets on the earthworm fauna of an area of grassland. *Pflanzenschutz-Nachrichten Bayer* **43**, 140-150.

Heimbach, F (1992) Effects of pesticides on earthworm populations: comparison of results from laboratory and field tests. In: *Ecotoxicology of Earthworms*. P W Greig-Smith, H Becker, P J Edwards and F Heimbach (eds.). Intercept, Hampshire. pp. 100-106.

Martin, T J (1993) The ecobiological effects of arable cropping including the non-target effects of pesticides with special reference to methiocarb pellets (Draza®, Mesurol®) used for slug control. *Pflanzenschutz-Nachrichten Bayer* **46**, 49-102.

Martin, T J; Forrest, J D (1969) Development of ®Draza in Great Britian. *Pflanzenschutz-Nachrichten Bayer* **22**, 205-243.

Reinecke, A J (1992) A review of ecotoxicological test methods using earthworms. In: *Ecotoxicology of Earthworms*. P W Greig-Smith, H Becker, P J Edwards and F Heimbach (eds.). Intercept, Hampshire. pp. 7-19.

THE HAZARD POSED BY METHIOCARB SLUG PELLETS TO CARABID BEETLES: UNDERSTANDING POPULATION EFFECTS IN THE FIELD

G PURVIS

Department of Environmental Resource Management, Faculty of Agriculture, University College, Belfield, Dublin 4, Ireland

ABSTRACT

The application of methiocarb baited molluscicide in the field is toxic to carabid beetle populations active at the time of application. The longer-term population impact of methiocarb use is, for most species, negligible as indicated by population recovery between repeated annual applications. One carabid species, *B. obtusum*, is exceptional in showing a persistent population depression following methiocarb use. The inability of this species to recover from pesticide-induced mortality is explained by the absence of a distinct dispersal phase in its lifecycle. Such dispersal ability is a characteristic of carabids which overwinter in arable fields as larvae and are subject to substantial mortality caused by routine soil cultivations. Dispersal ability, as indicated by inter-generational recovery of carabid populations following pesticide use, is the most appropriate criterion by which to assess pesticide hazard.

THE POTENTIAL HAZARD OF MOLLUSCICIDE PELLETS

Molluscicides formulated as baited pellets pose a significantly different hazard to non-target beneficial predatory arthropods in the crop ecosystem compared with more conventional pesticide spray applications. Spray formulations are likely to present a hazard to many susceptible beneficial groups active in the crop at the time of application or shortly afterwards through contact with residues left on the crop (Jepson & Thacker, 1990). Such non-target hazards are relatively indiscriminate but are often short-lived in the summer months, being largely dependent on the duration of residual toxicity before rapid recolonisation of the crop occurs (Unal & Jepson, 1991).

In contrast, the application of slug pellets represents a relatively discrete and localised placement of pesticide in the crop environment aimed at enticing slugs to consume molluscicide in a cereal-based, attractive, bait. To this end, pellets are formulated to persist and remain toxic to slugs for as long as possible following application. This persistence can range from 7 - 40 days depending on weather conditions, especially rainfall (T. J. Martin, pers. comm.). Slug pellets, therefore, pose little direct hazard to obligate predators or parasitoids. The cereal-based baited formulation presents an obvious and relatively persistent risk, however, to less exclusively predatory species which might be attracted by and consume pellets.

Methiocarb is generally regarded as the most effective active ingredient for molluscicidal baits (Kelly & Martin, 1989) and is a carbamate originally developed for its broad-spectrum insecticidal and acaricidal properties (Unterstenhofer, 1962). In laboratory and semifield arena tests, carabid beetles can be seen to eat more readily, slug pellets containing 4% methiocarb (Draza®) than pellets containing apparently less palatable molluscicides (Büchs *et al.*, 1989).

Although some carabid species differences in susceptibility were apparent in the latter study, methiocarb-based pellets were toxic to all carabids tested and, in some instances, mortality levels were greater in semifield than in small laboratory test arenas. Büchs *et al.*, (1989) explained this latter difference in terms of greater beetle hunger in semifield conditions.

MEASURING EFFECTS IN THE FIELD

Indirect hints from multifactor crop agronomy studies

In a study of the effects of a number of agronomic practices, including alternative methods of soil cultivation, disposal of straw residue and use of methiocarb-based slug pellets, Kendall *et al.*, (1986) reported significantly greater incidence of Barley Yellow Dwarf Virus (BYDV) on plots treated with double the normal rate of methiocarb pellets compared with untreated plots. This they argued, was probably a result of non-target effects on aphid predators, particularly Carabidae. However, no significant difference in the numbers of either Carabidae or Staphylinidae extracted by flooding soil blocks from methiocarb treated and untreated plots was found in this study.

Burn (1988) reported data from the Boxworth study which suggested that winter predation by polyphagous predators of *Drosophila* pupae glued to cards, was reduced on fields treated with methiocarb pellets and, additionally, that a more detrimental effect followed surface broadcast application compared with pellet incorporation in the soil. This effect, although observed on an unreplicated trial design was attributed to reduced incidence of *Bembidion obtusum* Serv., the adult carabid most active at the time of application.

Direct assessment of effects in the field

Kennedy (1988) used pitfalls to monitor carabid activity following experimental application of methiocarb pellets in late-autumn to replicated plots fenced laterally with polythene barriers but left open-ended to allow normal seasonal predator movements into plots from the field boundry. Beetle numbers trapped between methiocarb application and a normal mid-winter trough in total carabid activity, were substantially reduced on treated plots. A continued depression in the incidence of affected species was observed within treated plots well into the following spring after resumption of their normal activity on untreated areas.

Purvis and Bannon (1992), using a similar laterally barriered plot design, showed that annually repeated application of 4 % methiocarb pellets at the recommended rate of 5.5 kg/ha, broadcast on the soil surface or drilled into the seed bed, reduced total carabid incidence between autumn and late spring to less than 5% and 10 - 15% of untreated levels, respectively (Fig. 1). Only winter-active species, trappable on the soil surface at, or immediately following, application were consistently affected; these species being adult *Bembidion obtusum* and *Trechus quadristriatus* (Scrank) and larval instars of *Nebria brevicollis* (F) (Table 1). The latter is unusual amongst carabids in that the older larval instars actively forage on the soil surface of arable fields during the winter when they may be readily caught in pitfalls. Other common arable species not present in the fields at the time of application, e.g. *Agonum dorsale* (Pont.), *B. lampros* (Herbst.) and *B. aeneum* Germ., showed no consistent response to methiocarb use when they returned to the field from overwintering hibernation sites in the spring and early summer. *Pterostichus melanarius* (Illig.), a species overwintering

Table 1. Total carabid catches (Nov.-June) following repeated drilled and broadcast application of methiocarb in three successive crop seasons and a fourth year following the cessation of applications (data from Purvis & Bannon, 1992).

Species	Year	Mean seasonal catch			LSR	Pr. F (2,4 d.f.)
		Untreated	Drilled	Broadcast		
B. obtusum	1	420	85	35	1.9	P< 0.001
	2	317	53	35	1.8	P< 0.001
	3	135	31	6	1.4	P< 0.001
	4*	178	49	33	2.2	P< 0.01
N. brevicollis	1	835	348	230	1.1	P< 0.001
(larvae)	2	295	21	89	1.8	P< 0.001
	3	180	38	6	2.0	P< 0.001
	4*	117	147	112	1.3	N.S.
N. brevicollis	1	47	17	7	2.0	P<0.01
(adults)	2	46	11	9	2.2	P<0.01
	3	57	30	31	2.0	0.05<P<0.1
	4*	94	101	102	2.3	N.S.
P. melanarius	1	23	19	9	1.9	P<0.05
	2	21	15	16	2.1	N.S.
	3	275	338	282	1.4	N.S.
	4*	711	447	551	3.2	N.S.
T. quadristriatus	1	124	51	48	1.8	P<0.05
	2	138	64	85	1.3	P<0.01
	3	377	157	97	1.9	P<0.05
	4*	137	129	138	1.7	N.S.

in fields as larvae not active on the soil surface, was also seen to be unaffected by the applications when adults emerged from pupation from June onwards (Table 1). The appearence of these latter unaffected species in the early summer was in fact responsible for the eventual recovery of total carabid activity on treated plots following each application (Fig. 1). Affected species, however, continued to show depressed levels of activity in the spring for as long as adults of the affected generation remained active. In the case of *N. brevicollis*, when the subsequent adult generation emerged from pupation within plots in early summer, initial numbers caught clearly showed the continued effect of winter larval mortality following pellet applications (Table 1).

Eventual recovery of affected populations

Sampling in subsequent years of Purvis and Bannon's study prior to reapplication of pellets to the same plot areas, showed that the activity of adult *N. brevicollis* and *T. quadristriatus* on treated plots had fully recovered from the previous year's treatment. In the case of *B. obtusum*, however, reapplication of pellets in October/ November occurred prior to the appearance of a new adult generation in trap catches and so it was not possible to assess the recovery of this species at this time. In a fourth year, when pellets were not reapplied, activity of *N. brevicollis* and *T. quadristriatus* showed no evidence of previous pellet use whilst the incidence of *B. obtusum* remained obviously depressed (Table 1). When *B. obtusum* was excluded from consideration, total carabid activity in the final fourth year of this study on plots treated in previous years, was otherwise similar to that on untreated plots.

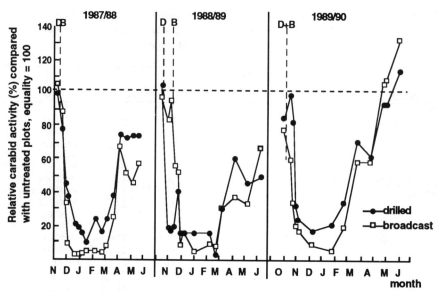

Fig. 1. Relative carabid activity on winter cereal plots treated annually with seed-drilled (D) and surface-broadcast (B) methiocarb compared with untreated plots (data from Purvis & Bannon, 1992)

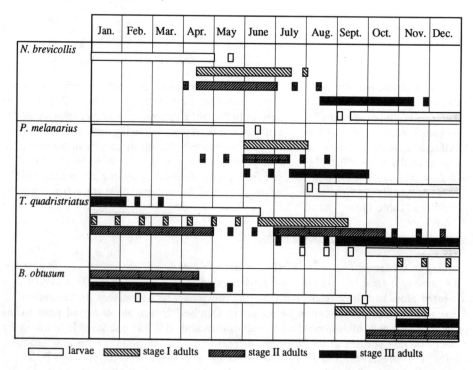

Fig. 2. Phenology of carabids overwintering in arable fields based on dissection of beetles trapped over three years - see text for explanation of stages (Fadl, unpublished data)

EXPLAINING DIFFERENCES IN POPULATION RECOVERY

Jepson (1988) quite logically argued that the ability of non-target populations to recover from pesticide applications was the most appropriate longer-term measure of their susceptibility to harmful pesticide effects. He highlighted dispersal ability as a basic ecological characteristic which determined the longer-term susceptibility of different beneficial groups to pesticide impact. In this regard, he categorised field resident carabids, excluding those species which disperse annually to and from field margins to hibernate overwinter as adults, as species with poor dispersal abilities and a high susceptibility to pesticide use. This view would suggest that winter populations affected by methiocarb use would be susceptible to longer-term population depression when clearly, with the exception of *B. obtusum*, this is not so. In contrast, den Boer (1985), on the basis of extensive ecological data, concluded that carabids were in fact, a generally highly dispersive, and 'r - selected ' group. Intuitively, this would seem to be necessary for populations to persist in an annually disturbed arable habitat. The singular susceptibility of *B. obtusum* to persistant population depression following methiocarb use suggests, however, that this species may be exceptional in this regard.

Carabid survival in the wider arable context - the effect of soil cultivation

Figure 2 illustrates the phenology of carabid species commonly found in arable fields during the winter in Ireland. These data are based on dissection of beetles from pitfall catches made in eight separate arable and grass-ley fields during a three year study (Fadl, unpublished). Female beetles were categorised as either; stage I - newly emerged with incomplete pigmentation and without ovary development; stage II - pre-reproductive with developing ovaries but no mature eggs present; stage III - gravid with mature eggs in the ovaries and stage IV - post-reproductive with the ovaries spent (see Fadl *et al* (1996) for further details). The indicated time of incidence of larvae in Figure 2, except for *N. brevicollis*, is deduced from the time of oviposition by stage III beetles and the subsequent appearance of stage I adults. On the basis of these data, it is obvious that overwintering larvae of species such as *N. brevicollis* and *P. melanarius*, both the larvae and adults of *T. quadristriatus* and adult *B. obtusum*, are subjected to routine autumn and spring soil cultivations in arable crops.

In a comparison of total seasonal pitfall catches of *P. melanarius* between June and August, Purvis & Fadl (1996) found that greater numbers were trapped in spring-sown cereal plots compared with autumn-sown plots on the same field (Table 2). However, catches of beetles emerging from pupation within 1 m^2 fenced arenas on these plots (see Purvis & Fadl, 1996 for details), clearly showed that spring, in comparison to autumn soil cultivation, caused a level of larval/pupal mortality of almost 80% (Table 2).

Table 2. Mean seasonal open plot trap catches of adult *P. melanarius* and mean pupal emergence density estimates from enclosed arenas on winter and spring-sown cereal plots (data from Purvis & Fadl, 1996)

	Winter-sown	Spring-sown	*LSR*	Pr. *F* (1,3 d.f.)
Open plot catch (June-August)	719.5	1022.5	1.28	P<0.01
Emergence density (nos. m^{-2})	13.8	2.5	3.29	P<0.05

This level of mortality is comparable in magnitude to the most detrimental documented pesticide effects on carabid populations. Open pitfalls failed to demonstrate the effect of spring cultivation, presumably because of rapid interplot movement immediately following adult emergence. Similarly, a comparison of total seasonal catches of *P. melanarius* in eight separate fields surveyed over a three year period and classified according to date of soil cultivation, failed to show any difference between field types (Fadl *et al*, 1996). However, catches of stage I females, newly emerged from pupation, showed that relatively few beetles were produced on spring cultivated fields compared with autumn cultivated or uncultivated new grass ley fields and largely support the view that soil cultivations, especially in late April/May, reduce survival of larval/pupal instars (Table 3a).

Table 3. Backtransformed least square means estimates for trap catches of *P. melanarius* in a survey of crops classified as autumn-cultivated (Oct./Nov.-winter cereals), early spring-cultivated (Mar./early Apr.-spring cereals), late spring cultivated (late Apr./May - beet, potatoes or maize) or uncultivated (grass ley) in the current and previous cropping season (see Fadl *et al*, 1996 for details)

Previous crop cultivation	a) Estimated newly emerged stage I female catch*				b) Estimated gravid stage III female catch*			
	Current crop cultivation time				Current crop cultivation time			
	Autumn	Early spring	Late spring	Uncult.	Autumn	Early spring	Late spring	Uncult.
Autumn	-	-	11	109	-	-	21	35
Early spring	76	-	-	-	43	-	-	-
Late spring	0.5	27	1.1	-	11	40	18	-
Uncultivated	122	-	3.6	1.7	77	-	283	191

* based on dissection of females

The subsequent distribution of gravid stage III females within the same crops, however, suggested that rapid post-emergence dispersal occurs between fields to allow populations to recover from this effect each year (Table 3b). Similar, as yet unpublished data were obtained for *N. brevicollis* in this survey of incidence on different field types and it seems probable that pre-reproductive dispersal following pupal emergence is an essential adaptation for the survival of exclusively larval overwintering species in seasonally cultivated arable fields.

T. quadristriatus overwinters in fields as both larvae and post-reproductive adults probably destined to resume reproduction in the spring (Fig. 2). Consequently, new adult beetles emerge within fields in almost every month of the year. This explains the rapid recovery of this species from methiocarb use which has little effect on larvae protected in the soil. It also explains why *T. quadristriatus* may actually take advantage of pesticide use to become rapidly more abundant following applications in the potential absence of competition from other species (Burn, 1992).

In marked contrast to all other arable field inhabiting carabids, *B. obtusum* remains active in fields throughout the winter as pre-reproductive adults which breed in the following spring (Fig. 2). This lifecycle strategy presumably avoids competition from other adult over-wintering *Bembidion* spp. which leave fields in the winter to hibernate in sheltered field boundries (Coombes & Sotherton, 1986), and enables *B. obtusum* to reproduce before these other species return to breed in the spring. Since the larval instars of *B. obtusum* occur in the summer months when no major soil disturbances occur, this species is pre-adapted to survive in cultivated fields without the need for an annual dispersal phase to make good the impact of soil cultivation on larval instars.

This strategy is only possible in a cultivated arable context because the adult instar is a relatively robust animal which is not harmed by routine cultivations. The disadvantage of such a strategy is the singular lack of an adaptation which allows population survival in the event of local field-scale mortality. Such a lifecycle uniquely predisposes *B. obtusum* to potentially long-term population depression following the use of pesticides at any time between adult emergence from pupation in August and the completion of egg laying in May (see also Burn, 1992). This probably explains the very patchy incidence of this otherwise highly successful species on individual fields (Purvis *et al.*, 1988 and Fadl, unpublished data). All other common carabids in arable fields are preadapted for long-term survival in the face of frequent local population disruption, either because of their need to make seasonal movements to and from adult overwintering sites at field margins or their widespread inter-generational dispersion following pupal emergence in the spring in order to make good substantial local annual mortality of larvae caused by routine crop husbandry.

CONCLUSIONS

Methiocarb molluscicide bait with obvious toxicity to carabids illustrates the need for a fundamental understanding of non-target population effects in the field before a realistic assessment of pesticide hazard can be made. Because of apparently very rapid post-emergence dispersion by those species which overwinter as larvae within arable fields and the on and off-field movements of adult hibernating species, most carabids exhibit robust inter-generational recovery following local population mortality. This capacity results from lifecycle adaptations which are otherwise necessary to ensure population survival in an environment routinely perturbed by many crop husbandry practices. The use of methiocarb in late autumn has, for the majority of affected carabids, a longer-term population consequence no more harmful than that of routine annual soil cultivation. In this context, the most meaningful way to gauge whether pesticide-induced mortality represents a real hazard to carabids in annual crops, is to assess the extent of inter-generational recovery between annual applications. This can be done most easily by measuring pupal emergence densities within treated sites. It is likely such studies will show that pesticides which do not have inter-seasonal persistence, including methiocarb, pose little actual hazard to carabids but that *B. obtusum* is susceptible to longer-term population effects because of its unique phenology.

ACKNOWLEDGEMENTS

The use of data in Figure 2 and Table 3 from the unpublished PhD studies of Mr. Abdoulla Fadl is greatfully acknowledged.

REFERENCES

Boer, P J den (1985) Fluctuations of density and survival of carabid populations. *Oecologia*. **67**, 322-330.

Büchs, W; Heimbach, U; Cznarnecki, E (1989) The effect of snail baits on non-target carabid beetles. In: *Slugs and Snails in World Agriculture*. I Henderson (ed). BCPC Monograph No. 41, pp. 245-252.

Burn, A J (1988) Assessment of the impact of pesticides on invertebrate predation in cereal crops. *Aspects of Applied Biology*. **17** (1), 279-288.

Burn, A J (1992) Interactions between cereal pests and their predators and parasites. In: *Pesticides, cereal farming and the environment*. P W Greig-Smith, G K Frampton, A R Hardy (eds), pp. 110-131. HMSO, London.

Coombes, D S; Sotherton, N W (1986) The dispersal and distribution of polyphagous predatory Coleoptera in cereals. *Annals of Applied Biology*. **108**, 461-474.

Fadl, A; Purvis, G; Towey, K (1996) The effect of time of soil cultivation on the incidence of *Pterostichus melanarius* (Illig.) (Coleoptera : Carabidae) in arable land in Ireland. *Annales Zoologici Fennici*. **33** (1), 207-214.

Jepson, P C (1988) Ecological characteristics and the susceptibility of non-target invertebrates to long-term pesticide side effects. In: *Field Methods for the Study of Environmental Effects of Pesticides*. M P Greaves; B D Smith; P W Greig-Smith (eds), BCPC Monograph No. 40, pp. 191-198.

Jepson, P C; Thacker, J R M (1990) Analysis of the spacial component of pesticide side-effects on non-target invertebrate populations and its relevance to hazard analysis. *Functional Ecology*. **4**, 349-358.

Kelly, J R; Martin, T J (1989) Twenty-one years experience with methiocarb bait. In: *Slugs and Snails in World Agriculture*. I Henderson (ed), BCPC Monograph No. 41, pp. 131-145.

Kendall, D A; Smith, B D; Chinn, N E; Wiltshire, C W (1986) Cultivation, straw disposal, and BYDV infection in winter cereals. *Proceedings, 1986 British Crop Protection Conference - Pests and Diseases* **3**, 981-987.

Kennedy, P J (1988) The use of polythene barriers to study the long-term effects of pesticides on ground beetles (Carabidae, Coleoptera) in small-scale field experiments. In: *Field Methods for the Study of Environmental Effects of Pesticides*. M P Greaves; B D Smith; P W Greig-Smith (eds), BCPC Monograph No. 40, pp. 385-390.

Purvis, G; Carter, N; Powell, W (1988) Observations on the effects of an autumn application of pyrethroid insecticide on non-target predatory species in winter cereals. In: *Integrated Crop Protection in Cereals*. R Cavalloro; K D Sunderland (eds), Rotterdam: A A Balkema, pp. 153-166.

Purvis, G; Bannon, J W (1992) Non-target effects of repeated methiocarb slug pellet application on carabid beetle (Coleoptera, Carabidae) activity in winter-sown cereals. *Annals of Applied Biology*. **121**, 401-422.

Purvis, G; Fadl, A (1996) Emergence of Carabidae (Coleoptera) from pupation: a technique for studying the 'productivity' of carabid habitats. *Annales Zoologici Fennici*. **33** (1), 215-223.

Unal, G; Jepson, P C (1991) The toxicity of aphicide residues to beneficial invertebrates in cereal crops. *Annals of Applied Biology*. **118**, 493-502.

Unterstenhofer, G (1962) Mesurol, a polyvalent insecticide and acaricide. *Pflanzenschutz - Nachrichten Bayer*. **15**, 177-189.

Session 6
Chemical Control

Session Organiser
and Chairman   Professor I D Bowen

KILL OR CURE? CONTROL OF AQUATIC MOLLUSC PESTS

E J TAYLOR, J S ARTHUR , I D BOWEN.
Dept. Pure and Applied Biology, University of Wales Cardiff, PO Box 915, Cardiff, CF1 3TL
United Kingdom.

ABSTRACT

Recent European Union funded research programmes have developed a wide range of
toxicity test methods based on the biological responses of freshwater
macroinvertebrates. These are for incorporation into the 7th Amendment for assessing
the hazards of chemicals which may potentially pollute freshwater environments.

Such tests also have a role within the design of chemical methods for controlling
aquatic pests. Firstly, they may be modified for use with target species to determine
the efficacy of chemicals for solving particular pest problems including those which
only require a sub-lethal control strategy e.g. inhibition of the feeding of crop pests.
Secondly, the tests quantify the effects of the chemicals on reference non-target
species providing a preliminary indication of the environmental impact of
treatments.

These two applications of toxicity tests have been integrated into a procedure for
screening plant extracts and proprietary chemicals for use against Pomacid and
Lymnaeid species of aquatic mollusc pests. The potential of short listed treatments is
discussed with respect to the biology of these snails and the pest problems which they
pose.

Aquatic toxicology employs aspects from all branches of biology but in the last twenty years
has evolved into a scientific discipline in its own right. It has developed to achieve the twin
goals of controlling and protecting organisms within managed and natural aquatic ecosytems.
Although it is possible, with the use of appropriate application factors, to predict chronic
effects from the results of acute lethal toxicity tests with fish and aquatic invertebrates
(Kenega, 1982) recently tests have been designed to enhance the certainty with which river
and lake ecosytems can be protected from pollutants. Three programmes sponsored by the
Commission of the European Union have promoted the design of new test systems:

Development and Validation of Methods for Evaluating Chronic Toxicity to Freshwater
Ecosystems (Environmental Programme CEC Contract No.EV4V-0110-UK (BA)).

Prediction of Toxic Effects in Freshwater Ecosystems - Validation of Laboratory Multi-
species Tests in Lake and River Mesocosms (Environmental Programme CEC Contract No.
EV5V CT91-0010).

Development of Physiological and Biochemical Toxicity Tests with Freshwater Species and
their Validation in the Laboratory and in Field Microcosms and Mesocosms. (Environmental
Programme CEC Contract No. EV5V-CT92-0200).

These projects have increased the suite of methods which, rather than assessing only the lethal effects of chemicals, are based on sub-lethal response criteria with a wide range of test species. The effects of four reference chemicals (copper, lindane, 3,4-dichloroaniline and atrazine) determined in laboratory tests have been compared with those derived from mesocosm studies. It was found that the 'no observed effect concentrations' provided by endpoints of sensitive laboratory tests could be protective of both pond and stream mesocosms.

Principles utilised in these projects have an application in assessing the suitability of various treatments for the control of aquatic mollusc pests. The test methods allow proposed chemical treatments to be screened with respect to their effect on different levels of the biological organisation of non-target species. In addition, where appropriate, the bioassays may be modified to determine the efficacy of managing pest species using sub-lethal control strategies. Thus the two purposes of aquatic toxicology (protection and control) can be united by the objectives of reducing both the environmental effects and the costs of pest management practices. The present research at UWCC into chemical control of snails, particularly *Pomacea spp.* (rice crop pests) and *Lymnaea spp.* (vectors of the liver fluke), draws from response criteria and protocols which proved useful in the CEC research.

Strategies for control of *Pomacea spp.* and *Lymnaea spp.*

Reviews highlighting the biology of these aquatic mollusc pests and the problems they cause are available in the literature (Litsinger and Estano, 1993; Halwart, 1994; Kela and Bowen, 1995). The strategy for control of these snails is influenced by the nature of the pest problem and the biology of the target organism. Traditionally the approach has been to apply lethal concentrations of chemicals, however this may not always be necessary. Snail species which are vectors of disease need to be eradicated in defined geographical areas and therefore the chemical must be applied to produce environmental concentrations which cause mortality or prevent the reproduction of individuals. For the latter strategy a lower amount of chemical may be added to the water than that required in the lethal dose. Exposure to sub-lethal concentrations of a chemical can reduce reproductive success of aquatic invertebrates directly by affecting reproductive behaviour or indirectly through reducing energetic resources available by inhibiting feeding and/or increasing metabolic demands (Maund *et al.* 1992). However, effecting a decrease in feeding rate *per se* may be sufficient to reduce the economic, health and social impact of snails which are crop pests. Thus for *Pomacea spp.* minimising or preventing their consumption of rice crop seedlings during the period the seedlings are susceptible to attack (up to 15 days after transplanting) would be appropriate.

Commercial products are currently used to control these snail pests. Obviously recommending an alternative chemical treatment would only be justified if it proved less damaging to the environment and/or more cost effective. At this point it is worth addressing a popular misconception that the optimum chemical is the one which is the most toxic to the snails. The treatment strategies (concentration, exposure period and frequency of application) have to be viewed in their entirety, inclusive of their effects on both target and non-target species. Then the most suitable chemical is that which proves to have the largest differential between the dose regime which solves the pest problem and that which would be an environmental hazard. Thus, if chemical A is efficacious at a total concentration of Y gl^{-1} and a hazard to the environment at $2Y$ gl^{-1} (2 fold differential in concentrations) whereas B has corresponding values of $10Y$ gl^{-1} and $100Y$ gl^{-1}, respectively (10 fold differential in concentration), then B is the better choice for minimising effects of chemical input on the ecosystem. However, the cost of producing the required amounts of each chemical would also be another factor to consider.

<u>Categories of chemicals being investigated at UWCC for their potential use in strategies to control *Pomacea spp.* and *Lymnaea spp.*</u>

1. Chemicals in plant extracts which are molluscicidal.
Plants may contain chemicals as a natural defence against consumption by molluscs. This provides researchers with a relatively 'instant' source of chemical which may prove useful in the management of snail pests. However, although derived from a natural source such chemicals are not necessarily 'environmentally friendly' as their toxicity to the snails may extend to a range of other aquatic organisms (Adewunmi and Marquis, 1981). The requirements and benefits of plant derived chemicals for use in controlling snails are well documented (Mott, 1987; Adewunmi, 1991) but generally only with respect to effecting mortality of target organisms.

2. Chemicals which increase intraspecific competition for *Lymnaea spp.* and *Pomacea spp.*
Intraspecific competition - the integration of the interactions within a population consisting of a single species - is an important factor in populations which tend to be self regulating. Territorial behaviour is indicative of intraspecific competition for space and may effectively control population size. For some species stress from crowding may lead to fighting and/or reduce the rate of reproduction. (Milne, 1961).
Under crowded conditions reproduction and growth of snails may be reduced not only as a result of physical encounters between individuals but also because of a conspecific reaction to aqueous concentrations of chemicals released by the snails (Wright, 1960; Berrie and Visser, 1963; Cazzaniga and Estebenet, 1988). This programme aims to confirm whether such a 'pheromone' exists for *Pomacea spp.* and *Lymnaea spp.* and if appropriate to identify the chemical/s involved. The chemical, if applied in the field at a concentration which falsely suggests a high population density of snails, may provide a method of reducing the growth and reproductive rate of snails and decrease the actual standing population of snails. Such a result would certainly reduce the impact of crop pests and the dose regime may be relatively harmless to other organisms.

3. Chemicals which increase burrowing behaviour of *Pomacea spp.*
Behavioural responses to pollutants have been used as the basis to several freshwater bioassays (Pascoe *et al.*, 1991). The burial of a number of snail species in response to extremely low concentrations of odours from predators and/or intraspecific snail juices (from damaged snails or homogenates) has been recorded (Snyder and Snyder, 1971). Invoking this response would effectively reduce the activity time of snails and consequently affect both feeding and reproduction. If the active ingredients which illicit this alarm response can be isolated and identified they may provide a very specific means of reducing the impact of crop pest species of snails.

4. Chemicals which affect the development and hatching of the eggs of each species.
Protecting a species from the effects of chemicals requires identification of the most sensitive, whilst still relevant, biological response criteria. The chemical control of a species warrants a similar approach in which the life-cycle event or physiological response which most lends itself

to cost effective and environmentally sound management is targeted. One such life-cycle event which may be the 'Achilles heel' for *Pomacea spp.* is the egg mass which is laid out of water. Currently mechanical methods e.g egg clappers are used to destroy the vivid pink conspicuous eggs of *P. canaliculata* (Awadhwal and Quick, 1991). However, chemical sprays may potentially provide a more efficacious means of reducing egg hatching. The chemical could be applied directly to the crops, which is likely to reduce the cost in comparison to treatments which require large volumes of chemical to be added to the water in order to affect the snails. In addition, the sprayed chemical although ultimately entering the aquatic environment will only be present at a relatively low concentration. In fact preliminary findings suggest that both development and hatching of eggs may be reduced by repeatedly wetting the eggs with water during their incubation. Obviously this would provide an extremely safe, readily available and cost effective means of controlling *Pomacea spp.* in rice paddies!

Screening programme

Figure 1. Programme for screening chemicals for use against aquatic snails.

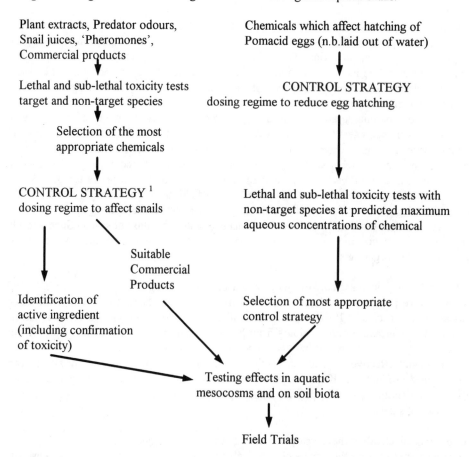

[1] Based on one or more of the following response criteria: mortality, reproduction, feeding, growth of individuals and populations and hatching of eggs laid in water.

One of the non-target reference species selected for use in the screening procedure is the freshwater amphipod *Gammarus pulex*. It was chosen because of its sensitivity to a wide range of toxicants (McCahon and Pascoe, 1988) which was endorsed by the results from tests performed during the previously mentioned CEC programmes. Defined protocols exist to investigate the various biological criteria listed in the footnote to Figure 1. A bioassay to measure the feeding of *G. pulex* (Taylor *et al.*, 1993) - developed during a CEC programme - has been modified to allow measurement of the feeding rate of *Pomacea spp.* and *Lymnaea spp.* The various protocols are detailed in Arthur *et al.* (1996). The feeding rates are determined using a time-response analysis of the feeding patterns of test organisms. A FT50 value - median feeding time- is calculated and this represents the time taken by the individuals in the treatment and control groups to consume 50 % of the food provided. It should be noted therefore that a larger FT50 value indicates a slower feeding rate. Table 1 shows the feeding of *G. pulex* and *Pomacea spp.* after they had been maintained in test solutions with different concentrations of the water extract of the Nigerian plant *Detarium microcarpum* or in control water (results for other plant extracts are detailed in Arthur *et al.*, (1996)).

Table 1. Median feeding time (FT50) of *Pomacea spp.* and *Gammarus pulex* after 24 h exposure to aqueous extracts of the plant *Detarium microcarpum* (95% confidence limits shown in parentheses). [*] indicates significant difference from control (p = 0.05).

Nominal Concentration of Aqueous Extract of *Detarium microcarpum* (mg/l)	Median Feeding Time (FT50) (mins)	
	Pomacea spp.	*G. pulex*
Control	29.5 (22.5-38.5)	46.1 (37.3-57.1)
10	52 (34.6-78.3)[*]	133.3 (85.5-207.9)[*]
30	60.2 (43.8-82.8)[*]	189.2 (129.8-257.7)[*]
50	15 % food eaten[1]	169.7 (139.5-222.4)[*]

[1] after 65 mins

It is evident that all the concentrations of plant extract reduced the feeding rate of both species in comparison to their respective control responses. Although further investigation is necessary, it is apparent that the active ingredient in *Detarium microcarpum* when applied at concentrations which reduce the feeding of *Pomacea spp.* may affect other organisms in a similar way. However, such a control strategy may be acceptable as it would only be a temporary measure (c.f. rice seedlings vulnerable to snail attack for a limited period) and may have minimal long term consequences for both target and non-target organisms.

In conclusion, this project will comprehensively evaluate both recognised and novel chemical treatments for the control of aquatic snail pests. Their efficacies and likely impacts on the aquatic environment will be compared to allow recommendation of an optimum control strategy based on an integration of suitable dosing regimes. It is apparent that the measurement of acute and chronic responses of the snails to chemicals will also yield substantial information about their autecology and behaviour.

REFERENCES

Adewunmi, C O (1991) Plant molluscicides: potential of Aridan, *Tetrapleura tetraptera*, for schistosomiasis control in Nigeria. *The Science of the Total Environment* **102**, 21-33.

Adewunmi, C O; Marquis, V O (1981) Laboratory evaluation of the molluscicidal properties of Aradin, an extract from *Tetrapleura tetraptera* (Mimosaceae) on *Bulinus globosus*. *Journal of Parasitology* **67**, 713-716.

Arthur, J S; Taylor, E J; Bowen, I D (1996) Optimising the chemical control of the rice pest *Pomacea canaliculata* (Lamarck). *These proceedings*.

Awadhwal, N K; Quick, G R (1991) Crushing snail eggs with a 'snail egg clapper'. *International Rice Research News*, **16**, 26-27.

Berrie, A D; Visser, S A (1963) Investigation of a growth inhibiting substance affecting a natural population of freshwater snails. *Physiology and Zoology*, **36**, 167-173.

Cazzaniga, N J; Estebenet, A L (1988) Effects of crowding on breeding *Pomacea canaliculata* (Gastropoda: Ampullariidae). *Comparative Physiology and Ecology* **13**, 89-96

Halwart, M (1994) The golden apple snail *Pomacea canaliculata* in Asian rice farming systems: present impact and future threat. *International Journal of Pest Management* **40**, 199-206.

Kela, S L; Bowen, I D (1995) Control of snails and snail-borne diseases. *Pesticide Outlook* **February 1995** 22-27.

Kenega, E E (1982) Predictability of chronic toxicity from acute toxicity of chemicals to fish and aquatic invertebrates. *Environmental Toxicology and Chemistry*, **1**, 347-358.

Litsinger, J A; Estano, D B (1993) Management of the golden apple snail *Pomacea canaliculata* (Lamarck) in rice. *Crop Protection* **12**, 363-370.

Maund, S J; Taylor, E J; Pascoe, D (1992) Population responses of the freshwater amphipod crustacean *Gammarus pulex* (L.) to copper. *Freshwater Biology* **28**, 29-36.

McCahon, C P; Pascoe, D (1988) Use of *Gammarus pulex* (L.) in safety evaluation tests - culture and selection of a sensitive life stage. *Ecotoxicology and Environmental Safety*, **15**, 245-252.

Milne, A (1961) Definition of competition among animals. *Symposia of the Society of Experimental Biology* **15**, 40-61.

Mott, K E (Ed.) (1987) Plant molluscicides.UNDP/World Bank/WHO Special Programme for Research and Training in Tropical Diseases, New York: John Wiley and Sons Ltd.

Pascoe, D; Gower, D E; McCahon, C P.; Poulton, M J; Whiles, A J; Wulfhorst, J (1991) Behavioural responses in freshwater bioassays. In: *Bioindicators and Environmental Management: Proceedings, Sixth International Bioindicators Symposium*, Dublin, *1990* D W Jeffrey and B Madden (eds), London: IUBS pp.245-254.

Snyder, N F R; Snyder, H A (1971) Defenses of the Florida apple snail *Pomacea paludosa*. *Behaviour* **40**, 175-215.

Taylor, E J; Jones, D P W; Maund, S J; Pascoe, D (1993) A new method for measuring the feeding activity of *Gammarus pulex* (L.). *Chemosphere*, **26**, 1375-1381.

Wright, C A (1960) The crowding phenomenon in laboratory colonies of freshwater snails. *Annals of Tropical Medicine and Parasitology*, **54**, 224-232

USE OF METALDEHYDE AS MOLLUSCICIDE IN MILKFISH PONDS

I G BORLONGAN, R M COLOSO
Aquaculture Department
Southeast Asian Fisheries Development Center
Tigbauan, Iloilo, Philippines

R A BLUM
Lonza Ltd., Basel, Switzerland

ABSTRACT

META R metaldehyde formulations were tested under laboratory and field conditions against brackishwater pond snails (*Cerithium* sp.). Under laboratory conditions the LC_{50} and LC_{99} 3 days after treatment ranged from 2 - 3.5 and 4.8 - 5.4 kg/ha, respectively. However, these levels proved ineffective when applied directly under actual pond conditions. In ponds with snail population of about 300/m², a higher application rate of 30 kg/ha is recommended. Application of META metaldehyde concentrations of 0 - 175 kg/ha did not affect milkfish juveniles (1-3 g body weight) 7 days after treatment. Results suggest that the META metaldehyde formulations were effective for pond snail control without detrimental effect on juvenile milkfish.

INTRODUCTION

Milkfish culture is an economically important industry in the Philippines. At present, one of the pressing problems faced by milkfish farmers is the control of brackishwater pond snails, particularly those belonging to the family Cerithidae. Snail infestation greatly affects milkfish pond production because the snails are serious competitors with the milkfish for algal food.

Brackishwater pond snails are small (less than 4 cm) and live in the intertidal zone on soft substrata. They can be found buried in the mud but can withstand long hours in more exposed conditions. Pond snails may also occur in estuarine habitats in high numbers. They are microphagous herbivores and graze on the pond bottom leaving tracks that disturb the growth of algae (Pillai, 1972).

Methods to control pond snails include handpicking from the pond bottom during the preparation of the pond. This practice is labour intensive and eggs and small juveniles are not removed. Heat treatment by burning rice straw (Trino et al. 1993) or calcium carbide (Wang 1989) has also been tried. However, the method poses some hazards from fire to farm workers. Botanical pesticides like tobacco dust (active ingredient, nicotine), derris root (rotenone), and teaseed cake (saponin) have been used with little or no negative environmental impact. However, supplies of these indigenous plant materials are limited and their use can be quite expensive. Organotin compounds like Aquatin and Brestan have been used to eradicate them but have been banned because of environmental concerns. Alternative methods to control the snails are needed.

Metaldehyde has been used in snail and slug control in agricultural crops. Its mode of action is highly selective on molluscs (disrupts the lining of the gut and stimulates excessive mucus secretion, leading to irreversible damage to mucus cells found in slugs and ultimately to death) with practically no effect upon non-target organisms such as soil microorganisms, parasitic beetles, honey bees, and earthworms. It has very low acute toxicity to water organisms such as algae, daphnids, and fish and, in water/sediment systems, it is rapidly biodegraded. Metaldehyde was also found to have low toxicity to birds such as quails and ducks (Camponovo, 1995).

In this study we determined the efficacy of metaldehyde in controlling brackishwater pond snails under controlled laboratory and field conditions. Its effect on milkfish juveniles was also studied.

MATERIALS AND METHODS

Study 1 (Controlled Laboratory Experiments)

The toxicity (LC_{50} and LC_{99}) of metaldehyde against brackishwater pond snails (*Cerithium* sp) was determined under laboratory conditions. Snails (about 1-1.5 g weight and 18 - 30 mm shell length) were handpicked from brackishwater ponds in Leganes, Iloilo, Philippines. They were fully acclimated to laboratory conditions prior to stocking. Three META metaldehyde (Lonza Ltd.) formulations (A, B, and C) were used. The experiments were conducted in circular basins (about 50 cm diameter) with a surface area of about 0.2 m^2. The basins were lain with soil (10 cm depth) taken from fishponds. The soil from the fish ponds was prepared and made snail-free by removal of snails, sun drying, pulverizing, and seiving. The basins were then filled with brackishwater (20-25 ppt) and a depth of 5 cm was maintained throughout the experimental period. Fifty snails were placed into each basin. The experimental design was completely randomized design (CRD) with 8 treatments and 4 replicates. Rates of metaldehyde application tested were 0, 5, 15, 25, 35, 45, 55, and 65 kg/ha. The response of snails to the various test concentrations was noted and mortality counts were done at 1, 2, and 3 days after treatment (DAT). Probit analysis (Finney, 1982) were used to determine the LC_{50} and LC_{99} values and their corresponding 95% confidence intervals for each exposure time.

<u>Study 2 (Pond Experiments)</u>

The effect of metaldehyde application against *Cerithium* sp. was also studied under pond conditions. The metaldehyde formulation C was used. Twenty-four (8 m x 4 m) ponds at the Iloilo State College of Fisheries (ISCOF), Barotac Nuevo, Iloilo, Philippines were used in the experiment. The ponds were prepared in such a way that they were free from snails at the start of the experiment. To facilitate counting of live and dead snails each pond compartment was bottom-lined with B-net and overlain with 10 cm of snail-free mud. A water depth of 5 cm in each pond was maintained. Snails from neighboring areas were collected and stocked into each pond at a density of 300 snails/m^2. The snails were acclimated in the ponds for 2 days prior to the application of metaldehyde. Two experiments were conducted. In the first trial, the metaldehyde application rates tested were 0, 5, and 15 kg/ha while in the succeeding experiment application rates tested were 0, 20, 30, 40, 60, 20 + 20, 30 + 30 and 40 + 40 kg/ha. In the first four test concentrations a single dose was applied at day 0 while in the latter three the applications were repeated at day 3. Three replicates per treatment were used in a completely randomized design (CRD). Mortality counts were done at 3 DAT and 6 DAT for the first trial and only at 6 DAT for the succeeding experiment by lifting the nets, removing the soil overlay and collecting all the snails into a basin. Mortality data were subjected to arcsin transformation and analyzed by analysis of variance (ANOVA) and Duncan's multiple range test (DMRT) using SAS computer software.

<u>.Study 3 (Milkfish bioassay experiments)</u>

The effect of metaldehyde formulations on milkfish were determined. Milkfish juveniles (1-2 g and 2-3 g, average body weight) from Tigbauan Main Station of SEAFDEC Aquaculture Department were used in the experiments. Milkfish of similar size ranges are preferred by the fish farmers for stocking in the grow-out ponds.

The metaldehyde formulations A, B, and C were tested on milkfish juveniles of 2-3g sizes in three separate tests. The concentrations tested were 0, 5, 15, 25, 35, 45, 55, 65, and 75 kg/ha. The treatments were replicated four times. Circular plastic basins (50 cm diameter, surface area of 0.2 m^2) were lain with clean pond soil (10 cm depth) and filled with brackishwater (20-25 ppt) to a depth of 5 cm. Metaldehyde was then applied and allowed to stand for 3 days. Under actual pond conditions, this intervening period would correspond to the treatment of snail infested ponds with metaldehyde. On the fourth day, the water level in the basin was increased to about 8 cm or approximately 15 liters of brackishwater. Milkfish were then stocked at 5 fish/basin. Mortality counts were done daily from 1-7 DAT. Fish were starved during the test.

In the trial using formulation C, a wider range of test concentrations was used- 0, 25, 50, 75, 100, 125, 150, 175 kg/ha. Milkfish juveniles used were smaller (1-2 g average body weight). All other procedures used were similar to those in earlier tests.

RESULTS AND DISCUSSION

Study 1.

Toxicity tests under controlled laboratory conditions of the different metaldehyde formulations against brackishwater pond snails (*Cerithium* sp.) showed that metaldehyde was effective against the snails. Lethal concentrations (LC_{50} and LC_{99}) of the formulations after 3 days of treatment ranged from 2-3.5 and 4.8 - 5.4 kg/ha, respectively (Table 1).

Table 1. Median lethal concentration (LC_{50}) and LC_{99} of different metaldehyde formulations against brackishwater pond snails, *Cerithium sp.*

Time	LC_{50} (kg/ha)	95% Confidence intervals Lower	95% Confidence intervals Upper	LC_{99} (kg/ha)
Formulation A				
1 DAT	26.5	18.6	27.2	75.0
2 DAT	5.1	0.31	7.7	69.0
3 DAT	3.5	-	-	5.4
Formulation B				
1 DAT	13.9	4.1	24.9	65.0
2 DAT	3.4	0.2	4.2	41.5
3 DAT	2.0	-	-	5.3
Formulation C				
1 DAT	11.5	4.9	17.7	60.0
2 DAT	2.9	0.3	4.1	27.6
3 DAT	2.2	-	-	4.8

Several workers (Mills *et al,* 1990, Triebskorn and Schweizer, 1990, Litsinger and Estano, 1993) had shown that Meta metaldehyde is effective under very wet conditions because it causes irreversible cell damage, particularly to the dermal mucous cells, with both cell membranes and organelles being affected. This refuted the previous hypothesis that the effects of metaldehyde are reversed by the slug's absorption of water from the environment (Godan, 1983). The efficacy of metaldehyde for the control of water snails in this study confirms this conclusion and that metaldehyde can be highly effective against aquatic species (Bowen and Mendis, 1995).

Study 2

Results in ponds suggest that the data from controlled laboratory experiments in basins may not be directly applicable under actual pond conditions (Table 2). Application rates of up to15 kg/ha appeared to be ineffective. The snails were quite active at 3 DAT and even lower mortalities were observed at 6 DAT. Higher counts on day 3 may include also moribund snails. The snails might have avoided the pellets by ceasing to feed or possessed detoxification mechanisms at sub-lethal levels of poison which enabled them to recover. Higher application rate was required in ponds to kill the snails quickly after exposure and leave no chances for recovery later (Table 3).

Table 2. Efficacy of metaldehyde (formulation C) against brackishwater pond snails (*Cerithium* sp.), Pond Expt. 1.

Dose (kg/ha)	% Mortality 3 DAT	6 DAT
0 (Control)	3[a]	5[a]
5	42[b]	16[a]
15	55[b]	14[a]

Means in a column with the same superscripts are not significantly different at p>0.05.

Table 3. Efficacy of metaldehyde (formulation C) against brackishwater pond snails (*Cerithium* sp.), Pond Expt. 2

Dose (kg/ha)	Ave. Mortality (%)	Normalized Value[1] (%)
0 (Control)	10[b]	0[b]
20	33[b]	23[b]
30	88[a]	76[a]
40	87[a]	77[a]
60	96[a]	86[a]
20 + 20	75[a]	65[a]
30 + 30	94[a]	84[a]
40 + 40	98[a]	88[a]

[1] Mortality of control groups subtracted
Means in a column with the same superscripts are not significantly different at p>0.05.

The average snail mortality data on the single application increased with increasing dose. A dose of 20 kg/ha killed only 30% of the population while a dose of 30 kg/ha eradicated about 90% and almost complete eradication was obtained at 60 kg/ha. Comparable values with that of 60 kg/ha single application were obtained at both 30 + 30 and 40 + 40 kg/ha. The data suggest that a single application of at least 30 kg/ha will be effective in treating pond snails. There is little benefit in making a follow-up application three days after the initial treatment or in increasing the dosage.

Study 3

No mortalities were observed in milkfish juveniles exposed to the different test concentrations of metaldehyde formulations up to 175 kg/ha (Tables 4 and 5).

The metaldehyde formulations were insoluble in water and the pellets tended to settle quickly on the surface of the soil. Milkfish juveniles are filter feeders and feed on the water column. Although the fish in the course of their swimming activity were observed to have some contact with the soil where the pellets were spread out, they were not attracted to the pellets and did not feed on them even if they were starved during the assay. The limited solubility of META metaldehyde in brackishwater and the non-palatability of META metaldehyde snail bait for milkfish may account for the non-toxicity of the formulation to non-target fish such as the milkfish in this experiment. Similar results were obtained by Tejada, et al (1992) in their toxicity experiments on carp and tilapia in irrigated lowland rice paddies. Future experiments should determine the acute toxicity of metaldehyde to milkfish by force feeding or intraperitoneal injection.

Table 4. Effect of metaldehyde (formulations A and B) on milkfish juveniles (2-3g)

Application rate (Kg/ha)	No. of fish per basin	Percent Mortality Days After Treatment (DAT)						
		1	2	3	4	5	6	7
0	20	0	0	0	0	0	0	0
5	20	0	0	0	0	0	0	0
15	20	0	0	0	0	0	0	0
25	20	0	0	0	0	0	0	0
35	20	0	0	0	0	0	0	0
45	20	0	0	0	0	0	0	0
55	20	0	0	0	0	0	0	0
65	20	0	0	0	0	0	0	0
75	20	0	0	0	0	0	0	0

Table 5. Effect of metaldehyde (formulation C) on milkfish juveniles (1-2g)

Application rate (Kg/ha)	No. of fish per basin	Percent Mortality Days After Treatment (DAT)						
		1	2	3	4	5	6	7
0	20	0	0	0	0	0	0	0
25	20	0	0	0	0	0	0	0
50	20	0	0	0	0	0	0	0
75	20	0	0	0	0	0	0	0
100	20	0	0	0	0	0	0	0
125	20	0	0	0	0	0	0	0
150	20	0	0	0	0	0	0	0
175	20	0	0	0	0	0	0	0

The results obtained in these experiments suggest that the META metaldehyde formulations were effective for pond snail control without detrimental effect on the milkfish. Future studies should look into ways of improving methods of application in ponds. Studies on the biodegradation rates of metaldehyde in brackishwater ponds are presently underway.

ACKNOWLEDGEMENTS

This work was a collaborative study between the Southeast Asian Fisheries Development Center (SEAFDEC) Aquaculture Department and Lonza Ltd., Basel, Switzerland. We thank Lucia Jimenez for technical assistance and the Iloilo State College of Fisheries (ISCOF), Barotac Nuevo, Iloilo, Philippines for the use of ponds.

REFERENCES

Bowen, I D; Mendis V W (1995) Towards an integrated management of slug and snail pests. *Pesticide Outlook* **6**:12-16.

Camponovo, F (1995) Environmental and Health Aspect of META Metaldehyde in the Context of Registration Requirements. A paper presented at the META Symposium. Visp, Switzerland. June 21, 1995.

Finney, D J (1982) Probit analysis. Cambridge Univ. Press, Cambridge, 318 p.

Godan, D (1983)   Pest Slug and Snails: Biology and Control. Springer-Verlag, Berlin.

Litsinger, J A ; Estano D B (1993)   Management of the golden apple snail *Pomacea caniculata* (Lamarck) in rice. *Crop Protection* **12:** 363 -370.

Mills, J D; McCrohan C R; Bailey S E; Wedgwood M A   (1990) Effects of Metaldehyde and Acetaldehyde on Feeding Responses and Neuronal Activity in the Snail, *Lymnaea stagnalis. Pesticide Science* **28:**89-99.

Pillai, T G (1972)   Pest and Predators in Coastal Aquaculture Systems of the Indo-Pacific Region. In: *Coastal Aquaculture in the Indo-Pacific Region,* T V R Pillay (Ed) Fishing News (Books), Ltd., London, 497 p.

Tejada, A W ; Cagampang S; Medina J R (1992)   Metaldehyde 6% Bait and Metaldehyde 50WP Paddy Fish Toxicity Trial. A Project Terminal Report, Calauan, Laguna, Philippines, 12 p.

Triebskorn, R; Schweizer H (1990) Impact of the molluscicide metaldehyde on the mucus cells in the digestive tract of the grey garden slug deroceras reticulatum, 8pp. In: Deuxieme Conference Internationale sur les Ravageurs en Agriculture, December 4-6, 1990, Versailles, France.

Trino, A T; Bolivar E C; Gerochi D D (1993) Effect of burning of rice straw on snails and soil in a brackishwater pond. *Journal of Tropical Agriculture* **11**(2):93-97.

Wang, S T C (1989) Pond on fire. *Aquaculture Techonology Talk*. 1st QTR. Publ,:1

NICLOSAMIDE, AN EFFECTIVE MOLLUSCICIDE FOR THE GOLDEN APPLE SNAIL (*POMACEA CANALICULATA* LAMARCK) CONTROL IN PHILIPPINE RICE PRODUCTION SYSTEMS

F V PALIS, R F MACATULA, L BROWNING

Crop Protection Division, Bayer Philippines Inc., San Juan, Metro Manila, Philippines

INTRODUCTION

The golden snail *Pomacea canaliculata* Lamarck, also known as golden apple snail and locally as golden kuhol, originated from South America and was introduced to Taiwan between 1979 and 1981 (Cheng, 1989). It was introduced in the Philippines for its potential as an alternate protein source (Adalla and Morallo - Rejesus, 1989). After intensive mass production in rural and urban communities, it became a destructive pest in 1986 as the hatchlings escaped and spilled away from rearing ponds to nearby lowland paddy fields via irrigation dykes and waterways. As the golden snail adapted to feeding on rice seedlings, various factors contributed to their continuous spread and infestation: it is very prolific, polyphagous and can survive buried in soil over long periods and has no known natural enemies locally.

The golden snail has since secured its status as a national pest in terms of distribution and potential for damage on rice seedlings (Rondon *et al.*, 1989).

In 1986, about 130,000 hectares of rice fields were affected and infestation increased to 400,000 hectares in 1989; currently, the infestation is estimated at over 800,000 hectares (Schnorbach, 1995).

Biological measures, mechanical/physical methods and cultural control are recommended ways of combatting the golden apple snail in Integrated Pest Management (IPM).

For chemical control of the golden apple snail, there are relatively few crop protection products available which have good efficacy, favourable user toxicity and satisfactory environmental impact (Schnorbach, 1995). In the absence of an acceptable molluscicide, the Department of Agriculture's anti-golden snail campaign seems stalled until an appropriated alternative is chosen as a component of an IPM recommendation.

Niclosamide, already recommended in the Philippines for schistosomiasis vector control by the Department of Health was evaluated as a candidate compound for golden apple snail control in 1990-91. Laboratory and field tests were conducted to evaluate the efficacy of 'Bayluscide 250 EC' (niclosamide 250 g/l EC) and other compounds (including botanicals) for the identification of an effective solution to the golden snail infestation. In 1991, Bayer Philippines obtained registration of niclosamide 250 EC and 'Bayluscide 70 WP' (niclosamide 70% WP) as recommended molluscicides for both direct seeded and transplanted lowland rice snail control from the Fertilizer and Pesticide Authority (FPA).

In consideration of the aforementioned concerns and anticipating a farmer-friendly use recommendation, various studies conducted under lowland rice condition were aimed to:

1. Evaluate the biological efficacy of niclosamide 250 EC for golden snail control.
2. Determine the effect of water depth at the time of treatment on the efficacy of niclosamide 250 EC for golden snail control.
3. Compare the efficacy of niclosamide 250 EC with other compounds and botanicals for golden snail control.
4. Determine if there are residues of niclosamide on paddy soil and in water after application.

MATERIALS AND METHODS

Experiment 1

In three separate trials, during 1991-1992 rice cropping season, niclosamide 250 EC was evaluated following FPA recommended procedures and approved agronomic practices for direct-seeded and transplanted lowland rice. Rice variety IR 66 was used in the study. The experiments were laid out using the randomized complete block design. Enlarged dykes were properly established to prevent snail migration and treatment seepage. Plots were kept saturated for four days. Irrigation water was introduced at 3-5 cm before snail introduction. Snails representing different sizes were introduced into the plots at a density of three snails per m². In the direct-seeded trials, snails were allowed to acclimatize for about an hour before treatment. Plots were rated on the degree of damage caused by the golden snails. In transplanted trials, snails were allowed to acclimatize overnight and missing hills were replanted before treatment application. Treatments involved were: niclosamide 250 EC at 1.0 and 1.5 l/ha and an untreated control.

Four quadrats (1.5 m x 4 m) were made in each plot (14 m x 8 m) to serve as sampling areas. Counting of snail mortality, number of eggmasses and crop damage assessment was done at 1, 3, 5, 7, 10 and 14 days after treatment application (DATA).

Experiment 2

The trial was conducted at Tropics Biological Research and development Center during the 1992 cropping to evaluate te efficacy of niclosamide applied at 1.0 and 1.5 l/ha 4 days after transplanting (DAT) at various irrigation water levels, namely at 0, 2, 4 and 6 cms depth. Each 2 m x 2 m plot had a surrounding enlarged earthen bund and was independently irrigated and drained as needed. Irrigation water was reintroduced based on treatment-specified depths at 4 DAT for each plot and treatments were applied within 2 hours after water introduction. The following data were taken: % dead snails at 6, 12, 24, 48 and 72 hours after application (HAA);
% crop damage at 24, 48, 72 and 96 HAA and 14 days after application (DAA); number of snail egg masses/plot at 24, 48 and 72 HAA.

Experiment 3

The experiment was conducted during the dry season of 1992. Each plot measured 12 m x 12 m and the treatments were replicated four times using a split plot design. The desired water level of 3-5 cm was maintained per plot per treatment throughout the observation period.

PRODUCT	RATE/ha
1. Madre de Cacao (*Gliridia sepium*)	200 kg (Bark)
2. Tobacco (*Nicotiana tabacum*)	
3. Tubli (*Derris elliptica*)	100 kg (Leaves)
4. Niclosamide 250 EC	40 kg (Roots)
5. Niclosamide 250 EC	1.0 l
6. Control	1.5 l
	(No Pesticide)

The leaves, bark and roots of tobacco, madre de cacao and tubli, respectively, were thoroughly dried and powdered. The application for the botanical plant materials was done by broadcasting. The chemical pesticides were applied y a knapsack sprayer at 160 l of spray volume per hectare.

After the last harrowing, the plots were cleared of golden snail. Treatments were applied a day before and 4 days after seeding (DAS). A day before application, thirty snails of different sizes or stages were collected and introduced at a density of three snails/m^2 per plot. Snail mortality was monitored 3 and 7 days after pesticide application.

Fourteen days after seeding, the degree of plant damage was determined based on 200 seedlings per plot. Egg masses laid by snails daily from introduction until 14 days after seeding were counted and recorded. Yield of rice was determined per plot at 14% moisture content.

Experiment 4

Supervised residue trials were conducted in collaboration with the National Crop Protection Center to determine the fate of niclosamide applied by foliar spray in a rice paddy ecosystem at the Central Experiment Station, University of the Philippines, Los Banos between March 27, 1991 to December 31, 1991. At 1 DAT niclosamide 250 EC was sprayed at the rate of 0.3 kg a.i./ha. Paddy water, maintained at 13 inches depth, was allowed to overflow into canals where cages containing five tilapia and five carp fishes were positioned at 0, 1, 5, 20 and 100 m away from the treated paddy. Fish mortality was recorded after 24 hours exposure at 0, 1, 3, and 7 days after initial exposure.

Random soil and water samples were collected at 0 (<3 hours), 1, 2, 3, 7, 14 and 23 days after application. At harvest, (127 DAT) leaves, stalk and rice grain samples were also collected. All samples were weighed and frozen prior to analysis, following recommended procedures for niclosamide residue determination at the Pesticide Residue laboratory of the National Crop Protection Center (NCPC). Three replicates of each substrate were used.

RESULTS AND DISCUSSION

Experiment 1

Niclosamide 250 EC applied at 1.0 and 1.5 l/ha provided excellent control of golden snail infesting both direct-seeded and transplanted lowland rice. Crop damage taken at 3 days after treatment was 34.4% in the untreated with negligible damage in niclosamide-treated plots both at 1.0 and 1.5 l/ha (Figures 1 & 2).

No egg masses were observed in niclosamide-treated plots as compared with 16-18.5 eggmasses per plot in the untreated. Similarly, rice yield in treated plots was 50% higher than in the untreated check (Figures 3 & 4).

Experiment 2

Excellent snail control was obtained in treated plots both in direct seeded and transplanted rice at 1.0 and 1.5 l/ha on plots with 4-6 cm water depth at the time of treatment.

Application of niclosamide on plots with 2 cm water depth at the time of treatment was better than application of the same product rates on plots without water (Table 1).

In the untreated plot crop damage was directly proportional to the increase in water depth. Significantly lower percent crop damage was noted at 1.0 and 1.5 l/ha of niclosamide with 2-6 cm water depth (Table 2).

Experiment 3

Niclosamide at 1.0 and 1.5 l/ha applied at one and four DAS showed significantly higher percent snail mortalities taken at 24 HAA than those plots treated with *G. sepium*, *N. tabacum* and *D. elliptica* (Figure 5).

On crop damage taken at 3 DAA, niclosamide applied 1 day before seeding (DBS) and 4 DAS gave lower seedling damage than the botanicals with 1.5-7.6% *N. tabacum* leaves at 200 kg/ha demonstrated lower seedling damage of 11.6-16.1% than the untreated control check with 31.4-50.8% crop damage (Figure 6).

The number of snail egg masses taken at 14 DAA was significantly lower on niclosamide treatments than on the botanicals (Figure 7).

Yield of rice was generally higher on niclosamide-treated plots than on both the botanicals and the untreated (Figure 8).

Experiment 4

There was a rapid decline in niclosamide residues in paddy water from an initial level of 0.96 mg/l to 0.02 mg/l at two days after application (Figure 9). Degradation followed pseudo-first order kinetics showing a half-life of 0.3 day; residues found three days after spraying (DAS) were below the detection limit of 0.0002 mg/l.

Niclosamide in paddy soil was below the detection limit of 0.03 mg/kg on the day of application. Maximum concentration of 0.1 mg/kg was attained at 1 DAS. However, residues were rapidly degraded to levels below the minimum detection limit of 0.03 mg/kg at 2 to 28 DAS (Figure 9). At harvest, niclosamide residues were below the detection limit of 0.03 mg/kg in rice leaves, stalk and grain indicating that the use of niclosamide in rice production does not lead to persistent residues in the various components of the rice paddy ecosystem.

The toxicity of niclosamide to two freshwater fish species, commonly reared for human consumption, was also assessed. Niclosamide was toxic to carp at the rate of 0.25 and 0.375 kg a.i./ha at the time of application where 100% mortality was observed from 0 to 100 m away from the treated paddy (Table 3). The practice of draining the paddy water takes place 60 days after transplanting and therefore the problem of fish toxicity is not usually a matter of concern. The dilution effect from unexpected torrential rains accompanying typhoons would be enough to lower the concentration to below toxic levels. Carp mortality was not observed at 1, 3 or 7 days after application of niclosamide under overflowing conditions, allowing the rapid decline of residues in paddy water to non-toxic levels.

Niclosamide was slightly toxic to tilapia as only 20% and 33% mortality was observed inside treated paddies on the day of application. A corrected percent mortality was observed for tilapia at 1 m from the treated paddy using 0.375 kg a.i.l/ha one day after application. This effect could be due to the release of previously-absorbed niclosamide from the soil matrix by desorption. Mortality observed in paddies treated at 0.250 kg a.i./ha was no different from the natural mortality observed on the introduction of fish. However, it is recommended that in areas where fish culture is practised, the fish should not be introduced until at least three days after niclosamide treatment.

CONCLUSION

Application of niclosamide 250 EC at 1.0 and 1.5 l product/ha provided excellent control of golden apple snails infesting both direct-seeded and transplanted lowland rice. Rice seedling damage and number of egg masses laid on niclosamide-treated plots were minimal.

Niclosamide 250 EC provided excellent control of golden snail at 1.0 and 1.5 l/ha with 4-6 cm water depth at the time of treatment. Seedling damage was directly proportional to the increase in water depth.

Niclosamide application at 1.0 and 1.5 l/ha at one and four days after seeding gave better efficacy than any of the botanicals tested.

Residues in paddy soil and water were below the detection limit. At harvest, residues were below the detection limit of 0.03 mg/kg in rice leaves, stalk and grain, indicating that the use of the product in rice production does not lead to persistent residues in the various components of the rice paddy ecosystem.

In areas where rice/fish culture is practised, introduction of fish in niclosamide-treated paddies must be made at least three days after molluscicide application.

REFERENCES

Adalla, C B; Morallo-Rejesus, B (1989) The Golden Apple Snail, *Pomacea* sp. A serious pest of lowland rice in the Philippines. In: *Slugs and Snails in World Agriculture*. I F Henderson (ed.). BCPC Monograph No. 41, pp. 417-422.

Calumpang, S M F; Medina, M J B; Tejada, A W; Medina J R (1994) Environmental impact of two molluscicides: niclosamide and metaldehyde in a rice paddy ecosystem. 4 p.

Cheng, E X (1989) Control strategy for the introduced snail *Pomacea canaliculata* in rice paddy. In: *Slugs and Snails in World Agriculture*. I F Henderson (ed.). BCPC monograph No. 41, pp. 69-75.

Mochida, O (1987) Pomacea snails in the Philippines. *International Rice Research Newsletter*. pp.12; 48-49.

Palis, F V; Macatula, R F; Sapin, A T; Dupo, H (1993) Efficacy of bayluscide 250 EC at various rates and water depth against Golden Snail (*Pomacea* sp.) in transplanted and direct seeded rice. *24th Annual Scientific Meeting of the Pest Management Council of the Philippines. College, Laguna, Philippines*. 49 p.

Palis, R G; Macatula, R F; Marchand, T; Dupo, H; Olanday, M and Estoy, G. (1994) Niclosamide (Bayluscide 250 EC): An Effective Molluscicide for Golden Apple Snail (*Pomacea* sp.) control in rice in the Philippines. *4th International Conference on Plant Protection in the Tropics*. March 8-31, 1994. 8 p.

Rondon, M B; Callo, D (1989) Distribution and mode of infestation of Golden Snail in rice farming. Paper presented at the Workshop on "*Environmental Impact of Golden Snail (Pomacea sp.) on Rice Farming System in the Philippines*". FMC-CLSU, Monoz, Nueva Ecija. 9-10 November 1989.

Schnorbach, H J (1995) The Golden Apple Snail (*Pomacea canliculata*): An increasingly important pest in rice and methods of control with Bayluscide. *Pflanzenschutz-Nachrichten Bayer*. pp. 313-346.

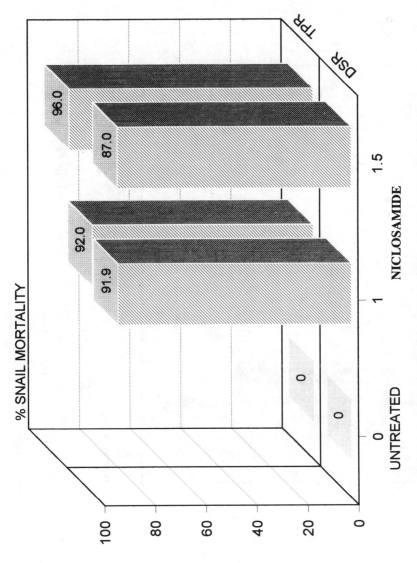

Fig. 1 EFFECT OF NICLOSAMIDE 250 EC AGAINST GOLDEN APPLE
SNAIL IN RICE. PHILIPPINES. 1991. WS.

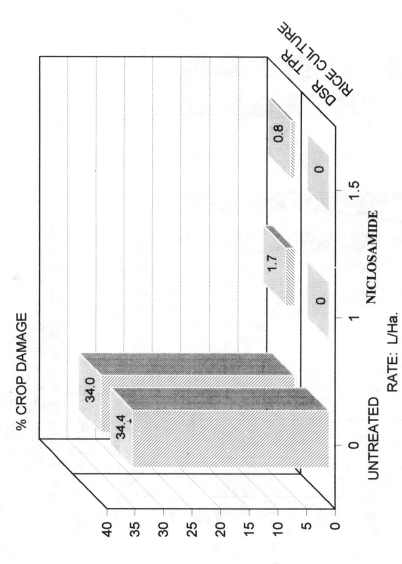

% CROP DAMAGE

Fig. 2 EFFECT OF NICLOSAMIDE 250 EC AGAINST GOLDEN APPLE
SNAIL IN RICE. PHILIPPINES. 1991. WS.

Mean of 3 trials.

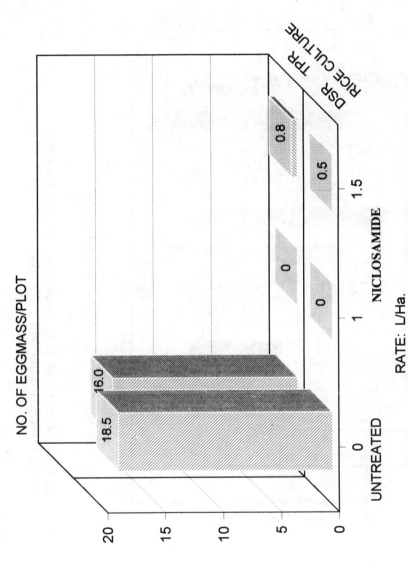

NO. OF EGGMASS/PLOT

RATE: L/Ha.

Fig. 3   EFFECT OF NICLOSAMIDE 250 EC AGAINST GOLDEN APPLE
SNAIL IN RICE. PHILIPPINES. 1991. WS.

Mean of 3 trials.

221

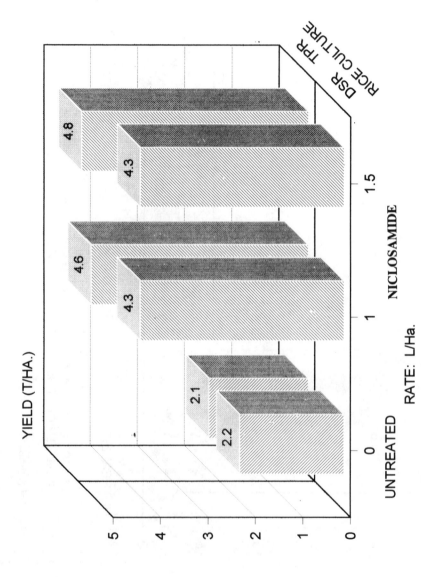

YIELD (T/HA.)

RATE: L/Ha.

Fig. 4 EFFECT OF NICLOSAMIDE 250 EC AGAINST GOLDEN APPLE
SNAIL IN RICE. PHILIPPINES. 1991. WS.

Mean of 3 trials.

Table 1. Effect of Niclosamide 250 EC Applied at Various Water Depths for Snail Control in Rice. TBRDC, Laguna, Philippines. 1992.

Treatment	Rate (L/Ha.)	Snail Mortality (24 HAT) [1] Water Depth (cm.)			
		0	2	4	6
Transplanted					
Untreated	0.0	0.0	0.0	0.0	0.0
Niclosamide	0.5	0.0	31.6	38.0	41.3
Niclosamide	1.0	3.6	63.3	86.0	63.0
Niclosamide	1.5	3.3	86.0	96.6	93.0
Direct Seeded					
Untreated	0.0	0.0	0.0	0.0	0.0
Niclosamide	0.5	0.0	23.0	38.3	31.3
Niclosamide	1.0	12.0	64.0	91.0	86.7
Niclosamide	1.5	19.3	70.3	90.0	100.0

[1] Average of 4 replicates; HAT = Hours After Treatment

Table 2. Effect of Niclosamide 250 EC Applied at Various Water Depth on Crop Damage. TBRDC, Laguna, Philippines. 1992.

Treatment	Rate (L/Ha.)	% Crop Damage (24 HAT) [1]			
		Water Depth (cm.)			
		0	2	4	6
Transplanted					
Untreated	0.0	1.3	10.7 a	19.0 a	33.3 a
Niclosamide	0.5	1.0 a	5.0 ab	2.6 b	5.3 b
Niclosamide	1.0	0.6 a	1.3 b	3.0 b	2.3 b
Niclosamide	1.5	4.0 a	1.0 b	0.0 b	1.0 b
Direct Seeded					
Untreated	0.0	16.7 a	43.3 a	53.3 a	73.3 a
Niclosamide	0.5	10.0 a	13.0 ab	43.0 a	73.0 a
Niclosamide	1.0	6.7 a	0.0 b	0.0 b	3.3 b
Niclosamide	1.5	10.0 a	0.0 b	0.0 b	0.0 b

[1] Average of 4 replicates; in a column means followed by the same letter are not significantly different at the 5% level by DMT; HAT = Hours After Treatment.

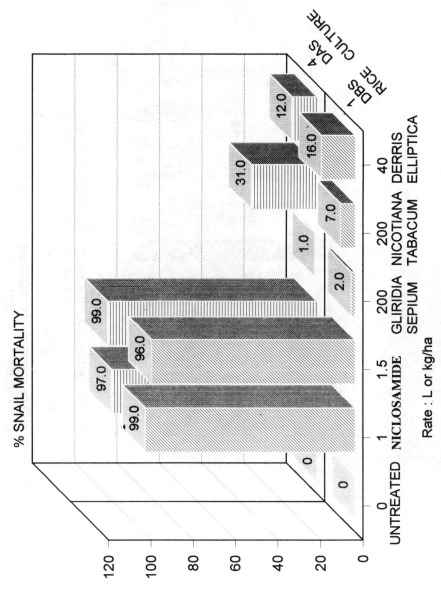

Fig. 5 EFFECT OF NICLOSAMIDE 250 EC AGAINST GOLDEN APPLE
SNAIL IN RICE. PHILIPPINES. 1991. WS.

225

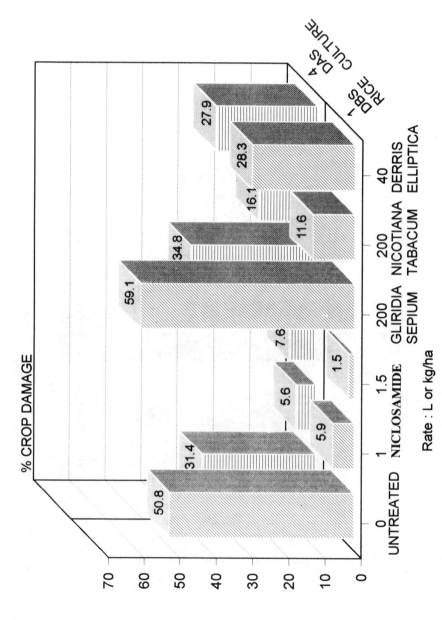

Fig. 6 EFFECT OF NICLOSAMIDE 250 EC AGAINST GOLDEN APPLE
SNAIL IN RICE. PHILIPPINES. 1991. WS.

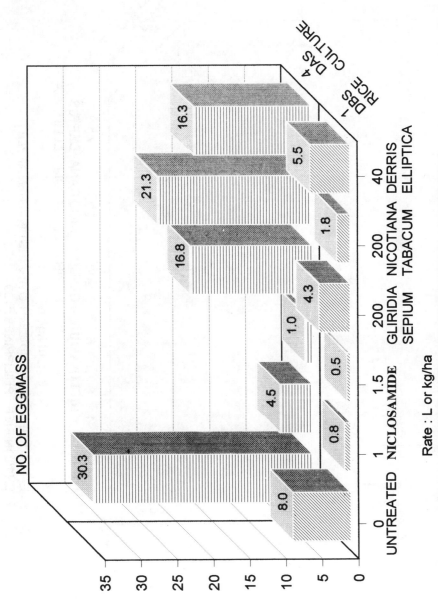

Fig. 7 EFFECT OF NICLOSAMIDE 250 EC AGAINST GOLDEN APPLE
SNAIL IN RICE. PHILIPPINES. 1991. WS.

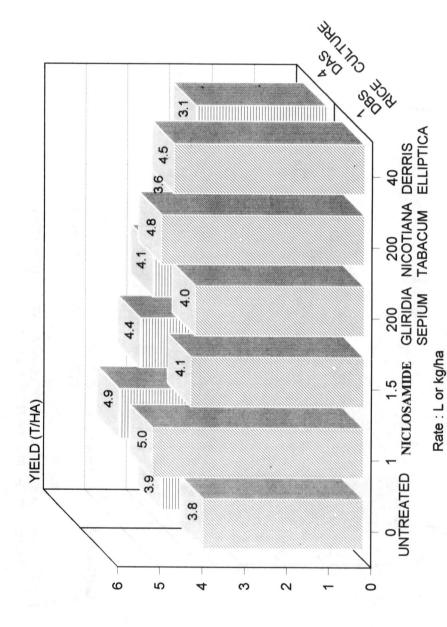

YIELD (T/HA)

RICE CULTURE
4 DAS
-1 DBS

UNTREATED   NICLOSAMIDE   GLIRIDIA   NICOTIANA   DERRIS
                          SEPIUM   TABACUM   ELLIPTICA

Rate : L or kg/ha

Fig. 8   EFFECT OF NICLOSAMIDE 250 EC AGAINST GOLDEN APPLE
         SNAIL IN RICE. PHILIPPINES. 1991. WS.

Figure 9. Niclosamide residues in paddy soil and water.

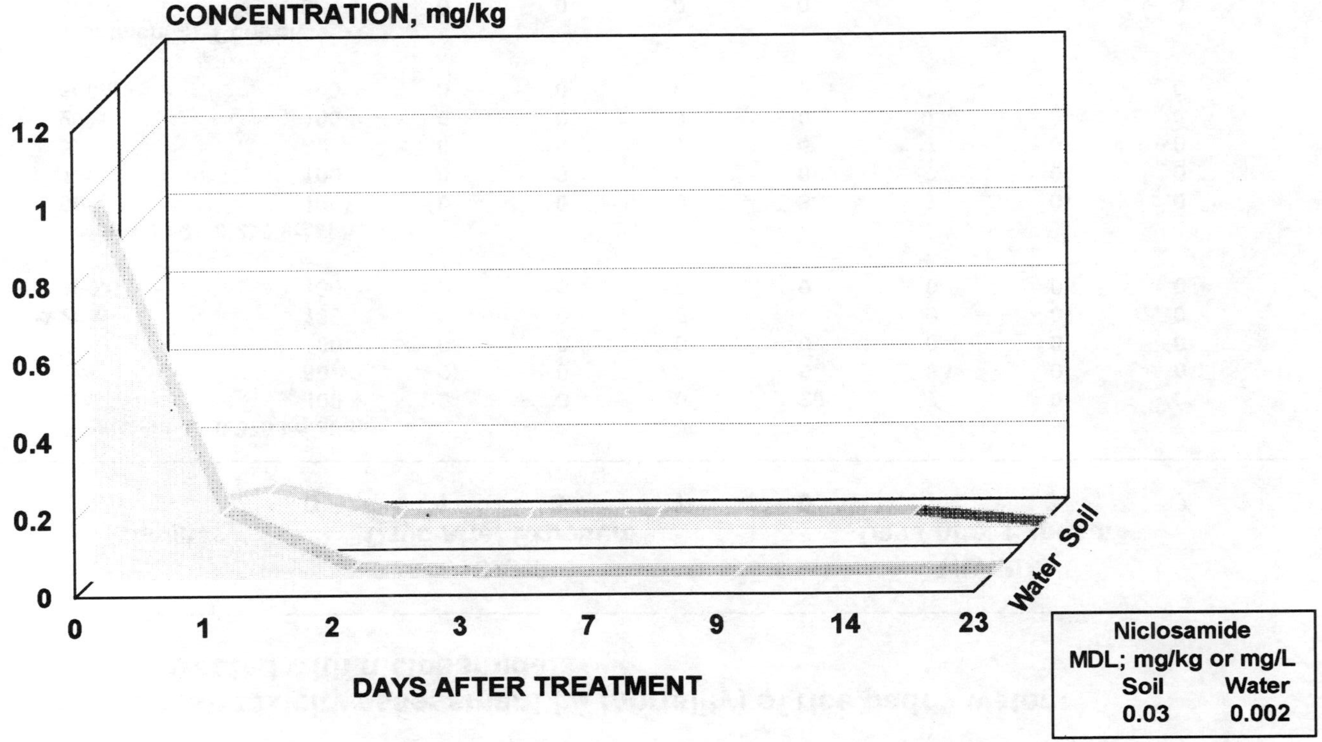

CONCENTRATION, mg/kg

1.2
1
0.8
0.6
0.4
0.2
0

0   1   2   3   7   9   14   23

DAYS AFTER TREATMENT

Water Soil

Niclosamide
MDL; mg/kg or mg/L
Soil   Water
0.03   0.002

229

Table 3. Fish toxicity assessment (% mortality) of rice paddy water treated with niclosamide.

Distance	CARP				TILAPIA			
	Days After Exposure				Days After Exposure			
	0	1	3	7	0	1	3	7
Treatment 1: 0.375 kg.ai/m								
0m	100	0	0	0	20	7	0	0
1m	80	0	0	0	0	33	0	0
5m	100	0	0	0	0	0	0	0
20m	100	0	0	0	0	0	0	0
100m	100	0	0	0	0	0	0	0
Treatment 2: 0.250 kg.ai/ha.								
0m	100	0	0	0	33	7	0	0
1m	100	0	0	0	0	7	0	0
5m	92	0	0	0	0	7	0	0
20m	100	0	0	0	0	7	0	0
100m	100	0	0	0	0	7	0	0
Treatment 3: Control								
0m	0	0	0	0	0	0	0	0
1m	0	0	0	0	0	7	0	0
5m	0	0	0	0	0	0	0	0
20m	0	0	0	0	0	0	0	0
100m	0	0	0	0	15	20	0	0

THE SCREENING OF NATURALLY OCCURRING COMPOUNDS FOR USE AS SEED TREATMENTS FOR THE PROTECTION OF WINTER WHEAT AGAINST SLUG DAMAGE

A L POWELL, I D BOWEN
School of Pure and Applied Biology, University of Wales Cardiff, PO Box 915, Cardiff, CF1 3TL, UK

ABSTRACT

Candidate compounds including phenolics, aromatics and monoterpenes were screened for use as seed treatments to protect winter wheat from seed hollowing by *D. reticulatum*. Compounds were tested for their molluscicidal and semiochemical effects and for phytotoxicity. Many compounds initially tested at 0.5% proved molluscicidal and repellent, however several were highly phytotoxic. Compounds showing favourable traits i.e. low phytotoxicity and high repellency were tested for their efficacy as feeding deterrents in seed dressings. Of the compounds tested only (-)menthol and thymol were highly effective, with (-)menthol completely inhibiting seed hollowing over the course of a week at 0.5%. Some compounds showed some feeding deterrency, but this decreased over time possibly due to the inherent volatility lowering the level of compound at the point of feeding to a concentration below a threshold level of deterrency.

INTRODUCTION

Monoterpenes have been shown to both repel and attract insects (Chapman *et al*, 1981) and a number are alarm and defence substances produced by insects (Gnanasunderam *et al*, 1981). They also act as feeding deterrents to slugs such as *Ariolimax dolichophallus* (Rice *et al*, 1978), with the bicyclic monoterpene (+)fenchone (Airey *et al*, 1989) and the acyclic monoterpene geraniol (Glen *et al*, 1989) having been isolated as allomones for *Deroceras reticulatum*.

Seed hollowing of winter wheat by *D. reticulatum* is a serious agricultural problem (Scott *et al*, 1984). The main control method utilized for crop protection against slugs is the broadcast application of pelleted baits containing either metaldehyde or methiocarb (Bourne *et al*, 1988). However, these compounds may affect a wide range of animals including birds and small mammals and the natural predators of slugs such as carabid beetles (Büchs *et al*, 1989). Chemoreception is important in gastropod food location (Croll & Chase, 1977, 1980) therefore, alternative control systems utilizing repellent or antifeedant semiochemicals may prove effective. In particular plant derived repellents of low toxicity applied as seed dressings could deter slug attack.

Many seed treatments of winter wheat have been investigated for their repellent effects upon *D. reticulatum* (Scott *et al*, 1977, Scott, 1981, Scott *et al*, 1984). However, due to the intimate contact of actives with the seed any phytotoxic effects of these actives may cause deleterious conditions for germination and seedling growth. Therefore, a desirable characteristic of a seed dressing is a wide margin between the effective repellent dose and the phytotoxic dose of the active (Griffiths, 1978).

Monoterpenes and soluble aromatics and phenolics have been shown to be toxic to vascular plants (Harborne, 1993). Studies on mechanisms of allelochemical inhibition of germination and growth have been reviewed by Einhellig (1986) and by Rice (1984). These allelopathic effects plus high volatility and short persistence may engender poor field development of monoterpene treated winter wheat and so render monoterpenes unsuitable as seed dressings (Airey *et al*, 1989). However, alternative regimes of crop protection could be developed exploiting the monoterpenes' semiochemical properties possibly using controlled release technology. This paper describes the screening of candidate aromatics and monoterpenes for semiochemical and possible molluscicidal effects upon *D. reticulatum*, together with integrated bioassays assessing both their phytotoxic effects upon winter wheat and their efficacies as seed dressings.

MATERIALS AND METHODS

Mature specimens of *D. reticulatum*, 0.6-0.8g were collected from outlying fields and gardens, housed at 5 ± 1°C in permanent darkness and starved for 48 hrs prior to experimentation.

Split substrate tests

Tests were conducted using methods outlined by Bowen & Antoine (1995). A 9cm diameter plastic petri dish was punctured with eight holes placed symmetrically about the circumference of the lid using a hot seeker. A 9cm Whatman 1 filter paper circle was cut in half, one half placed in the required concentration of test chemical made up in GPR grade methanol. This active half was placed in the petri dish and left for half an hour allowing the methanol to completely evaporate off. The other untreated half was then placed in the petri dish adjacent to the treated half and was considered as the control half. Both were then soaked in 1ml of distilled water. A replicate dish was prepared and a test slug placed in the control half of one dish with another slug introduced to the active half of the other dish and left for 24 hours at 13°C in complete darkness. Ten replicate dish pairs were prepared per treatment. A control dish was set up in tandem with a methanol only half previously left for half an hour for full evaporation and a slug placed in this active sector. A set of ten pairs of control dishes were prepared where all sectors were left untreated and soaked in distilled water.

After 24 hours it was noted which half the slug was occupying and whether it had died or not. Death was indicated by failure to respond to a 0.3Hz, 12V pulse of electricity from a CFP Model 8048 stimulator. The test animals were removed and the filter paper saturated in distilled water. Excess water was poured off and the filter paper covered in decolourising charcoal and left for five minutes. The charcoal was washed off and any mucus trail left by the slug picked up charcoal and left discrete black marks.

The following compounds were used in split substrate tests:

Cinnamyl alcohol

Salicylaldehyde

Vanillin

Thymol

(+)Limonene

αTerpinene

αTerpineol

(-)Menthol

(-)Menthone

(+)Carvone

Phytotoxicity tests

Winter wheat seeds, cv. Hereward, were coated with 0.5% a.i./wt seed in an adhesive sticker of 20% starch 20% gum acacia made up in distilled water. 120 seeds were coated with each active and placed on saturated filter paper in 0.07m² glass covered germination trays with thirty seeds per tray in six rows of five seeds at a rate corresponding to 266kg/ha. Four control trays were prepared in this way with untreated seeds and four trays were prepared with adhesive only treated seeds. Seeds were germinated at 24 ± 1°C, 16 hrs light. Germination was considered to have occurred when growth had reached the 07 stage of the decimal code (Tottman & Broad, 1987). Coleoptile growth was measured after seven days and percentage germination was also calculated after seven days. Coleoptile lengths were compared with controls using the Mann-Whitney non-parametric pairwise comparison as were calculated median percentage germinations. In all cases a significance level of p = 0.05 was used. Only compounds proving repellent in split substrate tests were used in phytotoxicity trials.

Terraria trials

Tests were conducted using methods outlined by Bowen & Antoine (1995). Four 0.07m² trays were prepared for each active as in the phytotoxicity tests, along with controls. Four specimens of *D. reticulatum* were introduced to each tray (representative of a level of 570,000 slugs/ha). Covered trays were held in total darkness at 13 ± 1°C. Numbers of seeds hollowed were counted per day for seven days. Median feeding times (FT50, the time at which 50% of seeds had been hollowed) were calculated for each treatment and controls, using a time response analysis of the cumulative percentage frequency of the seeds hollowed (Taylor *et al*, 1993). The data was computed using a FORTRAN program based on the methods described by Litchfield (1949). Calculated FT50s of treated seeds were compared with the control FT50. The percentages of seeds hollowed after seven days were also calculated. Only the less phytotoxic compounds were used in terraria trials.

RESULTS

Split substrate tests

Percentage contact mortality (n = 10) A. I. = 0.5%	
Control	0
Cinnamyl alcohol	100
Salicylaldehyde	0
Vanillin	0
Thymol	100
(+)Limonene	0
αTerpinene	0
αTerpineol	20
(-)Menthol	100
(-)Menthone	100
(+)Carvone	100

The control, Salicylaldehyde, (+)Limonene and αTerpinene were non-repellents. Cinnamyl alcohol, Thymol, αTerpineol, (-)Menthol, (-) Menthone and (+)Carvone were strongly repellent. Vanillin was weakly repellent.

<u>Phytotoxicity trials</u>

0.5% a.i./wt seed.

	Mean coleoptile length (cm) after seven days ($\pm$ SEM)	Median % germination after seven days
Control	10.4 (0.16)	100
Adhesive only	10.4 (0.20)NS	98.3 NS
Cinnamyl alcohol	0	0
Vanillin	5.3 (0.29)*	90 *
Thymol	4.6 (0.13)*	100 NS
αTerpineol	5.8 (0.21)*	100 NS
(-)Menthol	8.7 (0.22)*	98.3 NS
(-)Menthone	0	0
(+)Carvone	0	0

$p = 0.05$, NS = not significantly different from control, * = significantly different.

Cinnamyl alcohol, (-)Menthone and (+)Carvone completely inhibited germination.

<u>Terraria trials</u>

0.5% a.i./wt seed.

	FT50 (hrs) ($\pm$ 95% C. I.)	% seeds hollowed after seven days ($\pm$ SEM)
Control	51.26 (45.09 - 58.28)	95.8 (2.1)
Adhesive only	54.48 (48.89 - 60.71) NS	93.3 (1.9)
Vanillin	113.97 (93.45 - 139.00) *	70.0 (3.6)
Thymol	206.78 (183.65 - 232.82) *	35.8 (1.6)
αTerpineol	84.80 (73.01 - 98.31) *	80.8 (4.2)
(-)Menthol	0.0	0.0

* = Significantly different from control, NS = not significantly different.

The calculated FT50s of the control and the adhesive only treated seeds were not significantly different. (-)Menthol completely inhibited seed hollowing over seven days. No mortalities were seen in any treatment.

DISCUSSION

The results gained from split substrate tests show that the alcohols, both aromatic and monoterpenoid, and the ketones are contact molluscicides. These groups of compounds are also repellents. The aromatics salicylaldehyde and vanillin were not molluscicidal nor were the monoterpene hydrocarbons (+)limonene and αterpinene. However, vanillin was a repellent. A factor that may have affected the tests was the toxic fumigant effect of these volatiles that has been reported to affect slugs (Henderson, 1970). Lipophilic compounds have been reported to be effective contact molluscicides as they can cross the molluscan external epidermis (Briggs

& Henderson, 1987). Some hydrophilicity is required in order for the compound to pass the layer of protective mucus exuded by the slug (Deyrup-Olsen & Martin, 1982). The monoterpenes are lipophilic and mildly soluble (Weidenhamer *et al*, 1993) they may therefore cross the mucus/epithelial boundary of the slug integument. The alcohol and ketone monoterpenes could then disrupt the integrity of the external epithelium leaving the slug unprotected from dessication. However, these compounds are unsuitable for use as contact molluscicides due to their repellent properties as in the split substrate trials any contact with the repellent actives was forced.

In phytotoxicity trials the monoterpene ketones and cinnamyl alcohol completely inhibited germination at 0.5% a.i./wt seed. All other treatments, except the adhesive only, gave coleoptile lengths significantly different to the control. Only (-)menthol gave growth approaching that of the control. Soluble phenolics have been linked with allelopathy with compounds such as vannillic acid and cinnamic acid potent inhibitors of germination (Harborne, 1993). Reynolds (1987) linked inhibition of germination and growth with compounds lipophilicity and in trials showed that ketones were more phytotoxic than alcohols which were in turn more phytotoxic than hydrocarbons. It is interesting to note that in seed treatments that allowed germination, although coleoptile elongation was inhibited, percentage germination was on the whole not significantly different to the control.

The compounds that proved repellent and allowed germination and coleoptile growth were tested in terraria trials. It can be seen that the adhesive only treated seeds gave an FT50 not significantly different to the control. Vanillin, thymol and αterpineol all proved repellent seed dressings although all allowed seed hollowing over the seven days with thymol the most effective repellent. Due to the compounds volatility, levels of active present at the seed may drop to a level below which they remain effective repellents or anti-feedants. Only (-)menthol completely inhibited seed hollowing. Although volatile, levels of (-)menthol present at the seed remained repellent over seven days. (-)Menthol is known to affect molluscan neuronal activity (Haydon et al, 1982) and may affect the slugs olfactory processes. (-)Menthol also allowed the greatest coleoptile elongation over seven days of the repellent compounds and appeared the only compound suitable for possible use as a seed dressing. Further trials will be conducted utilizing alternative concentrations of actives. Different enantiomers may also be tested as this enantiomeric difference has proved important in behavioural trials using monoterpenes (Airey *et al*, 1989). However, high volatility conferring short field persistence and phytotoxicity may hinder the use of monoterpenes as seed dressings. Their repellent traits could be taken advantage of in a controlled release regime.

ACKNOWLEDGEMENTS

Many thanks to Prof I. D Bowen and Dr S. Antoine for their help and guidance.

REFERENCES

Airey W J; Henderson I F; Pickett J A; Scott G C; Stephenson J W; Woodcock C M; (1989) Novel chemical approaches to mollusc control. In 'Slugs and Snails in World Agriculture' (ed. I. Henderson), *British Crop Protection Council Monograph*, No. 41, pp. 301-307.

Bourne N B; Jones G W; Bowen I D (1988) Slug feeding behaviour in relation to control with molluscicidal baits. *Journal of Molluscan Studies*. **54**, 327-338.

Bowen I D; Antoine S (1995) Molluscicide formulation studies. *International Journal of Pest Management*. **41(2)**, 74-78.

Briggs G G; Henderson I F (1987) Some factors affecting the toxicity of poisons to the slug *Deroceras reticulatum* Müller (Pulmonata: Limacidae). *Crop Protection*. **6**, 341-346.

Büchs W; Heimbach U; Czarnecki E (1989) Effects of snail baits on non-target carabid beetles. In 'Slugs and Snails in World Agriculture' (ed. I. Henderson), *British Crop Protection Council Monograph*, No. 41, pp. 245-252.

Chapman R F; Bernays E A; Simpson S J (1981) Attraction and repulsion of the aphid *Cavariella aegopodii*, by plant odors. *Journal of Chemical Ecology*. **7 (5)**, 881-888.

Croll R P; Chase R (1980) Plasticity of olfactory orientation in the land snail, *Achatina fulica*. *Journal of Comparative Physiology*. **136**, 267-277.

Deyrup-Olsen I; Martin A W (1982) Surface exudation in terrestrial slugs. *Comparative Biochemistry and Physiology*. **72C** (1), 45-51.

Einhellig F A (1986) Mechanisms and modes of action of allelochemicals. In: *The Science of Allelopathy*. (A. R. Putnam and C. Tang eds.) pp. 171-188. Wiley Interscience.

Glen D M (1989) Understanding and predicting slug problems in cereals. In 'Slugs and Snails in World Agriculture' (ed. I. Henderson), *British Crop Protection Council Monograph*, No. 41, pp. 253-262.

Gnansunderam C; Young H; Hutchins R F N (1981) Defence secretions of New Zealand tenebrionids: I. Presence of monoterpene hydrocarbons in the genus *Artystona* (Coleoptera, Tenebrionidae). *Journal of Chemical Ecology*. **7 (5)**, 889-894.

Gould H J (1962) Tests with seed dressings to control grain hollowing of winter wheat by slugs. *Plant Pathology*. **11**, 147-152.

Griffiths D C (1978) Insecticidal seed treatments of cereals. *CIPAC Monograph No. 2*, pp. 59-73.

Harborne J B (1993) Biochemical interactions between higher plants. In *Introduction to Ecological Biochemistry*. 4th ed. pp. 243-263, Academic Press, London.

Haydon P G; Winlow W; Holden A V (1982) The effects of menthol on central neurons of the pond-snail, *Lymnea stagnalis* (L.). *Comparative Biochemistry and Physiology*. **73C** (1), 95-100.

Henderson I F (1970) The fumigant effect of metaldehyde. *Annals of Applied Biology*. **65**, 507-510.

Litchfield J T; Wilcoxon F (1949) A simplified method of evaluating dose-effect experiments. *Journal of Pharmaceutical Experimental Theses*. **96**, 99-113.

Reynolds T (1987) Comparative effects of alicyclic compounds and quinones on inhibition of lettuce fruit germination. *Annals of Botany*. **60**, 215-223.

Rice E L (1984) *Allelopathy*, 2nd ed. Academic Press.

Rice L R; Lincoln D E; Langenheim J H (1978) Palatability of monoterpenoid compositional types to the molluscan herbivore *Ariolimax dolichophallus*. *Biochemical Systematic Ecology*. **7**, 289-298.

Scott G C (1981) Experimental seed treatments for the control of wheat bulb fly and slugs. *Proc. 1981 British Crop Protection Conference. Pests and Diseases*. **2**, 441-448.

Scott G C; Griffiths D C; Stephenson J W (1977) A laboratory method for testing seed treatments for controlling slugs in winter wheat. *Proc. 1977 British Crop Protection Conf. - Pests and Diseases*, 129-134.

Scott G C; Pickett J A; Smith M C; Woodcock C M (1984) Seed treatments for controlling slugs in winter wheat. *Proc. 1984 British Crop Protection Conf. - Pests and Diseases*, 133-138.

Taylor E J; Jones D P W; Maund S J; Pascoe D (1993) A new method for measuring the feeding activity of *Gammarus pulex* (L.). *Chemosphere*. **26 (7)**, 1375-1381.

Tottman D R (1987) The decimal code for the growth stages of cereals, with illustrations. *Annals of Applied Biology*. **110**, 441-454.

Weidenhamer J D; Macias F A; Fischer N H; Williamson G B (1993) Just how insoluble are monoterpenes? *Journal of Chemical Ecology*. **19 (8)**, 1799-1807.

METAL CHELATES AS STOMACH POISON MOLLUSCICIDES FOR INTRODUCED PESTS, *HELIX ASPERA, THEBA PISANA, CERNUELLA VIRGATA* AND *DEROCERAS RETICULATUM IN AUSTRALIA*

C YOUNG

Dept. of Chemistry, University of Melbourne, Parkville, Victoria, 3052, Australia

ABSTRACT

Metal salts are contact poison for a range of terrestrial slugs and snails. However, in general, most metal salts are not effective as stomach poisons because the molluscs fail to eat sufficient metal compound to be lethal. Chelation of the metal, can in a number of cases, sufficiently increase the amount of metal compound consumed to be lethal. Obtaining an effective molluscicide using a chelate requires an appropriate concentration of active ingredient and a suitable bait formulation. In this paper results are presented using two bait formulations, a number of metal chelates and a range of concentrations of the chelate in the bait. Results are presented for *H. aspersa and T. pisana, C. virgata, Cochlicella* spp. *Limax maximus* and *D. reticulatum.*

BACKGROUND

The common garden snail, *Helix aspersa* and grey field slug *Deroceras reticulatum* are pests throughout temperate regions including south east Australia. The white snails, *Theba pisana* and *Cernuella virgata* are severe pest in Mediterranean countries, South Africa and areas of South Australia and Western Australia. (Smith and Kershaw, 1979, Baker, 1986). *Helix aspersa* and *Deroceras recticulum* cause damage by feeding and, in Australia, are usually controlled by use of bran/wheat based pellets containing either metaldehyde or methiocarb. They are severe pest in suburban gardens and in some horticultiural crops particularly as the major cities of Sydney, Melbourne and Adelaide have significant areas with heavy clay soils.

T. pisana feeds on seedling crops and legume-based pastures causing severe damage and occasionally total crop destruction in the sandy areas of South Australia (Baker, 1986). *C. virgata* feeds primarily on decayed organic matter (Pomeroy, 1969, Butler, 1976) and causes less damage by feeding. Both *T. pisana* and *C. virgata* cause significant damage to cereal crops because they climb on to the heads and stalks of cereals in late spring/early summer to aestivate. They clog machinery and contaminate grain at harvest.

The ecologies of the four molluscs considered here are very different. The common garden snail and the grey slug are, in general nocturnal feeders and in the daytime remain hidden on the underside of leaves, under rocks or in cracks in the soil. They flourish in moist conditions. On the other hand, *T. pisana* and *C. virgata* can survive long hot summer temperatures by aestivating on weeds, fence posts etc. In cold climates *T. pisana* hybernates in winter. Their ecology and life history in Australia has been extensively investigated (Baker, 1986, Baker, 1988, Baker and Vogelzang, 1988, Baker and Hawke, 1990). Heller (1982) has studied *T. pisana* in Israel under similar conditions.

MOLLUCICIDES

Molluscicides for use against slugs and snails can be divided into three types:

(i) contact-action molluscicides such as aluminium and copper sulfate crystals (Henderson and Martin, 1990) which are applied to the area inhabited by the snail or slug and are taken up passively when the snail or slug moves in this area.

(ii) irritant powder molluscicides such as silica grains which act by being taken up in the snail or slug locomotion mucus

(iii) stomach-action molluscicides (Henderson, 1969) such as metaldehyde and methiocarb pellets. Such molluscicides involve incorporation of the active ingredient in a bait. The greatest problem with such poisons is that a sufficient amount of poison must be ingested to ensure a lethal dose (Martin, 1991). In general metal salts contact poisons such as ferrous sulphate, copper sulfate and aluminium sulfate (Durdam, 1920, Anderson and Taylor, 1926) are ineffective in baits because the molluscs do not consume sufficient to be fatal. Most toxic compounds are also repellent and the interaction of toxicity with repellency prevents the ingestion of sufficient poison to kill the mollusc. With molluscicides of moderate toxicity, an optimum concentration within a narrow critical range is used in baits, high enough to ensure adequate dosage but not so high as to deter feeding at such low concentrations that a lethal dose is never ingested (Henderson et al. 1989).

This paper reports the toxicity of bran/wheat based pellets containing various amounts of metal chelate on the above mentioned snails and slug in a controlled "laboratory" situation. The results indicate that contact poisons with significant but relatively low toxicity can be used as the basis of successful stomach poisons. Some metal chelates offer considerable advantages over metaldehyde and methiocarb in that they are considerably less toxic to non-target organisms. Metal chelates are widely use as a source of trace elements in agriculture and horticulture. Simple metal salts are usually "locked up" when applied directly to the soil by reaction with hydroxide and phosphates in the soil. On the other hand chelates are often quickly assimilated by plants. The iron chelate used in the snail and slug pellets studied here poses no treat to the environment, indeed it is usually beneficial to the crop.

The effectiveness of all slug and snail baits is dependant on the feeding habits of the pests. In general, feeding activity peaks between 20-30 °C and decreases to a very low level below 2 °C. At temperatures above 30 °C feeding also decrease rapidly with an increase in temperature. The maximum feeding is also dependant on the humidity. In the case of metaldehyde, which depends partly, at least, on dehydration to kill the molluscs, the effectiveness is very temperature dependant.

Young (1996) has recently pointed out the molluscicidial activity of a group of substances collectively referred to as complexones (Anderegg, 1987). Work presented here involves iron complexones with the majority of work being undertaken on Fe EDTA, {Iron (III) sodium salt dihydrate of ethylenediaminetetraacetic acid}.

EXPERIMENTAL

Henderson and coworkers (1988, 1989, 1990) have made extensive investigations of the effect of chelated metal salt on molluscs both as stomach poisons or contact poisons. Their work has focused on slugs, principally *D. recticulum*. The work reported here is part of a continuing program on the evaluation of chelates as molluscicides. The work reported here is restricted to "laboratory" work. Field trials on molluscicides are time consuming and expensive to carry out if they are to be of any scientific value. The interested reader is referred to Byers et al. 1989 for a description of some of the difficulties encountered in field trials.

In the present work two types of "plot" were employed. In the first about 1 cm depth of sandy loam or potting mix was placed in a seed tray of dimensions 30 x 25 cms approximately. The top was covered with a 3 mm glass sheet of which about 70 per cent was covered with black polyethene film. The polyethene was attached with adhesive tape (on the outside), so that the snails or slugs were able to rest on a smooth surface out of direct sunlight. In the second a polycarbonate "food storage container" of 175 mm diameter and 80 mm height was employed, four air holes of 2mm diameter were made in the lid of each container. These containers were used as it was thought that, in the case of *T. pisana* and *C. virgata* eggs might be laid in the soil and if the seed trays were used extensive precautions were necessary to avoid introducing these snails to areas where they were not previously established. The smaller containers were used to study *D. recticulum* and the soil replaced by a layer of absorbent paper. This procedure was used because the slugs often buried themselves in the soil and were often difficult to find without disturbing the soil. It was often difficult to establish if buried slugs were alive or dead.

RESULTS and DISCUSSION

Bait formulation

Two different bait formulations were used, the first, (Low bran type) consisted of bran and wheat flour in the ratio of 1 part bran to 4 part wheat flour, together with small amounts of calcium stearate as a die lubricant, a binder and mould inhibitor. The second, (High bran type) consisted of wheat flour and bran in approximately equal proportions by weight together with a small amount of oat meal (to act as a lubricant), a small proportion of sugar (1-2%) and a mould inhibitor.

There appeared to be only a minor difference between the different bait formulations. In an experiment to compared the efficacy of Fe EDTA in different baits it appeared that the snails, *H. aspersa*, started to eat the high bran formulation with the small percentage of sugar earlier than the low bran formulation. However, the difference in the final kill rate was not statistical significant between the various bait formulations. The results of this trial are given in table 1.

Comparision with metaldehyde and methiocarb

A number of trials were carried out for a comparision between Fe EDTA metaldehyde and methiocarb. The results of two of these trials are given in table 2. These results indicate that metaldehyde is not particularly effective at lower temperatures. At higher temperatures metaldehyde is marginally more effective than either Fe EDTA or methiocarb pellets. This is consistent with metaldehyde acting as a dehydrating molluscicide. In general, lower kill rates are obtained during warm periods. This would be expected because at temperatures somewhat above 20°C feeding is reduced.

Comparision of different concentrations of Fe EDTA
            Results of using pellets with different concentrations of Fe EDTA are
given in table 3. The average daily maximum temperature in these trials was 25°C.
The results show that although kill rates are slightly higher with the higher
concentrations of Fe EDTA the effect above 6% is marginal. It is interesting to
note that even at 20%, the kill rate is high. This can only be the case if the pellets
are still palatable with this high concentration of Fe EDTA.

Comparision of different chelating agent
            The results of different Fe complexes are given in table 4. We have
compared the efficacy of pellets with the same amount of active ingredient. Fe
EDTA appears to be the most effective but this could be a result of using near the
optimium concentration of chelating agent in this case whereas the optimum
concentration for other chelates might not be near 8-10%.

Comparison of different species of slugs and snails
            Tables 3, 5-8 report the results of trial on *T. pisana, C. virgata, D.
recticulum, Cochlicella* spp., *Deroceras reticulatum* and *Limax maximus*. It can be
seen that in all cases the Fe EDTA based product was effective.

Comparison of different soil types
            In would be expected that soil type would have minimum influence on
the efficacy of methiocarb and Fe EDTA since these are non-dehydrating poisons. It
might be expected that possibly there would be some effect on metaldehyde pellets
because the dehydrating potential of the active would be counteracted by the
rehydration potential of the media. It would be expected that the moisture content
and hence, the rehydration potential of soils would vary considerably.

CONCLUSION
            Ferric EDTA offers the possibilty of a moluscicide that is of lower
toxicity and more environmentally friendlier than baits based on metaldehyde and
methiocarb.

ACKNOWLEDGMENTS
            I am indebted to friends, particularly Andrea and Stuart Henderson and
Daisy Taylor, for collecting snails and to my wife, Janet, for her help in the
experimental work. Work on development of Multigaurd Snail and Slug pellets was
carried out under a consultacy agreement with Multicrop (Aust.) Pty Ltd. I thank
the directors for permission to use some of the results obtained during the course of
the development in this paper.

REFERENCES
Anderegg, G (1987) Complexones in Comprehensive Coordination Chemistry, Vol. 2,
Sec. 20.3, ed G Wilkinson, R D Gillard and J A McCleverty.

Anderson, A W; Taylor, T H (1926) The slug pest. Bulletin of the University of Leeds,
Department of Agriculture, no 143, 1-14.

Baker, G H (1986) The biology and control of white snails (Mollusca:Helicidae),
Introduced Pests in Australia. CSIRO, Div Entomology Tech Pap 25.

Baker, G H (1988) The life history, population dynamics and polymorphism of Cernuella virgata (Mollusca:Helicidae). Aust. J. Zool., 36, 497-512.

Baker, G H and Vogelzang, B K (1988) Life history, Population dynamics and polymorphism of Theba pisana (Mollusca:Helicidae) in Australia. J. Appl. Ecol., 25, 867-887.

Baker, G H (1989) Damage, population dynamics, movement and control of pest helicid snails in Southern Australia., in BCPC mono. 41., Slugs and snails in world agriculture, 175-186.

Baker, G H; Hawke, B G (1990) Life history and population dynamics of Theba pisana (Mollusca:Helicidae) in cereal-pasture rotation. J. Appl. Ecol., 27, 16-29.

Butler, A J (1976) A shortage of food for the terrestrial snail Helicella virgata in South Australia, Oecologia, 25, 349-371.

Butler, A J; Murphy, C (1977) Distribution of introduced land-snails on Yorke Peninsula, South Australia., Trans. R. Soc., South Aust. 101, 91-98.

Byers, R A; Barratt, B I P; Calvin, D (1989) Comparison between defined-area traps and refuge traps for sampling slugs in conservation-tillage crop environments.,in BCPC mono. 41., Slugs and snails in world agriculture, 187-192.

Durham, H E (1920) Of slugs. Gardeners' Chronicle, 68, 85-86.

Glen, D M; Orsman, I A (1986) Comparison of molluscicides based on metaldehyde, methiocarb or aluminium sulphate. Crop Prot., 5, 371-375.

Heller, J (1982) Natural history of Theba pisana in Israel (Pulmonata:Helicidae), J. Zool., 196, 475-487.

Henderson, I F; Bullock, J I; Briggs, G G; Larkworthy, L F (1988) Aluminium(III) and Iron(III) complexes exhibiting molluscicidal activity, Aust Patent AU-B-22526/88.
Henderson, I F; Briggs, G G; Coward, N P; Dawson, G W; Pickett, J A; Bullock, J I; Larkworthy, L F (1989) A new group of molluscicidal compounds., in BCPC mono. 41., Slugs and snails in world agriculture, ed. I F Henderson 289-294.
Henderson, I F; Martin, A P (1990) Control of slugs with contact-action molluscicides, An. Appl. Biol., 116, 273-278.

Henderson, I F; Martin, A P; Parker, K A; (1990) Laboratory and field assessment of a new aluminium chelate slug poison. Crop Prot. 9, 131-134.

Henderson, I F; Parker, K A; Problems in developing chemical control of slugs. Aspects Appl. Biol., 13, 341-347.

Martin, A (1991) Molluscs as Agricultural Pests. Outlook Ag. 20, 167-174.

Pomeroy, D E (1969) Some aspects of the ecology of the land snail, Helicella virgata, in South Australia. Aust. J. Zool., 17, 495-514.

Smith, B J; Kershaw, R C (1979) Field guide to the non-marine molluscs of south eastern Australia. ANU Press, Canberra.

EXPERIMENTAL RESULTS

The application rate are Metaldehyde as the Defender label instructions, Methiocarb as the Yates Baysol label intructions, Fe EDTA (all formulations) as the proposed Multiguard Snail and slug pellet label (50 pellets per square metre). Differences at the 95% confidence level (P= 0.05) are denoted by different letters. The media was moist potting mix in all trials reported below and the feed sliced carrot unless noted otherwise. The results are for *Helix aspersa* unless denoted otherwise.

Table 1. Comparison of efficacy of different bait compositions and pellet size

(a)   Daily maximum temperature   24-35 °C
Number of snails dead after 7 days/Number of snails in plot

Control	0/6	0/6	0/6	0/6	total	0/24	a
Formulation 1 (low bran)2mm	5/6	6/6	3/6	5/6		19/24	b
Formulation 2 (low bran)3.5mm	4/6	5/6	6/6	5/6		20/24	b
Formulation 3 (low bran)2mm	5/6	3/6	3/6	6/6		17/24	b
Formulation 4 (high bran)3.8mm	5/6	5/6	5/6	5/6		20/24	b

(b)   Media - Moist sandy loam;Daily maximum temperature 25-35 °C; 8 days

Control	0/6	0/6	0/6	0/6	total	0/24	c
Formulation 1 (low bran)2mm	5/6	5/6	3/6	5/6		18/24	d
Formulation 2 (low bran) 3.5mm	4/6	5/6	6/6	5/6		20/24	d
Formulation 3 (low bran)2mm	5/6	3/6	3/6	6/6		17/24	d
Formulation 4 (high bran)3.8mm	5/6	5/6	5/6	5/6		20/24	d

Table 2. Comparison of efficacy for metaldehyde, methiocarb and Fe EDTA pellets

(a)   Daily maximum temperature 12-18 °C; High bran bait; 9 days

Control	0/6	0/6	0/6	0/6	0/6	0/6	1/6	0/6	0/6
	0/6	0/6	0/6			total 1/72 e			
Methiocarb	1/6	3/6	2/6	4/6	2/6	2/6	2/6	2/6	2/6
	4/6	4/6	2/6			total 30/72 f			
Metaldehyde	1/6	0/6	0/6	0/6	1/6	0/6	0/6	0/6	0/6
	0/6	1/6	1/6			total 4/72 e			
Fe EDTA	5/6	3/6	4/6	4/6	4/6	5/6	5/6	4/6	4/6
	5/6	5/6	1/6			total 49/72 g			

(b)   Daily maximum temperature 22-34 °C; Low bran bait; 6 days

Control	0/6	0/6	0/6	0/6	total	0/24	h
Metaldehye	4/6	4/6	4/6			12/18	i
Methiocarb	3/6	2/6	3/6	1/6		9/24	j
Fe EDTA	2/6	3/6	4/6			9/18	i,j

Table 3 Comparison of efficacy for Fe EDTA pellets containing various weight percentage of Fe EDTA
(a)   Daily maximum temperature 22-34 °C; Low bran bait; 6 days
Control 0/18; 1% Fe EDTA 3/18; 2.5% Fe EDTA 1/18; 6% Fe EDTA 8/18;
9% Fe EDTA 9/18; 12% Fe EDTA 11/18; 16% Fe EDTA 10/18; 20% Fe EDTA 13/18

(b)   Species *Theba pisana;* Daily maximum temperature 23-34 °C; Feed 20%bran + 80%wheat flour; High bran bait; 8 days
Control 0/40;   6% Fe EDTA 26/40; 9% Fe EDTA 23/40; 12% Fe EDTA 20/40;
20% Fe EDTA 25/40

Table 4. Comparison of efficacy of various chelating agents for *Helix aspersa*

Daily maximum temperature 18-24 °C; Low bran bait; 7 days

Control	1/6	0/6	0/6	0/6 total	1/24
10% Fe EDDHA	3/6	2/6	0/6	3/6	8/24
10% Fe DPTA	0/6	1/6	0/6	0/6	1/24
9% Fe EDTA	3/6	2/6	4/6	4/6	13/24

EDDHA is Ethylene-N,N'-bis(2-hydroxyphenylacetic acid), ferric-sodium complex

DPTA is Diethylenetriaminepentaacetic acid, ferric-sodium complex

Table 5. The efficacy of Fe EDTA pellets for *Cernuella virgata*

Species *Cernuella virgata;* Daily maximum temperature 23-35 °C;
Feed 20%bran + 80%wheat flour; Low bran bait

Control	0/5	0/5	0/5	total	0/15 k
Fe EDTA	2/5	4/5	3/5		9/15 l

Table 6. The efficacy of Fe EDTA pellets for *Cochlicella* spp.

Species *Cochlicella* spp.; Daily maximum temperature 12-18 °C;
Feed Lettuce; High bran bait; 7 days

Control	1/10	1/10	1/10	2/10	total 5/40 m
Fe EDTA	10/10	10/10	9/10	10/10	39/40 n

Table 7. The efficacy of Fe EDTA pellets for *Deroceras reticulatum*

Species Deroceras reticulatum; Daily maximum temperature 22-34 °C;
Low bran bait; 8 days; each slug in an isolated plot

Control   1/6   Fe EDTA   6/6

Table 8. The efficacy of Fe EDTA pellets for *Limax maximus*

Species *Limax maximus;* Daily maximum temperature 18-27 °C;
High bran bait; 7 days; each slug in an isolated plot

Control   0/6   Fe EDTA   6/6

Session 7
Biological Control

Session Organiser
and Chairman   Dr G R Port

FACTORS DETERMINING THE EFFECTIVENESS OF THE MITES *FUSCUROPODA MARGINATA* IN THE CONTROL OF THE SLUG PESTS *LAEVICAULIS ALTE*

S K RAUT

Ecology and Ethology Laboratory, Department of Zoology, University of Calcutta, 35 Ballygunge Circular Road, Calcutta 700 019, India

ABSTRACT

The slugs *Laevicaulis alte* are serious agri-horticultural pests in Oriental Africa, the Pacific islands, Indonesia and India. Various attempts are being made to control them through the use of biological agents. Since the mite *Fuscuropoda marginata* is an effective control agent for these slugs, an attempt was made to study the factors that determine the effectiveness of the mites to ensure maximum death of the slugs under consideration. Of the factors, soil moisture seemed to play the key role in predator-prey interactions. Temperature and predator-prey numbers also had great influence. At a temperature range of 25-35 °C, the mites attacked and killed slugs in maximum numbers when the soil moisture was in the range 46-65% of field capacity. Soil moisture regulates the amount and quality of mucous secretion by slugs and thereby the effectiveness of the mites in attacking slugs.

INTRODUCTION

Because of agri-horticultural pest problems in Oriental Africa, the Pacific islands, Indonesia (Godan, 1983) and India (Raut & Mandal, 1984, Raut & Panigrahi, 1990) various attempts are being made to control the slugs *Laevicaulis alte* (Godan, 1983, Panigrahi & Raut, 1994). Since control of pest organisms through the use of native biological agents is considered safe, and the mites *Fuscuropoda marginata* have been proved effective in reducing the number of *L. alte* through group predation (Raut & Panigrahi, 1991) it is essential to identify the factors that determine the effectiveness of these mites in the control of *L. alte*, prior to studies of the use of these predators as biocontrol agents on a pilot scale. Accordingly, some experiments were carried out in the laboratory, with different combinations of slugs and mites, at different soil moisture and temperature ranges, and the results are presented in this paper.

MATERIALS AND METHODS

With a view to noting the factors determining the effectiveness of mites (*F. marginata*) in killing slugs (*L. alte*), adult slugs were exposed to mites in the following combinations, in wooden boxes at seven different soil moisture ranges viz. 16-25, 26-35, 36-45, 46-55, 56-65, 66-75 and 76-85% of the field capacity of the soil to hold moisture.

Experiment I	:	10 slugs exposed to 10 mites
Experiment II	:	10 slugs exposed to 50 mites
Experiment III	:	10 slugs exposed to 100 mites
Experiment IV	:	50 slugs exposed to 500 mites in small and large boxes separately
Experiment V	:	100 slugs exposed to 500 mites in small and large boxes separately.

The boxes used in Experiments I, II and III, and the small boxes used in Experiment IV and V were of the same size (28 x 14 x 14 cm each). Each large box was 60 x 20 x 20 cm in size. Experiments I, II and III were carried out during July, September and November while Experiments IV and V were confined to July only.

Both mites and slugs were collected from their natural habitats at Sandeshkhali, 24-Parganas (North), West Bengal, India. They were cultured in the laboratory (Raut & Panigrahi, 1991) to maintain the stock. The slug and mite individuals used in the experiments were procured from this stock from time to time. Each wooden box was provided with a layer of loose soil, 5 cm thick, in the bottom. The required amount of water was added to the soil to give the chosen range of moisture, as mentioned earlier. Also, water as per need was added to the soil to maintain the required moisture range for a box, at intervals, throughout the experiments. Potato slices were regularly supplied to the slugs as food. The open mouth of the box was covered with cloth. In all cases the slugs were exposed to mites for a period of 10 days. After that the slugs were examined and the attacked individuals were counted and removed to a mite-free terrarium to note their fate, during the next 15 days, under similar soil moisture and temperature ranges. They were fed with lettuce during these 15 days. Three replicates were made for each experiment and the mean and standard deviations (SD) are calculated.

RESULTS

During the study period the room temperature ranged from 25-35°C in July, 19-31°C in September and 17-28°C in November. In all experiments, the mites did not attack the slugs at 16-25 and 76-85% soil moisture ranges (all soil moistures are expressed as a percentage of the field capacity of the soil to hold moisture).

Experiment I: The mites attacked 1-3, 1-2 and 1-3 slugs per trial at 25-35, 19-31 and 17-28°C irrespective of soil moisture. The number of attacked slugs that died ranged from 0-2 per trial. The mean number of the slugs attacked, that died or survived following attack is shown in Fig. 1a. The 10 mites, on average, attacked a maximum of 2.7 slugs of which 1.7 died, at 46-55 and 56-65% soil moistures and 25-35°C.

Experiment II: One to 10 slugs were attacked per trial under the various temperature and soil moisture ranges. In some cases, of the 10 attacked individuals, 8 died. Four to 16 mites in a group attacked each individual slug. The 50 mites attacked, on average, a maximum of 8.3 slugs, of which 7.7 failed to survive, at 56-65% soil moisture and 25-35°C (Fig.1b).

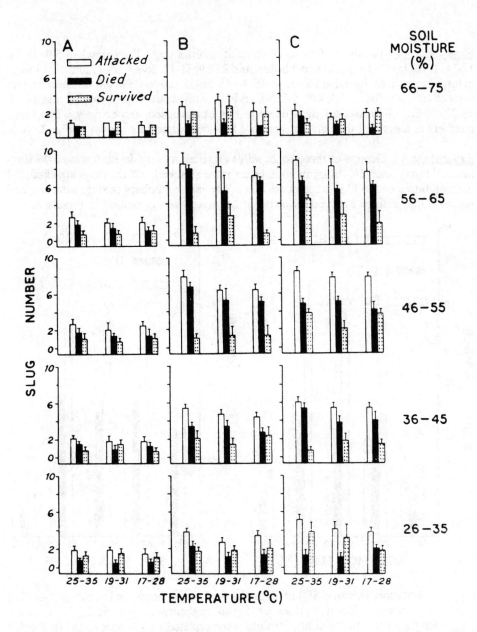

Fig.1. The number (mean ± SD) of slugs attacked by the mites, and numbers that died or survived following attack, at different soil moisture and temperature ranges in Calcutta, India. In the trials 10 slugs were exposed to 10 mites (A), 50 mites (B) and 100 mites (C).

Experiment III: Groups of three to 22 mites attacked each slug. In each trial, depending on conditions, 0-7 out of 1-9 attacked slugs, died. The mean number of slugs attacked and the fate of attacked individuals, varied with temperature and soil moisture (Fig.1c).

Experiment IV: Groups of four to 18 mites attacked a slug. They attacked 5-9, 11-14, 12-14, 10-17 and 3-4 slugs in small boxes, and 22-29, 34-42, 38-44, 35-41 and 12-14 slugs in large boxes in each trial at 26-35, 36-45, 46-55, 56-65 and 66-75% soil moisture ranges respectively. Of these, 1-3, 4-8, 5-7, 5-8 and 2-3 individuals in small boxes, and 18-19, 18-21, 17-22, 17-29 and 4-7 individuals in large boxes died, respectively. The mean numbers of slugs that died and/or survived following attack are shown in Figs.2a & b.

Experiment V: Groups of three to 21 mites attacked a slug. In each trial, 3-16 slugs in small boxes and 13-44 slugs in large boxes were attacked. Of the slugs attacked, 0-8 in small boxes and 4-30 in large boxes died. The mean numbers of slugs attacked and numbers that died or survived following attack have been presented in Figs.3a & b.

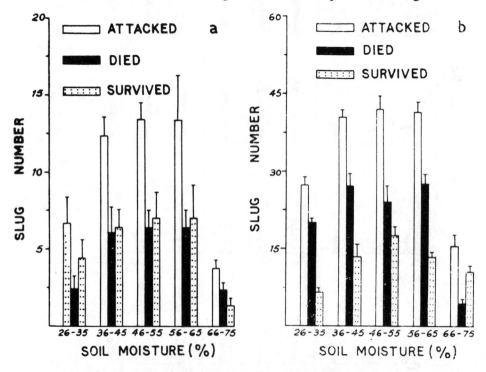

Fig.2 The number (mean ± SD) of slugs attacked by the mites, and numbers that died or survived following attack at 5 soil moisture ranges, at 25-35°C room temperature. In the trials, 50 slugs were exposed to 500 mites : (a) in a small box; (b) in a large box.

Results of ANOVA (Tables 1 and 2) clearly indicate that both density of mites and soil moisture have significant effects on the degree of attack on slugs and the likelihood of death of the attacked individuals. The results of X^2 tests indicate that the effect of soil moistures in small boxes is insignificant (X^2 = 7.19 for attack, 5.19 for death) but significant (X^2 = 16.61 for attack, 17.25 for death; both significant at 1% level) in large boxes. Also, results of t-tests suggest that mites in the large boxes had significantly greater attack rates (t = 6.12, significant at 1% level) and caused more slug deaths (t = 4.49, significant at 1% level) than mites in the small boxes. From the results of Experiment V it is evident that, irrespective of the size of the boxes, both attack (X^2 = 15.95, significant at 1% level for small box, and X^2 = 36.18, significant at 0.1% level for large box) and death (X^2 = 7.8, significant at 5% level for small box and X^2 20.78, significant at 0.1% level for large box) were highly influenced by soil moisture.

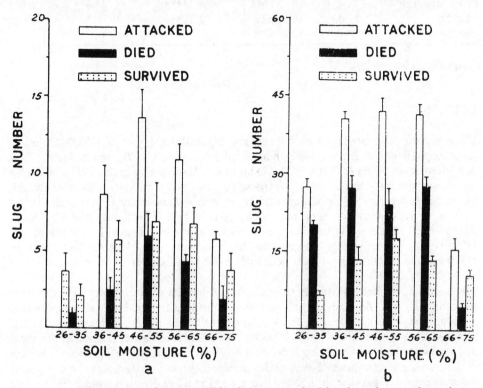

Fig. 3.   The number (mean ± SD) of slugs attacked by the mites, and numbers that died or survived following attack at 5 soil moisture ranges, at 25-35°C room temperature. In the trials, 100 slugs were exposed to 500 mites : (a) in a small box; (b) in a large box.

Table 1.   Results of ANOVA for studies of attack by 10, 50 and 100 mites on 10 slugs at different soil moistures.

Source of variation	df	SS	MS	F_{cal}
Density of mite	2	390.45984	195.22992	17.34*
Soil moisture	4	357.11587	89.278968	7.93*
Error	8	90.08693	11.260866	
Total	14	837.66264		

* significant at 1% level

Table 2.   Results of ANOVA of the death of slugs attacked by 10, 50 and 100 mites
           at different soil moistures.

Source of variation	df	SS	MS	F_{cal}
Density of mite	2	192.54756	96.27378	8.49*
Soil moisture	4	337.32743	84.331858	7.44**
Error	8	90.62517	11.328146	
Total	14	620.48016		

* significant at 1% level   ** 5% level

DISCUSSION

Mites are effective predators of various pest organisms (Hoy et al., 1983), including
nematodes (Royce & Krantz, 1991), house-fly larvae (Axtell, 1991) and molluscs (Turk
& Phillips, 1945, Viets & Plate, 1954, Godan, 1983, Raut & Panigrahi, 1991). The role
of mesostigmatid mites viz. *Macrocheles muscaedomesticae* (Macrochelidae),
Fuscuropoda vegetans (Uropodidae) and *Poecilochis monospinosus* (Parasitidae) in the
control of house-flies has been studied by Axtell (1991). Although the slug mites
Riccardoella limacum (Trombiidae) and *Fuscuropoda marginata* (Uropodidae) are
effective in reducing slug populations to a considerable degree, the effect of
environmental conditions on their potential to kill slugs has yet to be studied.

The results of the present studies clearly indicate that the effectiveness of *F. marginata*
in killing *L. alte* depends largely on soil moisture, irrespective of temperature.
Although, as a rule, increasing mite density resulted in increased numbers of slugs
attacked, data on the rate of attack by mites on slugs in small boxes contradict this idea.
The mites normally attack a slug individual on its dorsal surface, in a group of as many
as 22 individuals at a time. Since, under crowded conditions the slugs are compelled
to crawl over one another, it is likely that a good number of mites would die because
of being trapped by the mucus of the crawling slug. This is probably why the number
of slugs attacked was very low in small boxes in comparison to the number of such slugs
in large boxes. So, it appears that the predation potential of the mites is also a function
of space available to the slugs.

Of the slugs attacked by mites under any given condition, some usually survived. This
may be due to the capacity of different individuals to withstand such injury. Of course,
such resistance to attack is very much dependent on the ability of slugs to recover from
the wounds created by mite feeding (Raut & Panigrahi, 1991). As the degree of injury
is the main factor in determining the fate of attacked individuals, and the slugs usually
need to be fed on by the mites for three consecutive days to ensure death, interruption
of feeding on the second and/or third day would not only save the slugs from severe
injury but also from death. So, for field trials, attention should be given to releasing the
mites in such numbers that a slug individual would be likely to be fed on by the mites
for at least three days. Since the rate of attack determines the rate of death, and mites
attacked progressively fewer slugs both below and above the favourable soil moisture

ranges, with no attack at the lowest and highest soil moisture ranges tested in this study, it is likely that there is some sort of barrier, which may be thick mucus due to dry conditions or over-wet skin due to a very moist environment, that inhibits attack. This suggests that the release of mites in the field would need to be done at times of appropriate soil moisture in the slug habitat.

In the present experimental studies, only adult (sexually mature) slug individuals were considered and, in some cases 100% deaths followed attack by the mites. It is considered that more encouraging results would have been found in most cases if juvenile slugs were taken into account. This could be expected from the results of experimental studies made earlier by Raut & Panigrahi (1991). Since the effectiveness of attack by the mites is moisture dependent, and juvenile slugs are more sensitive to moisture, it is thought that attack on juvenile slugs would be more effective in causing slug death. However, further studies on the natural foods and food preferences of the mites, and their preference for feeding on slugs over other food items, in the presence of slugs of different sizes/age-groups, are very much needed. Obviously, attention should be given to the soil moisture conditions that encourage the mites to attack slugs in preference to other prey.

ACKNOWLEDGEMENTS

I am thankful to the Head of the Department of Zoology, Calcutta University for the facilities provided.

REFERENCES

Axtell, RC (1991) Role of mesostigmatid mites in integrated fly control. In: *Modern Acarology*, F Dusbabek & V Bukva (eds), Academia, Prague and SPB Academic Publishing bv, The Hague, Vol 2, pp. 639-646.

Godan, D (1983) *Pest slugs and snails, biology and control.* Springer-Verlag, Berlin, Heidelberg, New York, pp. 445.

Hoy, M A; Cunningham G L; Knutson L (eds) (1983) *Biological control of pests by mites.* Division of Agricultural Sciences, University of California, Berkeley, Special Publication 3304, pp. 185.

Panigrahi, A; Raut S K (1994) *Thevetia peruviana* (Family:Apocynaceae) in the control of slug and snail pests. *Memorias do Instituto Oswaldo Cruz*, Rio de Janeiro, **89**, 247-250.

Raut, S K; Mandal R N (1984) Natural history of the garden slug *Laevicaulis alte. Journal of Bengal Natural History Society*, New Series 3, 104-105.

Raut, S K; Panigrahi A (1990) Feeding rhythm in the garden slug *Laevicaulis alte* (Soleolifera : Veronicellidae). *Malacological Review*, **23**, 39-46.

Raut, S K; Panigrahi A (1991) The mite *Fuscuropoda marginata* (C L Koch) for the control of pest slugs *Laevicaulis alte* (Ferrussac). In: *Modern Acarology*, F Dusbabek & V Bukva (eds), Academia, Prague and SPB Academic Publishing bv, The Hague, Vol. 2, pp. 683-687.

Royce, L A; Krantz G W (1991) A new rearing method for nematophagous mites. In:

Modern Acarology, F Dusbabek & V Bukva (eds), Academia, Prague and SPB Academic Publishing bv, The Hague, Vol. 2, pp. 561-569.

Turk, F A; Phillips ST M (1945) A monograph of the slug mite *Riccardoella limacum* (Schrank). *Proceedings of the Zoological Society of London*, **115**, 448-472.

Viets, K; Plate H P (1954) Die okologischen (parasitologischen) Beziehungen zwischen Wassermilben (Hydrachnellae, Acari) und SuBwasser-Mollusken. *Zeitschrift fur angewandte Entomologie*, **35**, 459-494.

THE BIOLOGICAL CONTROL OF HELICID SNAIL PESTS IN AUSTRALIA: SURVEYS, SCREENING AND POTENTIAL AGENTS.

J B COUPLAND

CSIRO Division of Entomology, European Laboratory, Campus de Baillarguet, 34982 Montferrier sur Lez, France

ABSTRACT

The helicid snails, *Theba pisana*, *Cernuella virgata*, *Cochlicella acuta* and *C. barbara* are introduced snail pests of pasture and grain crops in Australia. Since 1990, surveys for potential biological control agents have been carried out in the western Mediterranean region where these snails are native. Over forty potential predators and parasites were discovered, ten of which showed enough promise for further study. Of these, several dipteran flies of the families Sarcophagidae and Sciomyzidae have shown the most potential as effective biological control agents.

INTRODUCTION

Four introduced species of Mediterranean snail, *Theba pisana*, *Cernuella virgata*, *Cochlicella acuta* and *Cochlicella barbara* (Helicidae) have become serious agricultural pests in South and Western Australia and are increasing problems in western Victoria and southern New South Wales (Baker, 1986, 1989). In South Africa, *Theba pisana* is an emerging problem and it is considered the most economically important pest snail species in Israel. In the United States there was a concerted effort at eradication of *T. pisana* from 1918 onward (Basinger, 1927) though the species still occurs in isolated locations near San Diego. All species cause severe damage to and occasionally total destruction of legume based pastures (e.g. annual medics, lucerne, clovers) and seedling arable crops (e.g. wheat and barley). Re-establishment of pastures in snail-infested areas is particularly difficult and stock reject both pasture and hay which is heavily contaminated with snails (due to slime).

All species except for *C. barbara* are also agriculturally important in southern Australia due to their habit of climbing onto the heads and stalks of cereals, beans, peas and increasingly grapes for the raisin industry, to aestivate in late spring. During harvest they clog machinery and contaminate the crop. The crop is then rendered either unacceptable or is downgraded. Export shipments of barley from South Australia and Western Australia have been rejected overseas, with Australia's reputation for good quality grain being damaged. Therefore these snails pose a serious threat to the export marketing of Australian cereals.

Since 1990, a biological control program against these introduced snails has been undertaken at the CSIRO Biological Control Unit in Montpellier, France, within the snails' native distributions. During this time large scale surveys for parasites and predators have been made, focussing primarily on south-western Europe. During these surveys many parasites and predators with potential as biological control agents were discovered. After initial screening, many were discarded as being to generalist while others were studied further. Final screening has further eliminated certain agents and the final agents have now been shipped to Australian quarantine.

MATERIALS AND METHODS

Surveys were conducted from 1990 to 1995 during all four seasons. During these surveys the following snail species were collected from over 200 sites in the region of Montpellier, France and from at least 300 sites in Italy, Portugal, Spain and Morocco during 1991-1995: *T. pisana, C. virgata, C. acuta, C. barbara, Helix aspersa, Trochoidea elegans, Cernuella explanata, Eobania vermiculata.*

Sites varied between pastureland, crop edges, waste land, road edges and littoral zones with sand dunes. Living and obviously parasitised snails were collected into plastic cages with gauze tops. Insects which emerged from these snails were then identified to species with the help of the British Museum of Natural History. Over 150,000 snails were sampled with over 40 different insects being reared from them. Of the various insects reared, Diptera were seen as having the most potential for possible release in Australia. The reason for this was that many of the beetles and other predators are well known to have a very broad host range and would not be specific enough for introduction. Thus most effort was made in screening the 21 dipteran species (Table 1) which were collected during the surveys.

Laboratory rearing conditions

Flies used for laboratory experiments were collected when they were first observed in the field and kept in clear plastic boxes (16cm x 14cm x 14cm) with moistened peat soil to maintain humidity. Flies were fed a mixture of honey, brewer's yeast, and skimmed milk powder and kept at 20° C under natural light conditions. Flies were found to mate and oviposit or larviposit readily under these conditions.

Host acceptance tests

Egg-laying Diptera such as the Sciomyzidae had their eggs collected from cages every two days after oviposition commenced. Eggs were placed on moistened filter paper until hatching. Active larvae were placed individually into test arenas which consisted of a petri dish containing moistened filter paper and a target snail. Target snails were collected from the wild at the time of larval hatch, to emulate the size and condition of prey snails. At least 30 replicate tests per fly species were made against each of nine locally collected snail species.

For larvapositing Diptera such as the Sarcophagidae, two gravid females were placed with 10 individuals of one of a range of 9 locally occurring helicid snail species in rearing cages (replicated 3 times) to determine the range of suitable hosts. Snails were checked every two days for signs of parasitism.

RESULTS

The 21 species of Diptera from 5 Families are shown in Table 1. Many of these species are known saprophages so no attempts were made to culture them, while with others this feeding behaviour became apparent after culturing began.

Sciomyzidae

The first flies to be cultured and screened were the Sciomyzidae as these are well known for their snail feeding habits (Berg, 1953), and one of the species collected during the survey was known to feed on *Theba pisana* (Knutson *et al.*, 1970). Collections of the sciomyzid species *Salticella fasciata, Coremacera marginata, Pherbellia cinerella; Trypetoptera punctulata, Dichetophora obliterata* and *Euthycera marginata* were maintained and larvae screened for specificity. Attention was focussed on *S. fasciata* as it was known to lay its eggs in the umbilicus of *Theba pisana* and this specificity would obviously make it an idea candidate as a biological control agent for this species. However both field and laboratory studies showed that this species was ineffective at killing the host snail and was more of a saprophage than a parasite (Hopkins & Baker, 1993; Coupland *et al.*, 1994).

Laboratory rearing was very difficult for *E. cribrata , C. marginata, T. punctulata* and *D. obliterata* although this confirmed that these species had exceptionally long preoviposition periods (up to 72 days) and very long larval and pupal development times (up to 189 days), a result which is consistent with Vala's (1989) contention that these species are univoltine. In contrast, the multivoltine species *P. cinerella* was easy to rear and had a generation time ranging from 24 - 40 days.

All of these sciomyzid species were able to kill several snail species, especially *C. marginata* and *P. cinerella* which were able to attack and kill every non-target species offered (Table 2) . These differences probably reflect differences in prey choices in the wild. It is known that some species of terrestrial sciomyzid are quite specific in either the species or genus which they attack, while others are not (Foote, 1977; Coupland *et al.*, 1994).

The wide prey range of these sciomyzids is important in assessing the appropriateness of these species as biological control agents of pest snails. Species with a wide prey range could possibly attack endemic non-target snails with disastrous consequences.

Another factor which is very important in assessing the potential of a biological control agent is both its habitat and climatic preference. Of the sciomyzids tested, most occur in forest or scrubland habitat with the exception of *P. cinerella* which is primarily a pasture species (Coupland & Baker, 1996). In Australia, the main target species occur in pastureland and grain crops. The most appropriate species would therefore be those which occur in similar situations in their native habitat. The only species which matches this criterion is *P. cinerella* and it must be considered the best candidate for possible introduction. Furthermore, its native range extends over areas which are very similar climatically to South Australia (northern Morocco and southern Spain) where the problems with helicid snails occur. The main drawback with *P. cinerella* is its wide prey range, it has been shown to attack and kill endemic Australian snails in feeding tests (D. Hopkins, pers. comm.). However, its strong preference for open pastures may reduce the likelihood of it overlapping with the habitats of endemic Australian snails, which do not occur in pastures.

Table 1. Fly species reared from field collected snails from 1991-1993. Snail species abbreviations: 1 = *Theba pisana*, 2 = *Cernuella virgata*, 3 = *Cochlicella acuta*, 4 = *Cochlicella barbara.*, 5 = *Helix aspersa*, 6 = *Trochoidea elegans*, 7 = *Cernuella explanata*, 8 = *Eobania vermiculata*. Percentage of flies collected from the various snail species.

Fly Species	Snail Species
Sarcophagidae	
Sarcophaga (Pierretia) nigriventris.	1(80%),2(15%),5(1%),8(4%)
Sarcophaga (Parasarcophaga) teretirostris	1(95%),2(5%)
Sarcophaga (Heliciophagella) hirticrus	1(95%),2(4%),5(1%)
Sarcophaga portschynskyi	1
Sarcophaga (Heteronychia) balanina	1
Sarcophaga (Kramerellia) anaces	1(95%),2(3%),7(2%)
Sarcophaga (Discachaeta) cucullans	1(95%),2(5%)
Sarcophaga (Heteronychia) filia	2
Sarcophaga (Heteronychia) uncicurva	1
Sarcophaga (Heliciophagella) maculata	1(90%),4(10%)
Sarcophaga (Heteronychia) graeca	3(98%),6(2%)
Calliphoridae	
Pollenia rudis	2
Melinda cognata	1(5%),2(95%)
Phoridae	
Spiniphora maculata complex	1(5%),2(5%),5(80%),8(10%)
Sciomyzidae	
Salticella fasciata	1(80%),2(20%)
Pherbellia cinerella	1(5%),2(90%),3(5%)
Coremacera marginata	1
Euthycera cribrata	1(20%),2(10%),8(70%)
Dichetophora obliterata	adults collected
Trypetoptera punctulata	adults collected
Anthomyidae	
Unidentified (perhaps *Homalomyia caticularis.*)	1(80%),2(20%)
Neoleria spp.	1(90%),2(10%)
Ephydridae	
Discomyza incurva	1(10%),2(10%),5(80%)

Sarcophagidae

The association between snails and sarcophagids has been known for a long time (Keilen, 1919, 1921). However, the nature of the relationship between snails and many sarcophagids is based largely on speculation, with few detailed studies having been carried out. During surveys made in the course of this study, 11 species of Sarcophagidae were reared from snails. Several of these were known to be saprophagous, while species such as *S. nigriventris* are known to parasitise a wide range of invertebrates (Disney & Cameron, 1975; Hopkins & Baker, 1993). However, species such as *S. penicillata*, *S. uncicurva* and *S. balanina* were found to be in close association with several of the target species. The first, *S. penicillata*, parasitises the conical snails *C. acuta* and *C. barbara* with occassional parasitism of *T. elegans,* another conical snail. It was never found to parasitise the globular snails *T. pisana* or *C. virgata* in the field or lab (Coupland & Baker, 1994). *Sarcophaga uncicurva* and *S. balanina* were found to parasitise only the globular snails *T. pisana* and to a lesser extent *C. virgata*. These sarcophagid species are multivoltine and have very fast generation times, averaging 18 days for *S. penicillata,* indicating the potential for up to 6 generations during a mediterranean summer. However in their native range, these species are heavily hyperparasitised (Coupland & Baker, 1994; Coupland unpublished data), though if released from these may have the ability to quickly increase in numbers in a relatively short period. Furthermore, their habit of attacking snails aestivating on plants, makes them particularly desirable agents. One of the main problems with target snail species is their habit of aestivating on the ears and stalks of cereals where they clog machinery and contaminate grain during harvest. Flies which attacked these particular snails and which caused the snail shell to fall during the flies' emergence would help alleviate this contamination.. Finally, the recorded distribution of all these species has to date only been in countries bordering the Mediterreanean region (Soós & Papp, 1986) a region with a similar climate to that in South Australia where the parasitoid would be introduced.

Table 2. Percentage of 9 locally occurring land snails killed by the first instar sciomyzid larvae.

Snail Species	C. marginata	E. cribrata	P. cinerella	T. punctulata	D. obliterata
Cernuella virgata	80%	77%	90%	0	0
Theba pisana	87%	73%	77%	0	0
Cochlicella acuta	90%	77%	93%	37%	23%
Cochlicella. barbara	83%	67%	93%	20%	20%
Troichodea elegans	77%	0	90%	10%	43%
Eobania vermiculata	77%	37%	10%	0	0
Pomatia elegans	47%	20%	3%	0	0
Lauria cylindaracea	23%	10%	43%	90%	80%
Discus rotundatus	73%	43%	67%	3%	93%

CONCLUSION

Of the 23 species of Diptera collected during surveys for potential biological control agents, 8 were studied in depth and screened for host specificity against the target snails. Of these the sciomyzid species *S. fasciata, C. marginata* and *P. cinerella*, along with the sarcophagid species, *S. penicillata, S. uncicurva* and *S. balanina* were sent to Australia for further host specificity screening. Since then work has stopped on the sciomyzids, but the sarcophagid species are now in the final phase of host-specificity testing.

ACKNOWLEDGEMENTS

This work was funded by the Australian Wool and Grains Research and Development Boards in grants to Dr. G. H. Baker. I thank Janine Vitou for valuable assistance in rearing the flies.

REFERENCES

Baker, G H (1986) The biology and control of white snails (Mollusca: Helicidae), introduced pests in Australia. *C.S.I.R.O. Technical Paper* No. 25, 31 pp.

Baker, G H (1989) Damage, population dynamics, movement and control of pest helicid snails in southern Australia. In: *Slugs and Snails in World Agriculture.* I F Henderson (ed.). BCPC Monograph No. 41, pp.175-185.

Basinger, A J (1927) The eradication campaign against the white snail (*Helix pisana*) at La Jolla. *California Monthly Bulletin, Department of Agriculture, State of California.* Vol. **26**. No. 2, pp. 51-76.

Berg, C O (1953). Sciomyzid larvae (Diptera) that feed on snails. *Parasitology,* **39**, 630-636.

Coupland, J B; Espiau A; Baker G (1994) Seasonality, longevity, host choice and infection efficiency of *Salticella fasciata* (Diptera: Sciomyzidae) a candidate for the biological control of pest helicid snails. *Biological Control.* **4**, 32-37

Coupland, J B; Baker G (1994) Host distribution, larviposition behaviour and generation time of *Sarcophaga penicillata* (Diptera: Sarcophagidae), a parasitoid of conical snails. *Bulletin of Entomological Research.* **84**, 185-189.

Coupland, J B; Baker G (1996) The potential of several species of terrestrial Sciomyzidae as biological control agents of pest helicid snails in Australia. *Crop Protection.* **14**, 573-576.

Disney, R H L; Cameron R A D (1975) Two further cases of parasitism by *Sarcophaga nigriventris* Meigen (Dipt., Sarcophagidae). *Entomologists Monthly Magazine,* **3**, 45.

Foote, B A (1977) Biology of Oidematops ferrugineus Diptera: Sciomyzidae, parasitoid enemy of the land snail Stenotrema hirsutum Mollusca: Polygyridae. *Proceedings of the Entomological Society of Washington.* **794**, 609-619.

Hopkins, D C; Baker G H (1993) Biological control of white and conical snails. In: *Pest Control and Sustainable Agriculture.* S A Corey; D J Dall & W M Milne (eds), CSIRO,Melbourne, pp. 246-249.

Keilen, D (1919) On the life-history and larval anatomy of *Melinda cognata* Meigen
(Diptera: Calliphorinae) parasitic in the snail *Helicella (Heliomanes) virgata* Da
Costa, with an account of the other diptera living upon molluscs. *Parasitology*. **11**,
430-455.

Keilen, D (1921) Supplementary account of the dipterous larvae feeding upon molluscs.
Parasitology. **13**, 180-183.

Knutson, L V; Stephenson J W; Berg C O (1970) Biosystematic studies of *Salticella
fasciata* Meigen, a snail killing fly (Diptera: Sciomyzidae). *Transactions of the
Royal Entomological Society of London*. **1223**, 81-100

Soós, A; Papp L (1986) *Catalogue of Palearctic Diptera*. Volume 12: Calliphoridae -
Sarcophagidae. 265 pp. Elsevier Science Publishers, Amsterdam

Vala, J C (1989) *Diptères Sciomyzidae Euro-Méditerranéens*. Faune de France, Fed. Fr.
Soc. Sci. Natur, Paris.

PROGRESS IN THE DEVELOPMENT OF ANTIBODIES TO DETECT PREDATION ON SLUGS - A REVIEW, PLUS NEW DATA ON A MONOCLONAL ANTIBODY AGAINST *TANDONIA BUDAPESTENSIS*

W O C SYMONDSON, M L ERICKSON

School of Pure & Applied Biology and School of Molecular & Medical Biosciences, University of Wales Cardiff, PO Box 915, Cardiff CF1 3TL

J E LIDDELL

School of Molecular & Medical Biosciences, University of Wales Cardiff, PO Box 911, Cardiff CF1 3US

ABSTRACT

Most of the invertebrate predators of slugs in Europe are polyphagous species. The analysis of prey preferences through laboratory trials cannot easily simulate field conditions, where prey choice is determined by a variety of factors, including dynamic changes in relative and absolute prey populations densities. Monoclonal antibody systems have therefore been developed for *post mortem* analysis of predator gut samples. They provide standard reagents for long term ecological studies, and progress in the development of a number of genus- and species-specific monoclonal antibodies against major pest slugs, including Arionidae and *Deroceras reticulatum,* is reviewed. Preliminary data is given for a new antibody against *Tandonia budapestensis*, which has proved in some ways to be too species-specific. Current work on conventional and engineered antibodies against *Arion distinctus* and *Arion intermedius* is reported.

INTRODUCTION

Predator-prey interactions involving polyphagous predators are notoriously difficult to study. Opportunistic feeders such as many ground beetles will, if confined in the laboratory, consume almost any invertebrate food offered to them, and this is reflected in the wide variety of prey consumed in the field (Larochelle, 1990). Laboratory feeding trials can tell you whether a certain species of predator is capable of consuming a particular size and species of mollusc for example (Symondson, 1989; Digweed, 1993), but provides little indication of prey preferences in the field. Prey choice and consumption rates in a crop environment are determined by a great many interacting factors, which cannot be adequately simulated in the laboratory. Certain carabid species, such as the Cychrini, may be behaviourally adapted to hunt molluscs (Wheater, 1989; Pakarinen, 1994), while others, such as the larger *Abax, Carabus* and *Pterostichus* sp., may be capable of taking slugs as a direct result of their size and feeding apparatus. However, all these genera feed on many other types of prey, and whether or not, and to what degree, they feed on slugs will also be determined by: the relative and absolute densities of different prey; concordance of predator and prey activity periods; how much time different prey spend on the ground or up in the crop; whether or not the predators climb the crop plants (Winder *et al.*, 1994); the crop architecture and particularly the stage of crop development (Symondson, 1993); direct and indirect effects of microclimatic conditions; the abilities of different prey to escape; the predators' ability to find different prey using visual, olfactory, gustatory or tactile cues, and many other factors.

Detecting predation on slugs in the field

The consumption of slugs by carabid beetles, in the laboratory, experimental plot and field, has been reviewed elsewhere (Symondson, in press). Indirect evidence suggests that where carabid numbers are increased in the field a reduction in slug numbers can follow (Altieri *et al.*, 1982). Conversely, where predators were excluded from field plots (Burn, 1988), or predator numbers reduced by insecticides (Grant *et al.*, 1982), slug numbers increased.

Although microscopic examination of the gut contents of field-collected predators can sometimes reveal the remains of slugs in the form of radular teeth or shells (Luff, 1974), these parts are not always consumed, particularly by predators that are primarily fluid-feeders. For this reason a number of studies have been conducted using polyclonal antisera to detect general slug proteins (Tod, 1973; Gruntal & Sergeyeva, 1989) and slug haemolymph proteins (Symondson & Liddell, 1993a; Symondson *et al.*, in press) amongst the gut contents of predators collected in the field. These antisera have revealed valuable information on general slug predation, and are particularly useful for identifying important predator species. With appropriate controls, they can give quantitative information on the effects of different agricultural practices on carabid-slug interactions (Symondson *et al.*, in press).

Monoclonal antibodies

Slugs are usually found, in any given field environment, in mixed populations comprising two or more species. The relative population densities of these species vary over time and between field sites, but the factors determining these observed distributions and species' ratios are little understood. Predation, however, could well be a factor. Within a crop, damage inflicted by slugs can depend upon the species present. For example, root crops, such as potatoes, are often severely damaged by Milacid slugs (e.g. Moens, 1980), which can feed deep beneath the soil surface. Wheat seeds and seedlings usually suffer greatest damage from *Deroceras reticulatum*, which mainly attacks plant material at or near the soil surface, and is most frequently the dominant pest species in terms of numbers and biomass (Runham & Hunter, 1970). It is therefore useful to know which predators are feeding upon which genera and species of slugs, and monoclonal antibodies allow you to do this.

Monoclonal antibodies were first developed by Köhler and Milstein (1975) with medical applications in mind. However, over the last decade invertebrate ecologists have started to use these techniques to investigate interactions between predators and their prey, or for pest identification (Symondson & Liddell, 1996a; Symondson & Hemingway, in press). Highly specific monoclonal antibodies have now been raised, for example, against the noctuid moth *Helicoverpa zea* (Greenstone & Morgan, 1989), the tracheal mites *Acarapis woodi* (Ragsdale & Kjer, 1989), eggs of the pink bollworm *Pectinophora gossypiella* (Hagler *et al.*, 1994), the grain beetle *Trogoderma granarium* (Stuart *et al.*, 1994) and larvae of the parasitoid *Microplitis croceipes* (Stuart & Greenstone, 1996).

Monoclonal antibodies against slugs

Although species-specific monoclonal antibodies have been raised against molluscan lectins of potential value in medicine (e.g. Schneider, 1985), none had, until recently, been produced for detecting predation. Monoclonal antibodies can provide uniform reagents in limitless quantity, allowing reproducibility of results over time. By contrast, no two polyclonal antisera will have the same properties, even when raised against the same antigen in the same individual mammal over a period of time (Tijssen, 1985). A general anti-slug monoclonal antibody, which reacts with all species of mollusc tested (Symondson *et al.*, 1995), has been used recently for an initial screen of the gut contents of several thousand *Pterostichus melanarius* collected from large-scale field experiments in arable fields (W O C Symondson, D M Glen, M L Erickson, S A Evans, C J Langdon, C W Wiltshire & J E Liddell, in prep.). Individuals that test positively will subsequently be tested with genus- and species-specific antibodies. This general antibody, however, designated DrW-2D11 (IgG isotype), was also used to investigate detection periods for slug remains in the carabid *P. melanarius* (Symondson & Liddell, 1995). It was found to be capable of identifying predation on slugs after 2.5 days at 16° C., and established that subsequent feeding on an alternative prey item, such as earthworm, will extend the detection period compared with starved controls. Although the length of the detection period (longer than that for any other monoclonal antibody used in predation studies to date) makes it very suitable for initial screening, this period is much shorter than that for a polyclonal antiserum (Symondson & Liddell, 1993c), and consequently fewer positive reactions would be expected.

A family-specific antibody (designated AaH-3E5, IgG isotype) was raised against *Arion ater* haemocyanin (Symondson & Liddell, 1993b). This antibody reacted strongly with all arionids tested, and has since been shown to react with *Geomalacus maculosus*, also in the Arionidae (W O C Symondson, unpublished data). As the reactions with all arionids was strong, the antibody can be used for detecting predation on arionid slugs in general. On sites where only one arionid species is present in significant numbers, as is the case with *Arion intermedius* on one of our current experimental sites (Symondson *et al.*, in press), this antibody is proving very useful. There were no cross-reactions with non-molluscan prey items, and in an enzyme-linked immunosorbent assay (ELISA) the antibody could detect less than 3.5 ng of antigen. The detection period, though short, was sufficient to identify predators that had consumed arionid slugs during the previous night.

Several species-specific monoclonal antibodies were selected against *D. reticulatum* (Symondson *et al.*, 1995), from which one has been chosen for screening field samples (Symondson & Liddell, 1996b). Reactions with all other species of slug, including *Deroceras caruanae* within the same genus, were not significant. The antibody (DrW-1G4, IgM isotype) could identify less than 12 ng of antigen as *D. reticulatum*, and remains of this species could be clearly identified as such for 38 hours after consumption by *P. melanarius*. In this case there were cross-reactions with the New Zealand Flatworm (*Artioposthia triangulata*), which fortunately is not found on any of our field sites in southern Britain, and the millipede *Polymicrodon polydesmoides*, which does not appear to be consumed by our principal target predator, *P. melanarius*. There were no cross-reactions with any potential prey found on our field sites. Monoclonal antibodies were also raised against slug eggs in a parallel study, and the results are reported elsewhere (Symondson *et al.*, 1995; Mendis *et al.*, this volume).

This paper now reports the development of a new monoclonal antibody against *Tandonia budapestensis* (Milacidae), and the results of preliminary specificity tests. Current work in progress, to raise antibodies against *Arion distinctus* and *A. intermedius,* is outlined in the Discussion.

MATERIALS AND METHODS

Monoclonal antibodies against *T. budapestensis* were created following the general methods described previously for other slugs (Symondson & Liddell, 1993b, 1996b). A mouse was immunised subcutaneously with 100 µl of a whole-body homogenate, diluted 1:20 in phosphate buffered saline (PBS) (approx 150 µg), mixed with an equal quantity of Freunds Complete Adjuvant. This was followed by a second immunisation after six weeks, substituting with Freunds Incomplete Adjuvant. An intravenous final boost was given four days later, comprising 200 µl of a 1:40 dilution of antigen (i.e. also approx. 150 µg).

Fusion was effected with polyethylene glycol, and was followed by cloning by limiting dilution and ascites production (Liddell & Cryer, 1991; Symondson & Liddell, 1993b). Screening, post fusion and during cloning stages, was by enzyme-linked immunosorbent assay (ELISA) of hybridoma supernatants. All supernatants at every stage were tested against both *T. budapestensis* and *D. reticulatum*, in order to select clones that did not cross-react with the latter species. Indirect ELISA tests were performed on plates coated over night with a 1:20,000 dilution of antigen in phosphate buffered saline (PBS) (i.e. the two slug species tested separately). The supernatants were used neat, and tested with a polyvalent goat anti-mouse Ig horseradish peroxidase conjugate (Sigma Chemical Company Ltd., Dorset), diluted 1:5000 in PBS-Tween 20. All other steps and reagents were as described in Symondson & Liddell (1996b). Isotyping of selected monoclonal antibodies was performed using the Serotec isotyping kit (Serotec Ltd., Oxford).

Reactions between the selected clone, TbW-1D8, and ten species of slug were tested by ELISA. Ten individuals of each species were tested in all cases except *Lehmannia marginata* (7), *Geomalacus maculosus* (6) and *Boettgerilla pallens* (4). Eight negative controls were included on each ELISA plate, comprising a 1:20,000 dilution of the gut contents of the carabid *P. melanarius* fed over night on the earthworm *Octolasion lacteum*. These controls provide a measure of the reaction of the antibody with heterologous proteins relevant to the subsequent use of this clone in the analysis of predator gut samples. A PBS only control was also included to monitor non-specific binding of antibodies and conjugates to the ELISA plate. All specimens were tested in duplicate on two separate ELISA plates. The ELISA tests where performed with ascitic fluid, diluted 1:10,000, and employed a sheep anti-mouse IgM horseradish peroxidase conjugate (Serotec), but in all other respects followed the same protocol as the screening tests.

RESULTS

The isotyping tests found that the selected clone was an IgM monoclonal antibody. ELISA results for tests with this antibody against ten species of slug are shown in Figure 1. The mean heterologous absorbance value (which also includes any non-specific binding to the plate) was 0.079 (S.E. 0.005). This background figure was subtracted from all data in Figure 1.

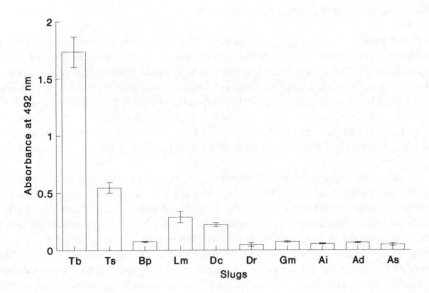

Figure 1. Absorbance readings for ten species of slugs tested with the TbW-1D8 monoclonal antibody. Tb = *Tandonia budapestensis*, Ts = *T. sowerbyi*, Bp = *Boettgerilla pallens*, Lm = *Lehmannia marginata*, Dc = *Deroceras caruanae*, Dr = *D. reticulatum*, Gm = *Geomalacus maculosus*, Ai = *Arion intermedius*, Ad = *A. distinctus*, As = *A. subfuscus*. Bars are ± S.E.

Statistical tests were not necessary to demonstrate that all species reacted significantly less strongly than the antigen species, *T. budapestensis*. There was a much lower level of reactivity with the closely related *T. sowerbyi* at a mean of 0.545 (S.E. 0.047), less than one third of the value for *T. budapestensis*. All other means were below 0.075, except for *L. marginata*, at 0.288 (S.E. 0.049) and *D. caruanae*, at 0.222 (S.E. 0.016).

DISCUSSION

Tests have still to be performed to determine whether this new antibody, designated TbW-1D8, can be used for identifying prey remains amongst the gut contents of predators. In particular, it is important to establish the post-feeding detection period for positive identification of prey as *T. budapestensis*, using a primary predatory species such as *P. melanarius*. However, as long as it can detect predatory events occurring during the previous night (i.e. a minimum detection period of 12 hours), then it should prove useful (Symondson & Liddell, 1993b). All antibodies raised to date against molluscs have been capable of doing this.

The low-level reactions with *D. caruanae* and *L. marginata* should not prove problematic. Neither of these species is found in large numbers in arable fields, and in any event manipulation of ELISA conditions plus the adoption of statistical controls over significance levels will eliminate any possibility of misidentification (Symondson & Liddell, 1993c, 1996b). It was interesting to note that there was no significant reaction with *B. pallens*, a milacid slug in the same family as *T. budapestensis*.

The reaction with *T. sowerbyi* could present difficulties when quantifying predation in an environment containing both this species and *T. budapestensis*. The pest status of these two species is similar (South, 1992), and in many ways it would have been an advantage to have had a less specific general anti-*Tandonia* antibody. For example, a predator gut sample that gives such a low reaction may contain 100% remains of freshly consumed *T. sowerbyi*, or it could, for example, contain 66% non-reactive prey material mixed with 33% freshly consumed *T. budapestensis*. Further clones from the same cell fusion are currently under test, including two IgG monoclonal antibodies showing a high degree of species specificity.

Other antibodies under development include clones raised against *A. distinctus*. Although at an early pre-monoclonal stage (June 1996) there are several cell lines producing antibodies against *A. distinctus* that do not cross-react with *D. reticulatum*, and further tests will soon follow to isolate specificities at the genus and species level. Phage antibodies are being produced against *A. intermedius* by genetic engineering techniques (Liddell & Symondson, 1996). These monoclonal antibodies are produced by ligating antibody heavy and light chain genes into a phagemid expression vector, creating recombinant gene libraries capable of generating many more antibody specificities than are possible by conventional hybridoma technology. Antibodies produced by these techniques can be used in ELISA tests in exactly the same way, to screen gut samples from field-collected predators. Production of antibodies by molecular techniques has a number of distinct advantages, principally the ability to create more specificities, and hence find antibodies that can distinguish between even very closely related species; the potential to produce monoclonal antibodies much more rapidly; the ability to dispense with expensive and labour-intensive tissue culture facilities and techniques.

Finally, predators that are found to be important natural enemies of slugs may play a part in the control of other pests within an arable ecosystem. In addition, species such as *P. melanarius* may switch between different prey species according to changing prey population densities and ratios. Indeed, polyphagous predators that remain within a crop when slug populations are low, subsisting on alternative prey, may perform a lying-in-wait strategy (Murdoch *et al.*, 1985), ready to prevent a resurgence of slug populations. Monoclonal antibodies are therefore currently being raised against four species of aphid, to assess the overall value of *P. melanarius* in the cropping system, and to monitor switching behaviour between slugs and aphids, and different species of both prey types, over time and in response to agricultural practices.

REFERENCES

Altieri, M A; Hagen, K S; Trujillo, J; Caltagirone, L E (1982) Biological control of *Limax maximus* and *Helix aspersa* by indigenous predators in a daisy field in central coastal California. *Acta OEcologia*. **3**, 387-390.
Burn, A.J (1988) Assessment of the impact of pesticides on invertebrate predation in cereal crops. *Aspects of Applied Biology*. **17**, *Environmental Aspects of Applied Biology* Part **2**, 173-179.

Digweed, S C (1993) Selection of terrestrial gastropod prey by cychrine and pterostichine ground beetles (Coleoptera: Carabidae). *Canadian Entomologist.* **125**, 463-472.

Grant, J F; Yeargan, K V; Pass, B C; Parr, J C (1982) Invertebrate organisms associated with alfalfa seedling loss in complete-tillage and no-tillage plantings. *Journal of Economic Entomology.* **75**, 822-826.

Greenstone, M H; Morgan, C E (1989) Predation on *Heliothis zea*: an instar-specific ELISA for stomach analysis. *Annals of the Entomological Society of America.* **84**, 457-464.

Gruntal, S Y; Sergeyeva, T K (1989) Food relations characteristics of the beetles of the genera *Carabus* and *Cychrus*. *Zoologicheskii Zhurnal.* **58**, 45-51.

Hagler, J R; Naranjo, S E; Bradley-Dunlop, D; Enriquez, F J; Henneberry, T J (1994) A monoclonal antibody to pink bollworm (Lepidoptera: Gelechiidae) egg antigen: a tool for predator gut analysis. *Annals of the Entomological Society of America.* **87**, 85-90.

Köhler, G; Milstein, C (1975) Continuous cultures of fused cells secreting antibody of predefined specificity. *Nature.* **256**, 495-497.

Larochelle, A (1990) The food of carabid beetles (Coleoptera: Carabidae, including Cicindelinae). *Fabreries.* Supplement **5**, 1-132.

Liddell, J E; Cryer, A (1991) *A Practical Guide to Monoclonal Antibodies.* Wiley, Chichester, 188 pp.

Liddell, J E; Symondson, W O C (1996) The potential of combinatorial gene libraries in pest-predator relationship studies. In: *The Ecology of Agricultural Pests: Biochemical Approaches.* W O C Symondson & J E Liddell (eds). Systematics Association Special Volume No. **53**, 347-366. Chapman & Hall, London.

Luff, M L (1974) Adult and larval feeding habits of *Pterostichus madidus* (F.) (Coleoptera: Carabidae). *Journal of Natural History.* **8**, 403-409.

Moens, R (1980) Le Problème des limaces dans la protection des végétaux. *Revue de l'Agriculture (Brussels).* **33**, 117-132.

Murdoch, W W; Chesson, J; Chesson, P L (1985) Biological control in theory and practice. *American Naturalist.* **125**, 344-366.

Pakarinen, E (1994) The importance of mucus as a defence against carabid beetles by the slugs *Arion fasciatus* and *Deroceras reticulatum*. *Journal of Molluscan Studies.* **60**, 149-155.

Ragsdale, D W; Kjer, K M (1989) Diagnosis of tracheal mite (*Acarapis woodi* Rennie) parasitism of honey bees using a monoclonal based enzyme-linked immunosorbent assay. *American Bee Journal.* **129**, 550-552.

Runham, N W ; Hunter, P J (1970) *Terrestrial Slugs.* Hutchinson, London, 184 pp.

Schneider, H.A.W. (1985) Properties of lectins from snails of the genus *Helix* probed by monoclonal antibodies. *Zeitschrift fur Naturforschling Section C. Biosciences* **40**, 254-261.

South, A (1992) *Terrestrial Slugs, Biology, Ecology and Control.* Chapman & Hall, London, 428 pp.

Stuart, M K; Greenstone, M H (1996) Serological diagnosis of parasitism: a monoclonal antibody-based immunodot assay for *Microplitis croceipes* (Hymenoptera: Braconidae). In: *The Ecology of Agricultural Pests: Biochemical Approaches.* W O C Symondson & J E Liddell (eds). Systematics Association Special Volume **53**, Chapman & Hall, London, pp. 300-321.

Stuart, M K; Barak, A V; Burkholder, W E (1994) Immunological identification of *Trogoderma granarium* Everts (Coleoptera, Dermestidae). *Journal of Stored Products Research.* **30**, 9-16.

Symondson, W O C (1989) Biological control of slugs by carabids. In: *Slugs and Snails in World Agriculture.* I. Henderson (ed.). *BCPC Monograph* **41**, pp. 295-300.

Symondson, W O C (1993) The effects of crop development upon slug distribution and control by *Abax parallelepipedus* (Coleoptera: Carabidae). *Annals of Applied Biology.* **123**, 449-457.

Symondson, W O C; Liddell, J E (1993a) The detection of predation by *Abax parallelepipedus* and *Pterostichus madidus* (Coleoptera: Carabidae) on Mollusca using a quantitative ELISA. *Bulletin of Entomological Research.* **83**, 641-647.

Symondson, W O C; Liddell, J E (1993b) A monoclonal antibody for the detection of arionid slug remains in carabid predators. *Biological Control.* **3**, 207-214.

Symondson, W O C; Liddell, J E (1993c) Differential antigen decay rates during digestion of molluscan prey by carabid predators. *Entomologia Experimentalis et Applicata.* **69**, 277-287.

Symondson, W O C; Liddell, J E (1995) Decay rates for slug antigens within the carabid predator *Pterostichus melanarius* monitored with a monoclonal antibody. *Entomologia Experimentalis et Applicata.* **75**, 245-250.

Symondson, W O C; Mendis, V W; Liddell, J E (1995) Monoclonal antibodies for the identification of slugs and their eggs. *EPPO Bulletin.* **25**, 377-382.

Symondson, W O C; Liddell, J E (eds) (1996a) *The Ecology of Agricultural Pests, Biochemical Approached.* Systematics Association Special Volume **53**. Chapman & Hall, London, 517 pp.

Symondson, W O C; Liddell, J E (1996b) A species-specific monoclonal antibody system for detecting the remains of Field Slugs, *Deroceras reticulatum* (Müller) (Mollusca: Pulmonata), in carabid beetles (Coleoptera: Carabidae). *Biocontrol Science & Technology.* **6**, 91-99.

Symondson, W O C (in press) Coleoptera (Carabidae, Staphylinidae, Lampyridae, Drilidae and Silphidae). In: *Natural Enemies of Terrestrial Molluscs.* G.M. Barker (ed.). CAB International, Oxford.

Symondson, W O C; Hemingway, J (in press) Biochemical and molecular techniques. In: *Methods in Ecological and Agricultural Entomology.* D.R. Dent & M.P. Walton (eds). CAB International, Oxford.

Symondson, W O C; Glen, D M; Wiltshire, C W; Langdon, C J; Liddell, J E (in press) Effects of cultivation techniques and methods of straw disposal on predation by *Pterostichus melanarius* (Coleoptera: Carabidae) upon slugs (Gastropoda: Pulmonata) in an arable field. *Journal of Applied Ecology.* **33**.

Tijssen, P (1985) *Practice and Theory of Enzyme Immunoassays.* Elsevier, Amsterdam, 549 pp.

Tod, M E (1973) Notes on beetle predators of molluscs. *Entomologist.* **106**, 196-201.

Wheater, C P (1989) Prey detection by some predatory coleoptera (Carabidae and Staphylinidae). *Journal of Zoology, London.* **218**, 171-185.

Winder, L; Hirst, D J; Carter, N; Wratten, S D; Sopp, P I (1994) Estimating predation of the grain aphid *Sitobion avenae* by polyphagous predators. *Journal of Applied Ecology.* **31**, 1-12.

EXPLORING AND EXPLOITING THE POTENTIAL OF THE RHABDITID NEMATODE *PHASMARHABDITIS HERMAPHRODITA* AS A BIOCONTROL AGENT FOR SLUGS

D M GLEN, M J WILSON, L HUGHES

IACR - Long Ashton Research Station, Department of Agricultural Sciences, University of Bristol, Long Ashton, Bristol BS18 9AF, UK

P CARGEEG, A HAJJAR

MicroBio Ltd, Unit 2 Centro, Boundary Way, Maxted Road, Hemel Hemstead, Herts, HP2 7SU, UK

ABSTRACT

Studies leading to the commercialisation, in 1994, of the slug parasitic-nematode *Phasmarhabditis hermaphrodita* as a biocontrol agent for use against slugs in the home garden market are described, together with ongoing investigations to exploit the potential use of this nematode for control of slug damage in a wide range of crops. The nematode is able to infect and kill all species of pest slug tested, as well as several snail species. However, larger species are thought to be susceptible only when young. Other invertebrates are not infected. The nematode is produced commercially in liquid fermenters. Infective larvae are harvested and, once formulated, they can be stored, under refrigeration, for several months. Infective larvae are applied to soil as a spray or drench at a rate of 3×10^9 ha^{-1}. Slug feeding activity is considerably reduced shortly after infection, thus providing rapid protection from slug damage. The nematode is effective in protecting crops from slug damage, when it is applied as a drench or sprayed to moist soil at, or shortly before, the stage when crops are susceptible. Efficacy in dry soil conditions can be improved by incorporating nematodes into the surface soil. Slugs avoid feeding or resting on areas of soil treated with nematodes, suggesting that it may be possible to target applications around individual plants or along crop rows. Despite its ability to kill a wide range of snail species, field studies indicate that it does not pose a threat to non-target molluscs living in field margins and hedgerows adjacent to arable areas treated with the nematode.

INTRODUCTION

Slug pests thrive in moist conditions, which they require for survival, activity, growth and reproduction and they are adapted to relatively low temperatures. A search was initiated in 1987, at IACR-Long Ashton Research Station, for a biological control agent which would be active against the full range of pest species in the cool moist conditions where slugs are most troublesome as pests. Early in 1988, the rhabditid nematode, *Phasmarhabditis hermaphrodita*, was discovered parasitising the field slug, *Deroceras reticulatum* (Wilson *et al.*, 1993a). Research on this nematode led to the launch in spring 1994 of a commercial product (Nemaslug*) for biocontrol of slugs in the home garden market (Glen *et al.*, 1994). This paper reviews and summarises current knowledge of the attributes and potential of this nematode as a biocontrol agent for slugs.

BIOLOGY AND INTERACTIONS WITH SLUGS

Phasmarhabditis hermaphrodita and a closely related species *Phasmarhabditis neopapillosa*, have been recorded from a number of locations in France and Germany, as non-feeding infective larvae (third stage larvae with mouth and anus closed, inside a retained second stage cuticle) living in the body cavity of slugs and snails, but later stages were not found in live slugs and snails (Maupas, 1900, Morand, 1988). These species were not considered to be parasites of slugs, but they were thought to show a degree of adaptation to life with slugs, in that infective larvae entered the bodies of living slugs where they remained until after the slug died, then grew and reproduced on the decaying slug cadaver, eventually producing infective larvae once again when the food supply became depleted (Mengert, 1953). Infective larvae are believed to survive in the soil ready to repeat the cycle by entering the body of slugs when the opportunity arises.

Our studies have shown that *P. hermaphrodita* is also able to live parasitically on slugs and kill them (Wilson *et al.*, 1993a,c). Infective larvae enter the shell cavity, beneath the slug mantle, probably via a small pore at the hind end of the mantle. Once inside the shell cavity, they are thought to shed their outer cuticle and release bacteria on which they feed and grow to hermaphrodite adults. At this stage, the shell cavity of infected slugs becomes swollen with fluid, causing a characteristic swelling of the mantle, and host feeding is strongly inhibited. The nematodes reproduce within the mantle cavity, which sometimes bursts as a result of the build up of pressure, then the slug dies. Dead slugs infected with *P. hermaphrodita* have never been found in plants or on the soil surface following applications of this nematode in the field. Thus, it is thought that slugs always move down into soil before they die as a result of nematode infection. Following death of the host, nematodes spread and reproduce over the entire cadaver. *Phasmarhabditis neopapillosa* has been found parasitising slugs at Long Ashton in much the same way as *P. hermaphrodita*. However, parasitism by *P. neopapillosa* has not been investigated in detail.

A soil-based bioassay has been used to investigate the relationship between the dose of nematodes applied to soil and their effects on slugs. In this bioassay, slugs are confined for five days in plastic boxes containing a clay or medium-loam soil moistened to field capacity (30% w/w water) with water alone or water containing different doses of nematodes. Thus, slugs are brought into contact with nematodes in soil, under conditions similar to the field. After this initial period of exposure, slugs are placed individually in petri dishes with moist filter paper and a disc of Chinese cabbage leaf, so that slug feeding and survival can be monitored. These bioassays have demonstrated that both feeding inhibition and mortality of the field slug, *Deroceras reticulatum*, are directly related to the dose of infective larvae to which slugs are exposed, with a given dose of nematodes causing stronger and earlier inhibition of feeding than mortality.

The optimum temperature for growth of *P. hermaphrodita* is *c.* 17°C and it is able to infect *D. reticulatum* at temperatures as low as 5°C (unpublished data). Thus, it responds to temperature in a similar manner to its slug hosts. For example, the optimum temperature for growth of *D. reticulatum* is 18°C and it can grow at temperatures as low as 5°C (South, 1992).

RANGE OF SPECIES AFFECTED BY *P. HERMAPHRODITA*

The soil-based bioassay described above has been used to demonstrate that *P. hermaphrodita* is able to kill all pest species of slugs tested, belonging to the three families containing the main pest species (Arionidae, Limacidae and Milacidae) (Wilson *et al.*, 1993a). A number of species of snails are killed by this nematode in laboratory bioassays (Wilson *et al.*, 1993c). Moreover, an isolate capable of killing helicid snails has been recorded from southern France (Coupland, 1995). We have developed a modified bioassay for testing the effects of nematodes on larger species of slugs and snails, including the garden snail, *Helix aspersa*. In this bioassay, 400 g of air-dried soil aggregates are placed in an even layer in the base of a plastic box (27 cm x 14 cm x 9 cm deep). The soil is moistened with 120 ml of water alone or water with suspended nematodes, applied from a wash bottle. Ten snails are placed on the soil in each box. Discs of Chinese cabbage are provided as food and replaced at regular intervals. The inside walls of the box are lined with 0.8 mm copper mesh to deter snails from leaving the soil. When three different size-classes of *H. aspersa* were exposed to the nematode in this way, mortality of the smallest size-class (<1 g) was directly related to nematode dose, but there was no evidence of nematode-induced mortality in larger snails (Fig.1). Similar results have been found for the large slug species, *Arion ater* agg. (unpublished data). Thus, where large slug and snail species are likely to cause damage, it would seem to be appropriate to apply the nematode as a biocontrol agent at the time when vulnerable juvenile stages are present.

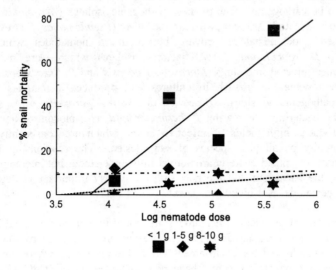

Figure 1. Mortality, after 21 days at 15°C, of three different size classes of *Helix aspersa* in relation to nematode dose, in a laboratory bioassay (10 snails per box, 2 boxes per dose, plus untreated controls).

Although the nematode is able to infect and kill a wide range of species of slugs and snails, other invertebrates, including earthworms and a range of insect species, have not been affected by exposure to high doses of the nematodes. Moreover, as the nematode is rapidly killed at 35°C (Wilson *et al.*, 1993b), it is not able to survive in homeothermic animals.

REARING, FORMULATION AND STORAGE

This bacterial-feeding nematode is cultured on nutrient-rich media, using techniques similar to those developed for entomopathogenic nematodes. For initial investigations of *P. hermaphrodita* as a biological control agent, it was cultivated in xenic cultures, containing an unknown mixture of bacteria (Wilson *et al.*, 1993b). Nematodes grown in xenic cultures were shown to be able to prevent slug damage and cause slug mortality, both in the laboratory (Wilson *et al.*, 1993a) and in mini-plot field experiments (Wilson *et al.*, 1994b). However, nematodes grown in this way are likely to be of unpredictable quality, because experiments have shown that the strain of bacterium used to culture *P. hermaphrodita* has a profound influence on its growth (Wilson *et al.*, 1995c) and ability to cause slug mortality (Wilson *et al.*, 1995b).

Satisfactory yields of the nematodes can be obtained from monoxenic cultures grown in liquid or solid-phase with several bacterial strains that have been found associated with the nematode (Wilson *et al.*, 1995c). However, only certain nematode-bacterial combinations are pathogenic to slugs and this is not related to the pathogenicity of the bacteria alone (Wilson *et al.*, 1995b). Nine bacterial isolates found associated with *P. hermaphrodita* were injected into the body cavity of the field slug *D. reticulatum*, but only two isolates were found to be pathogenic. One of these pathogenic isolates (*Aeromonas hydrophila*) did not support nematode growth, whereas the other (*Pseudomonas fluorescens*, isolate 140) supported good nematode growth, but produced nematodes which were not pathogenic to *D. reticulatum*. Two bacterial isolates which were not themselves pathogenic when injected into slugs (*Moraxella osloensis* and *P. fluorescens* isolate 141), supported good nematode growth in cultures and produced nematodes which were consistently pathogenic to slugs. Based on this work, *Moraxella osloensis* has been selected as the bacterium for rearing *P. hermaphrodita* as a biocontrol agent for slugs, because it produces high yields of infective larvae which are consistently capable of inflicting mortality on all pest species of slugs tested. The relationship between this nematode and its associated bacteria are not yet fully understood, but interactions between the nematode, the bacterium and the host slug are clearly important in determining the value of this nematode as a biocontrol agent.

In current commercial production, nematodes are reared in monoxenic liquid cultures with *M. osloensis* in fermenters. Yields of infective larvae are influenced by the composition of the growth medium and the environment within the fermenter, where yields in excess of 100,000 infective dauer larvae ml^{-1} have been achieved (Glen *et al.*, 1994). Infective dauer larvae are harvested from the fermentation medium by centrifugation followed by repeated washing in water, concentrated to give between 0.1×10^6 and 2.0×10^6 nematodes g^{-1} and mixed with calcium montmorillonite clay or finely powdered vermiculite to produce a water-dispersable friable formulation containing 0.05×10^6 to 1.8×10^6 larvae g^{-1} (wet weight), which is stored in bags of 8 μm high-density polyethylene. *Phasmarhabditis hermaphrodita* has been maintained in this formulation, under refrigeration, for 6 months. When required, the formulation is mixed with water and applied to soil using conventional spray equipment or a watering can fitted with a rose. The commercial formulation of *P. hermaphrodita* is sold in packs, each containing a minimum of 6 x 10^6 nematodes, with the recommendation that the product is suspended in water and applied to a total area of 20 m^2 of moist soil, using a watering can fitted with a rose.

EFFICACY AS A BIOCONTROL AGENT

Since 1989, the effects of nematodes applied to soil have been tested in a number of field trials in a range of crops at IACR-Long Ashton (Glen *et al.*, 1994; Wilson *et al.*, 1994a,b; 1995a,d,e; 1996) and, since 1993, in a number of independent tests, including Switzerland (Speiser and Andermatt, 1994).

Relationship between nematode dose and field efficacy

In initial trials at Long Ashton, the effects of the nematodes were compared with a standard product for slug control, methiocarb bait pellets ('Draza', Bayer Ltd), broadcast on soil at the recommended field rate of 5.5 kg product ha^{-1}. In some trials, the protection from slug damage improved steadily as nematode dose increased from 1 x 10^8 to 1 x 10^{10} ha^{-1}, whereas in one trial, there was no further improvement in protection at doses greater than 8 x 10^8 ha^{-1}. However, provided that the soil was moist at, and after the time of nematode application to soil, nematode doses of 3 x 10^9 ha^{-1} have provided reliable protection from slug damage as good as, or better than methiocarb. In some field trials, slug numbers have been substantially reduced (Wilson, *et al.*, 1994b, 1995c, 1996), whereas in others (Glen *et al.*, 1994; Wilson *et al.*, 1994a), no clear effects on slug populations have been detected, despite substantial reductions in slug damage.

Influence of timing of application on nematode efficacy

In two initial mini-plot field experiments, plant damage by slugs was recorded at weekly or bi-weekly intervals after application of nematodes to soil. Protection from slug damage increased over the first two weeks after application (Wilson *et al.*, 1994b). Since the most severe damage by slugs to crops, such as wheat and rape, is done soon after sowing, this suggests that it may be beneficial to apply nematodes to soil one or two weeks before susceptible crops are sown. However, field experiments have shown little or no benefit from nematode application 1 or 2 weeks prior to sowing a crop of cereals or rape, compared with application at the time of drilling (Wilson *et al.*, 1995e). In these experiments, application to moist soil appeared to be more important for achieving effective control of slug damage than the precise timing of application.

Because crops such as oilseed rape provide slugs with an excellent source of food and shelter, slug numbers usually increase considerably in these crops, with a consequent high risk of damage to following crops, such as winter wheat. We have tested whether it might be possible to use the nematode to prevent the build-up of slug populations in oilseed rape.

However, in two trials where the nematode was applied to oilseed rape at the time of drilling, soil sampling at intervals during the growing season revealed no effect on slug population increase in this crop (Wilson *et al.*, 1995e).

Value of incorporating nematodes into soil after application

Phasmarhabditis hermaphrodita, like its slug hosts, is very dependent on moisture for survival. Thus, nematodes are normally likely to be applied to moist soil where slugs are likely to cause damage. However, the surface of arable soils tends to dry rapidly following cultivation and, although there may be sufficient moisture below the surface for slug activity, nematodes applied to the soil surface under these conditions will become desiccated and die shortly afterwards. Poor performance of the nematode in some field trials has been associated with dry soil conditions (Wilson *et al.*, 1995d,e; 1996). We have investigated the possibility of incorporating nematodes into the surface 2-10 cm of soil in order to improve their reliability (Wilson *et al.*, 1996). When nematodes were applied to moist soil, incorporation had no effects on nematode efficacy. However, when nematodes were applied to a dry seed-bed immediately after drilling winter wheat, nematodes left on the soil surface were ineffective, whereas nematodes incorporated into the soil using a tined implement working to a depth of 5 or 10 cm, significantly reduced slug damage. Thus, shallow incorporation of nematodes was beneficial in dry soil conditions and had a neutral effect in moist soil. In many crops, nematodes could be readily incorporated into soil during seed-bed preparation. Different methods of incorporation now need to be tested.

Possibility of exploiting repellent effects of nematodes

In one field trial (Glen *et al.*, 1994; Wilson *et al.*, 1994a), it was noted that slug behaviour appeared to have changed on nematode-treated plots, such that more slugs were recorded beneath refuge traps placed on the soil surface of these plots, compared to untreated plots. Moreover, a higher percentage of slugs recorded beneath traps showed symptoms of nematode infection than slugs recorded in soil samples. We have since demonstrated in the laboratory that slugs avoid feeding and resting on soil treated with nematodes (Wilson *et al.*, 1995e). This finding suggests that it may be possible to apply nematodes only to soil immediately surrounding susceptible plants or along crop rows, thus, reducing the dose of nematodes required for effective protection. Opportunities for using this technique in crops such as wheat and rape are restricted because individual plants and crop rows are relatively close together. It is likely that the technique will be more suited to sugar beet and vegetable crops that are grown in widely spaced rows.

Effects on snails of conservation interest

Several snail species have been shown to be killed by *P. hermaphrodita* in the laboratory (unpublished data). However, in two field experiments where nematodes were applied to plots adjacent to field margins, no adverse effects on snails living in field margins were detected (Wilson *et al.*, 1995e). Experiments on vertical and horizontal movement of the nematodes in medium-loam soil and sand showed that the vast majority of nematodes moved less than 1 cm vertically or horizontally from the site of placement (unpublished data). These results indicate that application of nematodes to arable soil is unlikely to result in nematodes spreading to semi-natural habitats in sufficient numbers to affect slugs and snails living there.

Commercial trials

Since 1993, MicroBio Ltd have coordinated many European grower trials with the commercial formulation of *Phasmarhabditis hermaphrodita* (Nemaslug®). Initially, trials were set up in order to demonstrate efficacy against common slug species (predominantly *Deroceras reticulatum*). Attention has been paid to trial site and design, in order to quantify efficacy in comparison to chemical molluscicides sold on the home garden market. Table 1 summarises the data collated from one such trial. Less slug damage was recorded at the higher than the lower nematode dose. A single treatment with 3×10^9 nematodes ha^{-1} (3×10^5 nematodes m^{-2}) provided control equivalent to both metaldehyde-based formulations. This is the current recommended application rate.

The results in Table 1 are typical of the level of control recorded in a number of different trial locations across Europe, with a range of climatic conditions and soil types. However, reduced product efficacy has been observed when applied to heavy, waterlogged, soils. Here, it is likely that the nematodes were unable to penetrate the upper soil horizons, then quickly died upon soil drying.

Table 1. Field study to evaluate *Phasmarhabditis hermaphrodita* in comparison to two metaldehyde-based pellet formulations sold on the home garden market.

Treatment	Mean Percentage of wheat seeds hollowed	
	25 days after treatment	32 days after treatment
Untreated	20.6^a	33.2^a
Metaldehyde Formulation 1 (17.15 kg product ha^{-1})	6.4^{b-e}	11.6^{cd}
Metaldehyde Formulation 2 (18.35 kg product ha^{-1})	6.4^{b-e}	14.7^{bcd}
Nematodes (3×10^9 ha^{-1})	8.2^{bcd}	12.9^{bcd}
Nematodes (1×10^{10} ha^{-1})	3.7^{de}	5.4^d

Means in the same column followed by the same letter do not differ significantly
($P<0.05$)

More recent commercial trials have focused on both the fine tuning of the current label recommendation for the home garden market, and also the demonstration of efficacy in high value horticultural crops. Consistent slug control can be demonstrated in all but the harshest of environments. The application of *P. hermaphrodita* to mulched borders, or thatched lawns, resulted in a significant reduction in slug control. When entomopathogenic nematodes are used to control insect pests of turf grass, the turf is irrigated after application to prevent nematodes being trapped in the layer of thatch. At present, it is recommended that *P. hermaphrodita* is applied directly to moist, bare soil in order to maximise response. As previously mentioned, larger slugs and snails seem to be less sensitive to application of *P. hermaphrodita* than either their juvenile counterparts, or smaller species.

In the UK, T-tape irrigation (with low-pressure pipes on the soil surface) has become an increasingly popular method of irrigating mulched crops such as strawberries. This system has already been suggested (in horticultural press, and vine weevil workshops) for application of entomopathogenic nematodes for black vine weevil control (*Otiorhynchus sulcatus*) in strawberries. Table 2 summarises a T-tape grower trial, commissioned through the UK high street supermarket chain, Sainsbury's. This trial indicates that efficacy of *P. hermaphrodita* (applied at 3 x 10^8 ha^{-1} through double line T-tape) is comparable to chemical pelleted baits in terms of the percentage reduction in fruit damage. Numbers of carabid beetles remained unchanged following the use of *P. hermaphrodita*.

Table 2. Field study to evaluate the potential of T-tape irrigation to apply *Phasmarhabditis hermaphrodita* in commercially grown strawberries. Results are expressed as the percentage reduction in fruit damage compared to plots irrigated with water alone.

Treatment	Rate	Percentage reduction in fruit damage
Metaldehyde pellets (Metarex)	recommended (8 kg product ha^{-1})	59
Methiocarb pellets (Draza)	recommended (5.5 kg product ha^{-1})	28
Nematodes	8 x 10^8 ha^{-1}	51

Grower trials have demonstrated that control of plant damage can be sustained for 5-8 weeks following an application of *P. hermaphrodita*. A good example of this has been a series of UK field trials on Iceberg lettuce. Slug control is critical in this high value, often irrigated, edible crop. Growers routinely use 2-3 applications of pelleted baits throughout the crop's 9-10 week period of field growth. It is essential, to avoid crop contamination, that pelleted baits do not lodge within leaf bases, and that slugs are controlled before they move from the soil into the developing plant to both feed and seek refuge. These field trials concluded that a single application of *P. hermaphrodita* at the 11 leaf stage (approximately 4 weeks after planting) resulted in effective control of plant damage until harvest. A combination of the early inhibition of slug feeding, followed by slugs burrowing into the soil before death, makes *P. hermaphrodita* an ideal slug control agent for such a market.

CONCLUSIONS

Our results show that *P. hermaphrodita* is an effective, environmentally-sound means of controlling slug damage. The nematode is effective under a wide range of conditions against slugs causing damage in home gardens, horticultural crops and arable crops. Use of this nematode as a biocontrol agent does not have long-term effects on slug populations, thus, it will not threaten wildlife which depend on slugs as a source of food. Moreover, it does not pose a threat to snails of conservation interest living in field margins.

The main constraint on the use of this nematode to control slug damage in arable crops is one of cost. It is likely that, after the home-garden market, the nematode will next be used in high-value vegetable crops, such as brussels sprouts, and fruit crops, such as strawberries, where control of slug damage is unsatisfactory and where there is concern about possible residues of pesticides used for slug control. By allowing natural enemies to survive, the nematode provides the key to effective integrated control of a range of pests.

Through increased production capacity, and an on-going product research and development programme (formulation, application, etc), the harsh economic realities in using *P. hermaphrodita* in new markets will become less restrictive. Offering a high level of efficacy and persistence, along with an environmentally favourable profile, the future extension to the worldwide field use of this nematode is a distinct possibility.

ACKNOWLEDGEMENTS

The research described in this article was funded by the Agricultural Genetics Company Ltd, its subsidiary MicroBio Ltd and the Ministry of Agriculture, Fisheries and Food. Current research collaboration between IACR and MicroBio Ltd is funded under the LINK Programme 'Technologies for Sustainable Farming Systems'. IACR receives grant-aided support from the Biotechnology and Biological Sciences Research Council of the UK.

REFERENCES

Coupland, J B (1995) Susceptibility of helicid snails to isolates of the nematode *Phasmarhabditis hermaphrodita* from southern France. *Journal of Invertebrate Pathology*, **66,** 207-208.

Glen, D M; Wilson, M J; Pearce, J D; Rodgers, P B (1994) Discovery and investigation of a novel nematode parasite for biological control of slugs. *Brighton Crop Protection Conference - Pests and Diseases 1994*, **2**, 617-624.

Maupas, E (1900) Modes et formes de reproduction des nematodes. *Archives de Zoologie*, **8,** 464-642.

Mengert, H (1953) Nematoden und Schnecken. *Zeitschrift für Morphologie und Ökologie Tiere*, **41**, 311-349.

Morand, S (1988) Contribution á l'étude d'un systeme hôtes-parasites: nematodes associés á quelques mollusques terrestres. Doctorate Thesis, University of Rennes, pp. 335.

South, A (1992) *Terrestrial Slugs, Biology, Ecology and Control*, London: Chapman and Hall, 428 pp.

Speiser, B; Andermatt, M (1994) Biokontrolle von Schnecken mit Nematoden. *Agrarforschung*, **1**, 115-118.

Wilson, M J; Glen, D M; George, S K (1993a) The rhabditid nematode *Phasmarhabditis hermaphrodita* as a potential biological control agent for slugs. *Biocontrol Science and Technology*, **3**, 503-511.

Wilson, M J; Glen, D M; George, S K; Butler, R C (1993b) Mass cultivation and storage of the rhabditid nematode *Phasmarhabditis hermaphrodita*, a biocontrol agent for slugs. *Biocontrol Science and Technology*, **3**, 513-521.

Wilson, M J; Glen, D M; Pearce J D (1993c) Biological Control of Molluscs. *World Intellectual Property Organisation, International Patent Publication Number:* WO 93/00816.

Wilson, M J; Glen, D M; George, S K; Pearce, J D; Wiltshire, C W (1994a) Biological control of slugs in winter wheat using the rhabditid nematode *Phasmarhabditis hermaphrodita*. *Annals of Applied Biology*, **125**, 377-390.

Wilson, M J; Glen, D M; Wiltshire, C W; George, S K (1994b) Mini-plot field experiments using the rhabditid nematode *Phasmarhabditis hermaphrodita* for biocontrol of slugs. *Biocontrol Science and Technology*, **4**, 103-113.

Wilson, M J; Glen, D M; George S K; Hughes, L A (1995a) Biocontrol of slugs in protected lettuce using the rhabditid nematode *Phasmarhabditis hermaphrodita Biocontrol Science and Technology*, **5**, 233-242.

Wilson, M J; Glen, D M; George S K; Pearce, J D (1995b) Selection of a bacterium for the mass production of *Phasmarhabditis hermaphrodita* (Nematoda: Rhabditidae) as a biocontrol agent for slugs. *Fundamental and Applied Nematology*, **18**, 419-425.

Wilson, M J; Glen, D M; Pearce, J D; Rodgers, P B (1995c) Monoxenic culture of the slug parasite *Phasmarhabditis hermaphrodita* (Nematoda: Rhabditidae) with different bacteria in liquid and solid phase. *Fundamental and Applied Nematology*, **18**, 159-166.

Wilson, M J; Glen, D M; George, S K; Pearce J D; Rodgers, P B (1995d) The potential of the rhabditid nematode *Phasmarhabditis hermaphrodita* for biological control of slugs. *Journal of Medical Applied Malacology,* in press.

Wilson, M J; Hughes, L A; Glen, D M (1995e) Developing strategies for the nematode, *Phasmarhabditis hermaphrodita*, as a biological control agent for slugs in integrated crop management systems. In: *Integrated Crop Protection: Towards Sustainability?* R G McKinlay & D Atkinson (eds). BCPC Monograph No. 63, pp. 33-40.

Wilson, M J; Hughes, L A; Hamacher, G M; Barahona, L D; Glen, D M (1996) Effects of soil incorporation on the efficacy of the rhabditid nematode, *Phasmarhabditis hermaphrodita*, as a biological control agent for slugs. *Annals of Applied Biology*, **128**, 117-126.

Session 8
Integrated Pest Management

Session Organiser
and Chairman   Dr D M Glen

AN INTEGRATED PEST MANAGEMENT (IPM) APPROACH TO THE CONTROL
OF THE BROWN GARDEN SNAIL, *(HELIX ASPERSA)* IN CALIFORNIA CITRUS
ORCHARDS.

N J SAKOVICH

University of California Cooperative Extension Ventura County, California, USA.

ABSTRACT

The brown garden snail *(Helix aspersa)* can be a major pest in citrus orchards
throughout California. Not only are the leaves damaged, but high yield losses
may occur. Chemical controls do not always work and are expensive. New
sustainable methods are being tried with a much success. The predatory snail,
(Rumina decollata), is being reared and released in many orchards for the
control of the brown garden snail. In addition, copper bands wrapped around
the trunks of trees provide an excellent barrier to deny snail access into the
tree. An integrated pest management approach is working well in controlling
this pest.

INTRODUCTION

The brown garden snail, *(Helix aspersa* Muller), was intentionally introduced from France into
several areas of California between 1850 and 1860. By 1900, the snail was present throughout
many of California's agricultural districts. It has been a problem to citrus growers ever since
(Basinger, 1931).

The brown garden snail is an important pest in citrus orchards, however the degree of damage
they cause is mostly dependent upon weather conditions and irrigation habits. During years
of drought, growers may see little or no damage. The snail is relegated to a rather minor pest.
During years of high rainfall, especially with warm spring rains, snails can easily become the
number one pest. Counts of feeding snails within a given tree have reached 1,000. Crop losses
can be 40% and sometimes reach 90-100% (Sakovich and Bailey, 1985). Large amounts of
money are spent on snail baits, often with minimal results.

During these heavy rainfall years, the mature trees themselves usually do not incur significant
damage. However, when snails feed on the foliage of one- or two-year old trees, major damage
can occur as these trees become completely defoliated. In rare cases, snails have also been
found to chew on the outer bark of the young trees.

The critical concern however, is with the fruit. Snails will create tiny holes as they feed on the
flavedo, leaving the white albedo exposed. This fruit is either left in the orchard or, if picked,
will be eliminated at the packinghouse. As a consequence of extensive feeding by one or more
snails, large holes will be created in the fruit, often hollowing out the fruits' interior.

Light feeding may actually cause more serious damage. This is because the lightly damaged fruit is often overlooked by pickers and packinghouse sorters. These extremely small wounds are perfect entry points for the *Penicillium* fungi, a major decayer of citrus fruit. The contaminated fruit is then packed and exported. It may remain on board ship for as long as two weeks. Upon arrival, the cartons are opened and the fruit will be completely covered with the blue or green mold. Not only is this shipment lost, but future shipments are also suspect by the now angry customer.

Over the past thirty years, the brown garden snail has actually become an increasing problem. This is due mainly to the advent of preemergence herbicides and along with them, the practice of noncultivation. Under noncultivation, no longer is equipment pulled through the orchard, disrupting the soil. Therefore, snail nests, and adults to a lesser degree, are not destroyed (Morse and Sakovich, 1986). Also, with a greater use of low volume irrigation, the frequency of water application has increased, creating a better environment for the snail.

CONTROL

Control may be accomplished through chemical, nonchemical or biological means. Some growers incorporate all three of these methods, termed Integrated Pest Management (IPM), to keep snails in check.

Chemical

Mesurol (methiocarb) is a good molluscicide and had been used in California until recently when registration was not renewed. It may still be registered in some states. Presently, metaldehyde is the only chemical registered in California for the control of snails in citrus. Dosage and frequency vary depending upon the severity of the problem, which in turn is generally dependent upon the amount of rainfall and morning fog.

Placement is very important for effective control. If upon ingestion of a sublethal dosage of metaldehyde, the snail reaches shelter (shade and high humidity), the snail may survive. Therefore, placing the bait directly under the tree has not always been successful. But placing the bait halfway between two rows is sometimes too great a distance for a majority of the snails to travel (Sakovich and Bailey, 1985). Therefore, placing baits out from underneath the dripline (outermost reach of the tree), but still close to it seems best. It is also preferable to apply these baits just after an irrigation or rain. However, it is not always possible to get equipment into the orchard when the ground is wet. The advantage to the next bait to be discussed, *Deadline 40*, is that it can be applied before a rain or irrigation without loss in efficacy.

Deadline , from Pace International LP, is a metaldehyde product in liquid form, having the consistency of white liquid glue. This material is very efficacious in controlling snails (Morse and Sakovich, 1986). In addition, the manufacturer claims it has an attractant which has been borne out by observations. Special machinery, or at least some modifications are needed to apply this highly viscous material to the orchard floor. Some growers have purchased custom-built applicators which are pulled by ATV-vehicles. As the machinery is pulled through the orchard, strips of liquid Deadline are laid on the ground close to the dripline of the tree. A major benefit of the Deadline liquid is that it is very persistent and will not dissolve or wash away during rainy weather.

In addition to Deadline 40 , a dry bait called Deadline Bullets is also available. This formulation has also proved very efficacious in University of California tests, and is used by growers who chose not to use the special equipment designed for the liquid Deadline . The problem mentioned above, concerning snail recovery from a sublethal dosage of metaldehyde, does not seem to be true when using the Deadline products.

Baits are expensive, especially in years of heavy rainfall when high dosages are required along with multiple applications. Growers are also aware that at any given time a pesticide may be withdrawn from use. This, along with worker safety issues, has driven many growers to seek alternatives to pesticides.

Mechanical barriers

Skirt pruning is the removal of the lower-most branches, generally those which touch the ground or are quite close to it. This type of pruning, in conjunction with some type of barrier placed around the trunk of the tree, can be effective in keeping snails out of the trees. Even skirt pruning alone, during relatively dry years, has helped to minimize snail damage. At one time resistance had been met to removing skirts and with them much fruit. Research however, conducted over a seven year period in valencia oranges, has shown that skirt pruning does not decrease tree yields. After four years, cumulative yields were virtually equal between skirted and nonskirted trees. In fact, in the second year after pruning, yields were higher in the skirt-pruned trees (Sakovich, 1984).

With mature trees, which are skirted for the first time, yields will obviously decrease that first year. However, skirted trees do compensate by setting more fruit higher up in the tree. In addition, fruit located on the skirts is often badly bruised and must be culled. If growers begin their skirt pruning practices when the trees are young, as they now do, there is no initial fruit loss. First time pruning of older trees is usually done with a chain saw and loppers. Subsequent pruning is often done mechanically.

Because there are several other advantages to skirt pruning, this procedure has become a very popular operation. Other advantages include: (1) better ant and fuller rose weevil control, (2) lower incidences of phytophthora brown rot and gummosis, (3) an easier inspection of the irrigation system and (4) frost protection - allows for better movement of cold air (Fisher et. al., 1983).

Once trees are skirt-pruned, usually at a height of 45-60 centimeters, the trunks provide the only access way into the tree. A barrier must be applied. Research conducted at the University of California, Riverside found copper to be an excellent repellent to snails (Sakovich and Bailey, 1985). The exact mechanism is unknown. Perhaps a slight electrical shock or simply the copper is extremely irritating to the foot of the snail.

There are basically two ways of applying the copper as a barrier. The trunks can be painted or sprayed with a slurry of copper bordeaux mixture. The width of the copper band should be at least six inches and it is applied between ground level and the first branching. Since this material is relatively soluble, rain and irrigation water can wash it away after about one year. A small percentage of white latex house paint, or a good spreader/sticker may be added to increase the persistence of the copper.

For a more permanent barrier, a copper foil, 'Snail-barr,' is wrapped around the trunk (Sakovich and Bailey, 1985). One end of the foil is stapled to the tree, then the foil is wrapped around and fastened with a nickel plated paper clip. Several extra inches of foil are left to allow for expansion of the trunk. These barriers have been found to last more than five years. However in some areas, birds and other small animals tear the bands off the trees, presumably seeking food such as insects which hide between the foil and trunk. Inspections must be made to see that the bands are in place and working properly.

Biological control

The decollate snail *(Rumina decollata)* is a predatory snail which attacks, kills and consumes the brown garden snail (Fisher et. al., 1980). Many growers are now distributing this snail in their groves. Properties where this predator has become well established either harbor low numbers of brown snails or none at all. Depending upon how much an individual is willing to spend, the decollates may be distributed throughout the entire grove, or a handful placed under certain core trees. These core trees will then act as a nursery as the decollates multiply. Later they may be moved to other parts of the orchard. Supplemental food (rabbit pellets or left over salad) and cover (old fertilizer bags) can be provided to enhance chances of establishment. Extra, light irrigations also increase the chance of a quicker colony establishment (Sakovich et. al., 1984).

Decollate snails, however, do not provide an overnight cure. In many areas it will take 4-6 years (dependent upon how many are originally released) to establish a large enough population to sufficiently dominate the pest snail. The decollates feed mainly on the young snails. To obtain quicker control, it is recommended to first bait the orchard, then upon waiting 60 days, the decollates can be released. Decollates are susceptible to snail bait.

Ducks have been used for many years in other countries for snail and weed control. Recently, growers in California have established their "herds" of ducks. The ducks have done an excellent job of controlling snails and a fair job in controlling weeds. There is of course extra care and labor involved in handling the ducks. They must be penned within a sturdy enclosure whenever they are not being watched to protect them from predators such as dogs and coyotes. Usually in the morning, a man will take them out for as little as half an hour to scavenge for food. They are then penned for the rest of the day. Once the snails are gone, however, the grower now has a problem of what to do with the ducks. Therefore, it has become the practice for several growers to join together and share the ducks.

The best approach to controlling snails in a citrus orchard is IPM. When dealing with high populations of snails, they must first be baited in order to knock down the pest population. This needs to be done early in the season before large numbers of snails are found in the trees. Once in the trees, they are very difficult to remove. As mentioned, ducks may also be used at this point instead of chemical baits. Then, simultaneous copper banding and decollate releases are to be done. The copper bands will keep the snails out of the trees for the immediate future. This will allow time for the decollate population to increase and begin feeding on the brown snail. After approximately 4-6 years, the decollate population will have grown sufficiently to keep the brown garden snail in check. At this time, maintenance on the barriers may cease. If more time is needed for decollate establishment, maintenance will continue on the bands. As time goes on the decollates may be collected and transferred to other groves or given to neighbors.

REFERENCES

Basinger, A J (1931) The European brown garden snail in California. _University of California Agricultural Experiment Station Bulletin_ **515,** 22 pp.

Fisher, T W; Bailey, J; Sakovich N J (1983) Skirt pruning, trunk treatment for snail control. _Citrograph_ 68 **(12)** 292-297.

Fisher, T W; Orth, R; Swanson, S (1980) Snail against snail. _California Agriculture_, Nov-Dec 1980, 18-20

Morse, J G; Sakovich N J (1986) Control of _Helix aspersa_ on citrus and avocado. _Journal of Agricultural Entomology_ **3** 342-349.

Sakovich, N J; Bailey, B (1985) Skirt pruning and tree banding as snail controls. _Citrograph_ 70, **(13)** 18-21.

Sakovich, N J (1984) New methods of snail control in Southern California. _1984 International Citrus Congress, Sao Paulo, Brazil_, 486-487.

Sakovich, N J; Bailey, J B; Fisher, T W (1984) Decollate snails for control of brown garden snails in Southern California citrus groves, _University of California Cooperative Extension Bulletin_ **21384** 4pp.

SOWN WILDFLOWER STRIPS IN ARABLE LAND IN RELATION TO SLUG DENSITY AND SLUG DAMAGE IN RAPE AND WHEAT

T FRANK

Zoologisches Institut der Universität, Baltzerstr. 3, CH-3012 Bern, Switzerland

ABSTRACT

In 1994 and 1995, slug density and slug damage were estimated in sown wildflower strips and adjacent fields (oilseed rape and winter wheat). Investigations began as soon as seedlings appeared above ground. Slug density and slug damage in rape were much higher in parts of fields near the wildflower strips than in other areas of fields. Within a distance up to 2 m from the wildflower strips, complete crop failures occurred. Areas treated with metaldehyde could be protected from such crop losses. In contrast to rape, in wheat fields bordering on wildflower strips, increased slug damage was never observed adjacent to wildflower strips. It is suggested that this difference in the pattern of damage between rape and wheat was due to *Arion lusitanicus*. Adults of this species were active when rape was susceptible to damage, but they had died by the time that wheat was at risk.

INTRODUCTION

Wildflower strips are linear semi-natural biotopes sown inside fields with mixtures consisting of about twenty-five herbaceous species (Heitzmann, 1994). The creation of sown wildflower strips has been shown to encourage epigeic beneficial arthropods (e.g. Lys & Nentwig, 1992; Frank & Nentwig, 1995) but nothing was known until recently about the effects of sown wildflower strips on adjacent fields in terms of slug damage. Sometimes increased slug damage close to wildflower strips was observed by farmers but there has been no published information on such observations until now. Therefore, slug studies in sown wildflower strips and adjacent fields were started, in 1994, on a farm with high slug densities, in order to observe maximum slug damage in rape and wheat fields bordering on sown wildflower strips. This study examines slug damage and slug density at increasing distances from sown wildflower strips. Moreover, it investigates whether slug damage can be prevented by using metaldehyde pellets.

MATERIAL AND METHODS

The investigation took place in autumn 1994 and 1995 on a farm in Belp, south of Bern, Switzerland (Fig.1). The investigation was started as soon as seedlings of oilseed rape and winter wheat appeared above ground, and lasted for the following five weeks, during the phase when young crops are most vulnerable to slug attack. Crops and wildflower strips were sown on fields with reduced chemical input with wet soils, providing suitable conditions for slugs. In the outer parts of each rape field metaldehyde pellets were used, whereas no slug pellets were applied in the central parts of the field (Fig. 1). Slug pellets (5% metaldehyde at 8 kg product / ha) were broadcast on the soil surface once in 1994 and three times in 1995 due to severe slug problems in that year (Table 1a). In wheat, in contrast to rape, slug pellets were never used. There was a difference in the method of establishing rape

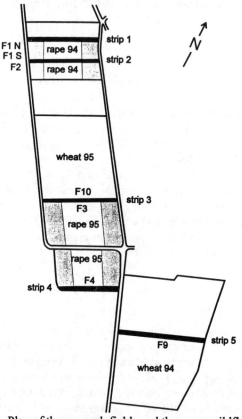

Figure 1. Plan of the research fields and the sown wildflower strips (coloured dark grey) studied in 1994 and 1995. Dotted areas in rape = field areas where metaldehyde was used.

Table 1a. Details of the research fields.

Field	Crop studied	Sowing date	Foregoing crop	Application of metaldehyde (5% at 8 kg / ha)	Cropped area (ha)
F1	oilseed rape 94	6.09.94	winter wheat	20.09.94	0.5
F2	oilseed rape 94	6.09.94	winter wheat	20.09.94	0.5
F3	oilseed rape 95	6.09.95	winter wheat	6., 21., 28. 09.95	1.3
F4	oilseed rape 95	6.09.95	winter wheat	6., 21., 28. 09.95	0.8
F9	winter wheat 94	10.10.94	oats	-	3.0
F10	winter wheat 95	19.10.95	sugar beet	-	2.5

Table 1b. Details of the sown wildflower strips.

Wildflower strip	Sowing date	Length (m)	Width (m)
strip 1	28.04.94	138	3
strip 2	28.04.94	140	3
strip 3	3.04.95	145	4
strip 4	3.04.95	110	6
strip 5	3.05.04	162	3

in the two years. In 1994, the field was ploughed and harrowed before sowing rape, whereas in 1995 rape was direct-drilled into wheat stubble. All wildflower strips were sown in spring of the year in which they were studied (Table 1b).

Slug density was estimated using bait stations containing wheat bran, cat food and water in a ratio of 1:1:5 by weight. Each bait station consisted of a petri dish (140 mm diameter) filled with 20-25 g bait. In each wildflower strip, eight petri dishes with bait were placed 1 m from the edge of the adjacent field. In the adjacent fields, bait stations were located at 1, 3 and 5 m from the wildflower strips with eight stations at each distance. Eight further bait stations were placed in the middle of each field, 18 to 45 m from the wildflower strips (depending on field size). In each wildflower strip and within each distance from the strips, the bait stations were located 7 m apart. In rape, slug density was estimated only in the central parts of the field where no slug pellets were used. Slug numbers were recorded weekly on wet evenings with high slug activity. Petri dishes with bait were placed on the soil surface at dusk; four hours later slugs on the dishes were counted and identified to species. After that, the petri dishes with baits were removed from the field.

Every day after estimating slug numbers, slug damage was evaluated using a defoliation scale from zero to four: 0 = no damage; 1 = 1-25% defoliation; 2 = 26-50% defoliation; 3 = 51-75% defoliation; 4 = 76-100% defoliation. Slug damage was evaluated 1, 3 and 5 m from the wildflower strips and in the middle of fields without (rape, wheat) and with application of metaldehyde pellets (rape). At each distance from the wildflower strips, slug damage was evaluated at ten randomly chosen places. Defoliation scores were based on slug damage on five (rape) or ten plants (wheat) in each place. Means were calculated for each distance and week to give an approximate score of damage caused by slugs.

For data analysis, slug numbers were transformed to square roots. Slug damage scores were analyzed using untransformed values. ANOVA was used to evaluate whether there were significant differences in slug density or slug damage between the different distances on each date. Tukey's HSD-test was used to determine significant differences.

RESULTS

Rape

Slug damage and slug densities in 1994 in field F2 are representative of all wildflower strips and adjacent rape fields in 1994. In the part of field F2 where no slug pellets were used, damage was not significantly different at different distances from wildflower strip 2 during the first two weeks after appearance of the rape seedlings above ground (Table 2). The same situation was also observed in fields F1 N and F1 S. However, from the third week onwards, slug damage was significantly higher at 1 m from wildflower strip 2 than at 3 to 18 m from this strip. This damage at 1m increased from week three to week five and developed into a complete crop failure. Severe slug damage took place in all three rape fields and reached up to two metres from the wildflower strips. The areas of fields in which slug pellets were used

Table 2. Slug damage score (Mean±SE) in rape fields in 1994 (F2) and 1995 (F3),without and with application of metaldehyde, at different distances from wildflower strips. Different letters mean significant differences within each date (Tukey; P<0.01). n.s. = not significant.

F2	Without metaldehyde			
Date	1m	3m	5m	18m
20.09.94	0.90±0.16 n.s.	0.52±0.15 n.s.	0.28±0.08 n.s.	0.70±0.14 n.s.
28.09.94	1.24±0.14 n.s.	0.96±0.10 n.s.	1.14±0.11 n.s.	0.82±0.70 n.s.
4.10.94	2.60±0.24 a	1.54±0.13 b	1.38±0.15 b	1.34±0.12 b
12.10.94	3.28±0.24 a	1.78±0.15 b	1.32±0.11 b	1.22±0.09 b
18.10.94	3.68±0.16 a	1.36±0.14 b	1.22±0.08 b	1.24±0.11 b
F2	With metaldehyde			
Date	1m	3m	5m	18m
20.09.94	0.56±0.12 n.s.	0.46±0.09 n.s.	0.48±0.07 n.s.	0.48±0.10 n.s.
28.09.94	0.76±0.08 ab	0.58±0.09 ab	0.40±0.10 ac	0.90±0.09 b
4.10.94	1.08±0.18 n.s.	0.92±0.11 n.s.	0.68±0.05 n.s.	1.08±0.12 n.s.
12.10.94	1.42±0.25 n.s.	1.00±0.07 n.s.	0.94±0.08 n.s.	1.10±0.08 n.s.
18.10.94	1.36±0.24 a	0.86±0.07 ab	0.62±0.07 bc	1.06±0.09 ab
F3	Without metaldehyde			
Date	1m	3m	5m	45m
22.09.95	2.03±0.32 a	0.80±0.19 b	0.74±0.13 b	1.14±0.21 b
29.09.95	2.98±0.45 a	1.38±0.14 b	0.88±0.16 b	2.88±0.28 a
5.10.95	3.43±0.20 a	1.48±0.19 b	1.08±0.22 b	3.14±0.27 a
13.10.95	3.64±0.22 a	2.50±0.20 b	1.36±0.12 c	3.98±0.02 a
17.10.95	3.53±0.24 a	2.08±0.21 b	1.14±0.07 c	3.96±0.03 a
F3	With metaldehyde			
Date	1m	3m	5m	45m
22.09.95	1.46±0.30 n.s.	0.82±0.57 n.s.	0.68±0.53 n.s.	1.48±0.78 n.s.
29.09.95	1.40±0.19 ab	0.72±0.54 a	0.44±0.45 a	2.42±1.08 b
5.10.95	1.78±0.25 a	0.86±0.59 ab	0.48±0.47 b	1.20±0.69 ab
13.10.95	2.28±0.80 a	1.16±0.68 b	0.94±0.61 b	1.82±0.89 ab
17.10.95	1.72±0.26 a	0.82±0.57 b	0.72±0.54 b	1.32±0.73 ab

confirmed the extent of slug damage in 1994. The most important effect of the application of metaldehyde pellets was that slug damage at 1 m from the wildflower strip was much lower than in untreated areas. Although slug damage at 1 m from the wildflower strip was slightly higher than at 3 to 18 m in areas treated with metaldehyde, these differences between 1 m and other distances were almost always not significant. Sample points 3 to 18 m from the wildflower strip in areas treated with slug pellets always had slightly less slug damage after five weeks than at corresponding distances where no slug pellets were used.

The number of slugs trapped in 1994 corresponded very well to the damage in rape, thus, in

Table 3. Number of slugs (Mean±SE, square root scale) in untreated parts of rape fields in 1994 (F2) and 1995 (F3) and in adjacent wildflower strips 2 and 3. Different letters mean significant differences within each date (Tukey; P<0.01).

F2					
Date	strip 2	1m	3m	5m	18m
19.09.94	1.99±0.21 a	1.34±0.26 a	0.37±0.26 b	0.00±0.00 b	0.00±0.00 b
27.09.94	2.52±0.19 a	1.88±0.17 a	0.48±0.24 b	0.12±0.12 b	0.00±0.00 b
3.10.94	2.07±0.16 a	2.82±0.14 b	0.62±0.18 c	0.00±0.00 c	0.25±0.16 c
11.10.94	2.04±0.29 a	1.29±0.32 ab	0.12±0.12 c	0.30±0.20 bc	0.25±0.16 bc
17.10.94	2.41±0.17 a	2.38±0.22 a	0.30±0.20 b	0.12±0.12 b	0.00±0.00 b
F3					
Date	strip 3	1m	3m	5m	45m
21.09.95	2.25±0.10 a	1.47±0.26 ab	0.48±0.24 b	0.68±0.20 bc	1.41±0.26 ac
28.09.95	2.20±0.28 a	1.25±0.25 ab	0.98±0.24 b	0.90±0.21 b	1.55±0.26 ab
4.10.95	2.15±0.23 a	2.26±0.19 a	0.85±0.20 b	0.72±0.23 b	1.58±0.27 a
12.10.95	1.83±0.14 a	1.97±0.12 ab	1.90±0.23 a	0.97±0.34 a	1.96±0.38 b
16.10.95	1.86±0.14 ab	2.19±0.17 ab	1.83±0.20 ab	1.14±0.32 b	1.73±0.33 a

Table 4. Slug damage score (Mean±SE) in wheat fields in 1994 (F9) and 1995 (F10) at different distances from wildflower strips. Different letters mean significant differences within each date (Tukey; P<0.01). n.s. = not significant.

F9				
Date	1m	3m	5m	25m
28.10.94	0.17±0.05 n.s.	0.09±0.02 n.s.	0.21±0.05 n.s.	0.32±0.08 n.s.
2.11.94	0.46±0.07 n.s.	0.34±0.05 n.s.	0.62±0.07 n.s.	0.66±0.08 n.s.
7.11.94	0.48±0.06 a	0.27±0.05 a	0.74±0.05 b	0.97±0.03 b
14.11.94	0.77±0.06 ab	0.61±0.04 a	0.88±0.04 bc	1.06±0.04 c
23.11.94	0.95±0.05 ab	0.78±0.04 a	0.89±0.07 ab	1.03±0.04 b
F10				
Date	1m	3m	5m	18m
10.11.95	0.08±0.03 n.s.	0.04±0.02 n.s.	0.01±0.01 n.s.	0.02±0.02 n.s.
14.11.95	0.11±0.03 n.s.	0.07±0.03 n.s.	0.06±0.03 n.s.	0.05±0.03 n.s.
17.11.95	0.15±0.04 a	0.08±0.02 ab	0.07±0.01 ab	0.03±0.01 b
24.11.95	0.10±0.02 n.s.	0.09±0.03 n.s.	0.06±0.02 n.s.	0.04±0.02 n.s.
29.11.95	0.29±0.06 a	0.09±0.03 ab	0.12±0.03 ab	0.07±0.03 b

the region of wildflower strip 2, many more slugs were collected from the baits 1 m from the strip than at 3 to 18 m from the strip (Table 3). The same observation was also made in fields F1 N and F1 S during all five weeks of the investigation.

In parts of field F3 without slug pellets in 1995, slug damage was significantly higher at 1m than at 3 and 5 m from the wildflower strip 3 from the first week onwards and this developed into a complete crop failure (Table 2). The same situation was also observed in field F4 (data not presented). In the centre of field F3 (45 m from the wildflower strip 3), slug damage was conspicuously high, with defoliation scores which were approximately as high as those at 1 m. This result was in contrast to rape in 1994. In 1995, in the areas of field F3 where slug pellets were applied, slug damage at 1m was significantly greater than at 3 and 5 m from the fourth week on (Table 2). However, the defoliation score at 1m was not very high (2.28 ± 0.80 maximum), therefore in the part of the field treated with slug pellets complete crop failures were not observed. As in 1994, there was less slug damage at distances 3 m and 5 m from the wildflower strips 3 and 4 and in the field centres where metaldehyde pellets were used than at the corresponding distances from wildflower strips in the parts of fields without slug pellets. In field F3, high slug numbers were found in wildflower strip 3 and at 1 m from this strip. In weeks four and five, slugs were also abundant at 3 m from this strip. In the centre of field F3, numbers of slugs were high too, especially in the last two weeks of observation (Table 3).

Wheat

In both years, slug damage in fields F9 and F10 was very low with extremely low defoliation scores (Table 4). Furthermore, in both fields slug damage at 1 m from the wildflower strips was never significantly higher than at 3 m. In field F9 in 1994, an opposite tendency was observed with significantly higher slug damage at 25 m than at 1 m from the wildflower strip 5 in the third and fourth weeks. In field F10 in 1995, slug damage at 1 m was significantly higher than at 18 m in weeks three and five. However, these differences occurred at a very low level of defoliation. In both years, high numbers of slugs were recorded in the sown wildflower strips. Whereas in the centre of field F9 in 1994 high slug numbers were observed, slugs were less abundant in the other parts of the field. Additionally, no decrease in numbers was observed from 1 m to 5 m from the wildflower strip. In contrast to 1994, slug densities in 1995 tended to a decrease from 1m towards the field centre.

DISCUSSION

In both years, slug damage in rape was much higher close to the sown wildflower strips than in other parts of the field. Within a distance up to two metres from the wildflower strips, almost all young rape plants were eaten by slugs in parts of the field where no slug pellets were used. This complete crop failure was recorded in all rape fields studied. Similar observations in rape fields without applications of slug pellets along sown wildflower strips have also been made in other regions in Switzerland (Högger, personal communication). Slug damage in rape corresponded very well with slug numbers recorded at bait stations. In 1994, *Arion lusitanicus* comprised 75% of all slugs collected in wildflower strips and adjacent rape. This species was very abundant in the wildflower strips and at 1m from the strips but was virtually absent at distances of 3 m or more from the strips. It appeared that individuals of *A. lusitanicus* moved between the wildflower strips and the adjacent rape probably because of the nutritional attractiveness of young rape plants. Poor food conditions increase the dispersal of *Arion ater* (Hamilton & Wellington, 1981) and this appeared to be true for *A. lusitanicus* too, since they left the wildflower strips during the night in order to feed on young rape plants, obviously a more attractive food than the older plants in the wildflower strips. Since *Deroceras reticulatum*, the second commonest slug species in Belp, did not show a distribution pattern like *A. lusitanicus*, it is assumed that the large amount of damage up to two metres from the wildflower strips was caused primarily by *A. lusitanicus*.

In 1995, both species mentioned above, were more abundant at 1 m than at 3 to 5 m from the wildflower strips with higher densities of *D. reticulatum* in general. Thus, both species appeared to be responsible for the complete crop failure of rape in 1995 close to the wildflower strips. The complete crop failure in the centre of field F3 in 1995, independent of the wildflower strips, was apparently primarily caused by *D. reticulatum* because this species was by far the most abundant one there. *Arion fasciatus* also reached relatively high densities there from the fourth week onwards, so this species was also probably responsible for the enormous damage in the centre of field F3. This severe damage was most likely due to the direct-drilling of rape seeds into wheat stubble. Direct-drilling meant minimal soil disturbance and conservation of soil moisture. This treatment in 1995 provided optimal conditions for slugs resulting in much higher densities of *D. reticulatum* in rape in 1995 than in 1994. Similar observations were also made elsewhere (Mabbett, 1991; Kendall et al., 1995).

Since metaldehyde is known to be an effective molluscicide (e.g. Bailey & Wedgwood, 1991; Henderson et al., 1992; Ester & Nijenstein, 1996) the lower slug damage in parts of fields where metaldehyde was applied was not unexpected. However, it was important to know that complete crop failures along sown wildflower strips could be prevented by using 5% metaldehyde pellets at 8 kg / ha.

In contrast to rape, slug damage in wheat was much lower and crop losses never occurred. Although large numbers of slugs were recorded at bait stations in wildflower strip 5 in 1994, slug numbers at 1 m from the strip were not higher than at 3 or 5 m. This was the reason why slug damage in 1994 at 1 m from the strip was not higher than at other distances. The significantly higher damage in wheat in 1994 at 25 m than at 1 m from the strip was probably due to the relief of the ground surface which was slightly lower at 25 m than the rest of field F9. This area of the field provided favourable conditions for slugs in terms of soil humidity resulting in higher densities of *D. reticulatum, A. fasciatus and Deroceras laeve* there than in other parts of the field. Although at 1 m from wildflower strip 3 at the edge of field F10 in 1995, slugs were more abundant than at 3 and 5 m, slug damage was not significantly different. This could be due to the fact that *A. lusitanicus* was conspicouosly less abundant in late October and November, the time when wheat was susceptible to damage, than at the time when rape was susceptible. *A. lusitanicus* laid its eggs primarily in August and September. After oviposition most adults died. Therefore, dispersal of adult *A. lusitanicus* between wildflower strips and the adjacent wheat crop was very rare. Thus, *A. lusitanicus* was unable to cause high damage in wheat close to the wildflower strips.

This study showed the potential extent of slug damage to rape bordering on sown wildflower strips associated with high densities of *A. lusitanicus*. According to farmers and gardeners, this species has increased rapidly during the last decade. Therefore, increasingly severe problems with *A. lusitanicus* in vegetables and arable crops must be expected in the future. This may be especially true for rape because concentrations of glucosinolates, known to protect rape from slugs (Glen et al., 1990; Giamoustaris & Mithen, 1995), continue to decline in rape cultivars.

ACKNOWLEDGEMENTS

I am thankful to the farmer, M. Gygli, for providing his fields and to C. Högger for helpful comments.

REFERENCES

Bailey, S E R & Wedgwood, M A (1991) Complementary video and acoustic recordings of foraging by two pest species of slugs on non-toxic and molluscicidal baits. *Annals of applied Biology*. **119**, 163-176.

Ester, A & Nijenstein, H J (1996) Control of field slug (*Deroceras reticulatum* (Müller)) by seed-applied pesticides in perennial ryegrass assessed by laboratory tests. *Journal of Plan Diseases & Protection*. **103** (1), 42-49.

Frank, T & Nentwig, W (1995) Ground dwelling spiders (Araneae) in sown weed strips and adjacent fields. *Acta Oecologica*. **16** (2), 179-193.

Giamoustaris, A & Mithen, R (1995) The effect of modifying the glucosinolate content of leaves of oilseed rape (*Brassica napus ssp. oleifera*) on ist interaction with specialist and generalist pests. *Annals of applied Biology*. **126**, 347-363.

Glen, D M; Jones, H; Fieldsend, J K (1990) Damage to oilseed rape seedlings by the field slug *Deroceras reticulatum* in relation to glucosinolate concentration. *Annals of applied Biology*. **117**, 197-207.

Hamilton, P A & Wellington, W G (1981) The effects of food and density on the movement of *Arion ater* and *Ariolimax columbianus* (Pulmonata: Stylommatophora) between habitats. *Researches on Population Ecology*. **23**, 299-308.

Heitzmann, A (1994) Die Vegetationsdynamik in angesäten Ackerkrautstreifen in Abhängigkeit verschiedener Saatmischungen. *Zeitschrift für Pflanzenkrankheiten und Pflanzenschutz*. Sonderheft **XIV**, 75-83.

Henderson, I F; Martin, A P; Perry, J N (1992) Improving slug baits: the effects of some phagostimulants and molluscicides on ingestion by the slug, *Deroceras reticulatum* (Müller) (Pulmonata: Limacidae). *Annals of applied Biology*. **121**, 423-430.

Kendall; D A; Chinn, N E; Glen, D M; Wiltshire, C W; Winstone L; Tidboald, C (1995) Effects of soil management on cereal pests and their natural enemies. In: Ecology and Integrated Farming Systems, D M Glen, M P Greaves & H M Anderson (eds), Chichester: Wiley, pp. 83-102.

Lys, J A & Nentwig, W (1992) Augmentation of beneficial arthropods by strip-management. 4. Surface activity, movements and activity density of abundant carabid beetles in a cereal field. *Oecologia*. **92**, 373-382.

Mabbett, T (1991) Straw incorporation trials reveal arable slug damage will increase. *Agriculture International*. **43** (11), 304-306.

THE POTENTIAL FOR COMMON WEEDS TO REDUCE SLUG DAMAGE TO WINTER WHEAT

R T COOK*, S E R BAILEY, C R McCROHAN
University of Manchester, School of Biological Sciences, Oxford Road, Manchester M13 9PT, UK

* now at Kingston University, Dept of Life Sciences, Kingston upon Thames, Surrey KT1 2EE, UK

ABSTRACT

Lack of plant diversity in monocultures may exacerbate slug damage to crops simply because the crop is the only available food. Provision of an alternative food source during the early, vulnerable stages of a wheat crop might reduce levels of damage caused by slugs. Weeds are a readily available potential food source for slugs. This paper describes experiments which examine this possibility with the field slug, *Deroceras reticulatum*.

Slugs exhibited a range of preferences for the leaves of 12 weed species, and the presence of highly palatable weeds reduced the consumption of wheat leaves. In the field, damage to wheat seeds and seedlings was reduced in the presence of weeds, although they were not as effective as metaldehyde pellets over the short duration of each test. A laboratory study indicated that a slug's foraging behaviour can be affected by the nature of the food encountered first during a night. While weeds could provide some protection to a crop after metaldehyde pellets have ceased to be effective, the main practical difficulty is to ensure that palatable species of weeds dominate in a field.

INTRODUCTION

The field slug, *Deroceras reticulatum*, is a pest of winter wheat in the UK. The slugs eat the seeds and seedlings below ground, and graze flag leaves after emergence which reduces grain yield (Kemp & Newell, 1987). Chemical control, principally the application of poison bait pellets, is not always reliable, especially in wet conditions which accelerate the degradation of the pellets and facilitate recovery of slugs from poisoning. Furthermore, the pellets are toxic to non-target organisms. Studies into integrated approaches to pest slug control have identified cultural methods, such as soil cultivation, seed depth and soil compaction, as factors which can affect levels of slug damage (Glen *et al.*, 1989, 1990).

There has been little research into the effect of alternative food on crop losses caused by molluscan pests. Airey (1988) reported that the presence of lettuce leaves in pots containing potato tubers significantly reduced slug damage to the tubers. Raut & Ghose (1983) found that the presence of non-crop plants reduced damage by *Achatina fulica* to a range of crop plants. Damage is mostly caused by slugs already present in the soil rather than by large numbers immigrating into a field. This, and the limited mobility of slugs, requires the alternative food source to be distributed throughout the crop, rather than in strips between crop rows. Many pest species, including *D. reticulatum*, are known to feed on fresh, green plant material (Hunter, 1968), and exhibit distinct preferences between the leaves of different species (Duval, 1971; Mølgaard, 1986). Foraging usually occurs on or near the soil surface, penetrating deeper only when sufficiently large spaces exist between clods of earth (Stephenson, 1975). Common weeds could represent a readily available and inexpensive alternative food for slugs with a consequent reduction in damage to the buried wheat seeds.

This paper describes experiments which examine slug preferences for 12 species of common weed, the potential for the presence of palatable weeds to reduce slug damage to wheat, and the foraging behaviour of *D. reticulatum* in the presence of wheat seeds and weeds.

MATERIALS AND METHODS

Preferences for weeds

Field slugs (0.4g to 0.9g) were placed individually in clear plastic containers (14 x 7.5 x 5 cm) without food for 24 h before each experiment. The containers were stored in an incubator set at 15^0C with a 12 h photoperiod. The lids of the containers had a 2 cm diameter hole, covered with cotton gauze, to allow circulation of air. The base of each container was covered with a 1 cm deep layer of moist sterilised loam. Each slug was provided with a choice of three wheat leaves (cv Hereward at Zadok's stage 11; seeds from PBI, Cambridge) and three leaves of one of 12 weed species (seeds from John Chambers, Kettering). The weeds tested are listed Table 1. The wheat and weeds were all grown from seed in a glasshouse (John Innes No. 2 compost; temperature $15\text{-}20^0$C; 16 h photoperiod). The weed plants were one month old when used in the experiment.

The areas of the leaves were measured with an image analyser before and after presentation to the slugs. Each test lasted 72 h, with leaves being replaced every 24 h. Approximately equal areas of wheat and weed leaf were given to each slug, with wheat leaves alternating in a row with weeds. Fifteen replicates were carried out for each weed species, and any slug which ate neither weed nor wheat over the three nights was excluded from the analysis. The relative preference for each weed species was recorded as an acceptability index as follows:

$$\frac{\text{Area of weed leaf eaten}}{\text{Total area of weed and wheat eaten}}$$

The indices were arc-sine transformed to radians to correct for the tendency for proportional data to form binomial, rather than normal, distributions.

Field experiment: potential for weeds to reduce slug damage to wheat

During October and November 1994, six open-topped aluminium arenas (70 x 50 x 30 cm) were dug into the soil which had been ploughed and harrowed, but not rolled. The arenas had an electrified barrier to prevent the escape or entry of slugs over the top. Slugs already in the soil within an arena were removed daily by collecting any found under a plastic sheet covered with silver foil over a three day period. Wheat seeds (cv Hereward) were sown at a density of 325 m^{-2} and a depth of 5 mm to ensure that they were accessible to slugs. Four treatments were tested: wheat seeds only; wheat seeds and weeds; wheat seeds and metaldehyde pellets; wheat seeds, metaldehyde pellets and weeds. A fifth arena, the control, contained wheat seeds and weeds, but no slugs. Before introduction to an experimental arena, all slugs were kept without food for 24 h in the sixth arena to acclimatise them to the arena conditions. One month old weed plants were positioned in seven rows between the seed rows. One each of dandelion (*Taraxacum officinale*), fat hen (*Chenopodium album*) and red dead-nettle (*Lamium purpureum*) were placed randomly in each row. Thirty metaldehyde mini-pellets (Bio Slug Mini-Pellets, Pan Britannica Industries Ltd.) were scattered on to the soil surface in the appropriate arenas. Eight adult slugs were placed in each arena, except the control. After 72 h the wheat seeds were examined for damage. All treatments were repeated a total of seven times. This experiment was subsequently repeated using wheat seedlings at Zadok's growth stage 10.

Laboratory experiment: feeding behaviour of *D. reticulatum* in the presence of wheat seeds and weeds

A wooden arena (35 x 35 x 30 cm) was half-filled with moist sterilised loam, and kept at 15^{0}C and a minimum of 90% RH, with a 12 h photoperiod. The arena's inner walls were painted with copper paint which is repellant to slugs. Watch glasses, painted black, acted as refuges for slugs in each corner. Slugs were exposed to either 32 wheat seeds only, 32 wheat seeds and 12 dandelion leaf discs, or 32 wheat seeds and 12 groups of 3 chickweed leaves. Each treatment was carried out twenty times. Each food item was attached to a miniature crocodile clip connected to a miniature microphone. Vibrations caused by a slug rasping or biting on food were detected by the microphone, from which signals were fed into a computer programmed to identify the particular microphone and record biting activity. Any signal occurring within 1.5 sec of another was ignored to prevent more than one signal being recorded for each bite. A 'meal' was defined as a feeding phase separated by at least 10 min from another. The dandelion leaf discs were 1.8 cm in diameter. Chickweed leaves were much smaller, and 3 were attached to each probe. Leaf material was attached to a thin strip of acetate with a small amount of gelatine to provide extra rigidity.

A single slug (0.4g to 0.8 g) was placed in a similar arena without food for 24 h before each test to acclimatise the slug. It was then placed under a refuge in the test arena from 17:00 to 10:00.

RESULTS

Preferences for weeds

Distinct preferences for weed species were exhibited by the slugs (Anova: P <0.001), dandelion being the most palatable and groundsel the least (Fig. 1). Four species were eaten in significantly greater quantities than wheat leaves, while five were relatively unpalatable (Table 1). The presence of unpalatable weeds resulted in slugs consuming more wheat than

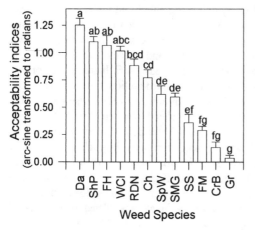

Fig. 1   Preferences for weed species. Shared letters = no significant differences. Ch=chickweed; CrB =creeping buttercup; Da=dandelion; FH=fat hen FM=field madder; Gr=groundsel; RDN=red dead-nettle; SS=sheep's sorrel; ShP=shepherd's purse; SMG=smooth meadow grass; SpW= field speedwell; WCl=wild clover. Bars=SE. N=12 to 15.

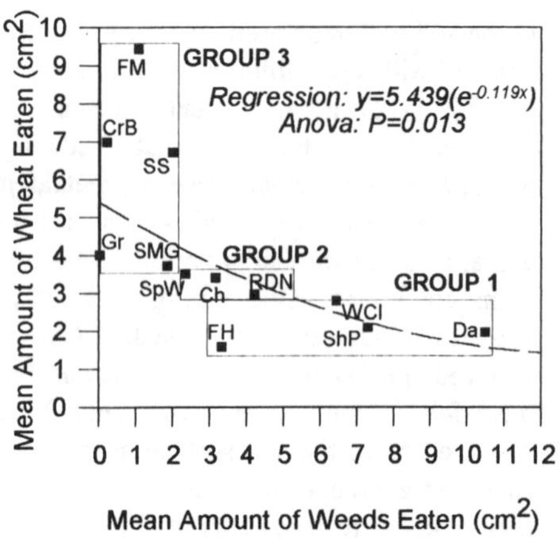

Fig. 2   Amount of wheat eaten *vs.* amount of weed eaten. (See Fig.1 for species, Table 1 for groups).

slugs which were offered more palatable weed species (Anova: P < 0.001. Fig. 2). Linear and non-linear regressions were carried out to establish the line of best fit for the data in Figure 2. The slope of the line and the distribution of weeds in Figure 2 suggest that slugs given less palatable weeds (groups 2 & 3) consumed less in total (wheat and weeds) compared to slugs given palatable species, the exceptions being field madder, sheep's sorrel and fat hen. If they were compensating fully for eating less weed by increasing wheat consumption, the line would have a slope of nearer -1.

Field experiment: potential for weeds to reduce slug damage to wheat

Undamaged seeds recovered increased from 33.1% ± 5.4 s.e. in the arenas with seeds and slugs to 55.1% ± 3.8 s.e. when weeds were present (Fig. 3a; t-test: P < 0.01). Damaged wheat seedlings declined from 39.8% ± 4.7 s.e. to 23.6% ± 3.2 s.e. in the presence of weeds (Fig. 3b; t-test: P = 0.017). However, the application of metaldehyde resulted in levels of damage no different from those in the controls over the 72 h duration of each test. The number of undamaged seeds recovered from the controls was less than 100% because it was not always possible to find all the seeds, and there were occasional signs of damage to seeds caused by unidentified organisms.

Feeding behaviour of *D. reticulatum* in the presence of wheat seeds and weeds

The presence of dandelion leaves significantly reduced the number of seeds damaged by the slugs compared to arenas with seeds only and seeds and chickweed (Fig. 4. Kruskall- Wallis

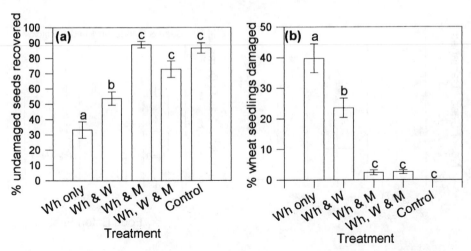

Fig. 3 Percentages of (a) wheat seeds undamaged and (b) wheat seedlings damaged in field expts (Wh = wheat; W = weeds; M = metaldehyde. Bars = SE. Shared letters indicate no significant difference)

test: P=0.002). The lower palatability of chickweed was reflected by the significantly fewer number of bites per meal (median 33, inter-quartile range (i-q) 27.5-53.5) and slower bite rate (6.6 bites per minute (bpm), i-q 5.9-8.2) compared with the other two food types (665 bites (i-q 365-862) and 14 bpm (i-q 12.4-16.0) for wheat seeds; 355 bites (i-q 263-476) and 17 bpm (i-q 13.7-17.9) for dandelion. Kruskall-Wallis: P<0.001 for both bites per meal and bite rate). The slugs took a median of 4 meals per night (i-q 3.0 - 6.0). Most slugs ate the first item encountered, although the size of the first meal on chickweed was only 58 bites (i-q 23-81), compared with 1391 (i-q 1273-1883) for wheat seeds and 534 (i-q 432-691) for dandelion. Figure 5 shows that the nature of the food first eaten affected the slugs' subsequent behaviour. When a dandelion disc was eaten first, the number of wheat seeds

Table 1. The preferences of common weed leaves compared to winter wheat leaves (Paired t-tests: ★ = 0.05>P>0.001; ★★ = P < .001)

Preference Group	Species	Significance
GROUP 1 (preferred to wheat)	Dandelion (*Taraxacum officinale*)	★★
	Shepherd's Purse (*Capsella bursa-pastoris*)	★★
	Wild Clover (*Trifolium repens*)	★★
	Fat Hen (*Chenopodium album*)	★
GROUP 2 (no preference)	Red Dead-Nettle (*Lamium purpureum*)	-
	Field Speedwell (*Veronica agrestis*)	-
	Chickweed (*Stellaria media*)	-
GROUP 3 (wheat preferred)	Smooth Meadow Grass (*Poa pratensis*)	★★
	Sheep' Sorrel (*Rumex acetosella*)	★★
	Field Madder (*Sherardia arvensis*)	★★
	Creeping Buttercup (*Ranunculus repens*)	★★
	Groundsel (*Senecio vulgaris*)	★★

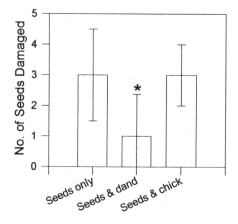

Fig. 4 Median number of seeds damaged per slug per night in lab arenas. Dand=dandelion; Chick=chickweed. Bars=inter-quartile (iq) range. *=significant difference.

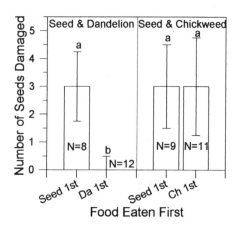

Fig. 5 Median number of seeds damaged per slug per night according to which food eaten 1st. 'X' axis= food 1st eaten Bars=iq range. Shared letters indicate no significant difference.

eaten in the night was significantly reduced compared to when a seed was eaten first (Mann-Whitney test: P<0.001). This was not evident when chickweed was eaten first. Figure 6 shows that, after a slug had eaten a wheat seed first in arenas containing seeds and dandelion (Fig. 6a), it subsequently took a greater number of bites from seeds than dandelion. However, after dandelion had been eaten first, the situation was reversed. There was no such reversal in arenas with seeds and chickweed, where seeds were always eaten in greater quantities (Fig. 6b).

DISCUSSION

Preferences for weeds

The ability of slugs to discriminate between the leaves of different plants is well known (Dirzo, 1980; Duval, 1971). All the plants tested in this paper, except for field madder and meadow grass, had soft leaves which slugs generally find acceptable unless they contain deterrent secondary compounds (Dirzo, 1980; Mølgaard, 1986). Grasses are generally unacceptable to slugs, probably because of the presence of silica bodies (phytoliths) in the leaves (Dirzo, 1980).

The tendency for slugs given the less palatable weeds to consume less in total may be because they were unable to fully compensate for the reduced intake of weeds by increasing wheat leaf consumption. Alternatively, consumption of species containing deterrent compounds might reduce the motivational state of the slugs to feed. The relatively large amounts of wheat eaten by slugs given field madder may therefore be a result of leaf toughness, rather than secondary chemicals, deterring slugs from eating this weed.

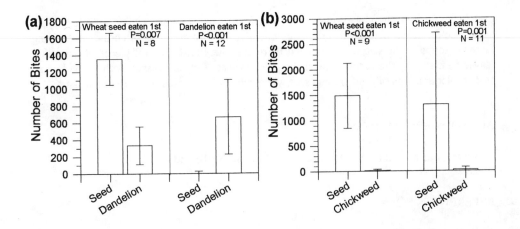

Fig. 6 Median number of bites per slug on (a) wheat seeds & dandelion, and (b) wheat seeds & chickweed in a night according to which item was eaten 1st (1st meal excluded). Bars=iq range.

Nevertheless, slugs given less palatable weeds ate significantly more wheat than slugs given palatable species.

Field experiment: potential for weeds to reduce slug damage to wheat

The weeds provided a degree of protection to the wheat, albeit to a limited extent with the species and densities used. The experiment was designed to test the relative effectiveness of the weeds when metaldehyde pellets were at their most potent. After a few days poison pellets tend to degrade, especially in wet weather which also facilitates the recovery of slugs from poisoning (Bourne *et al.*, 1990; Kemp & Newell, 1985). The presence of palatable weeds after the poison pellets have ceased to be effective could provide a degree of protection comparable with that shown in Fig. 1(b), where damage to wheat seedlings was almost halved in arenas with weeds compared with arenas with wheat only.

Feeding behaviour of *D. reticulatum* in the presence of wheat seeds and weeds

The importance of the presence of palatable weeds is highlighted by this study; damage to wheat seeds was significantly reduced in the presence of dandelion, but the less palatable chickweed had no effect.

The slugs exhibited a capacity for short-term learning. Early consumption of dandelion resulted in slugs taking very few bites from wheat seeds, with most seeds being ignored. When dandelion was not eaten, or eaten later, the slugs were more likely to eat a wheat seed, and generally consumed more wheat in a night. A meal on chickweed at any time in a night had no effect on wheat seed consumption. The data suggest that young dandelion leaves are preferred to wheat seeds and that, when a slug is aware of a highly palatable food source, it tends to ignore seeds in favour of that palatable food. The number of bites per meal on dandelion was lower than on wheat seeds, which appears to contradict the apparent preference for dandelion. However, after a meal on dandelion, slugs often returned to the same disc for subsequent meals and, consequently, the disc area was often the determining factor in terms of the size of a meal. Furthermore, slugs often bite off pieces of dandelion leaves by clamping the radula against the jaw. Wheat seeds are rasped with the radula so that smaller amounts of food are ingested with each rasp compared to a bite of a leaf.

ACKNOWLEDGEMENTS

This work was funded by BBSRC grant PG34/800 to SERB and CRM.

REFERENCES

Airey, W J (1988) The influence of alternative food on tuber damage by slugs. *Journal of Molluscan Studies.* **54**, 131-138.

Bourne, N B; Jones G W; Bowen I D (1990). Feeding behaviour and mortality of the slug *Deroceras reticulatum* in relation to control with molluscicidal baits containing various combinations of metaldehyde and methiocarb. *Annals of Applied Biology.* **117**, 455-468.

Dirzo, R (1980) Experimental studies on slug-plant interactions I. The acceptability of thirty plant species to the slug *Agriolimax caruanae*. *Journal of Ecology.* **68**, 981-998.

Duval, D M (1971) A note on the acceptability of various weeds as food for *Agriolimax reticulatus* (Müller). *Journal of Conchology.* **27**, 249-251.

Glen, D M; Milsom N F; Wiltshire C W (1989) Effects of seed bed conditions on slug numbers and damage to winter wheat in a clay soil. *Annals of Applied Biology.* **115**, 197-207.

Glen, D M; Milsom N F; Wiltshire C W (1990) Effect of seed depth on slug damage to winter wheat. *Annals of Applied Biology.* **117**, 693-701.

Hunter, P J (1968) Studies on slugs of arable ground III. Feeding habits. *Malacologia.* **6**, 391-399.

Kemp, N J; Newell P F (1985) Laboratory observations on the effectiveness of methiocarb and metaldehyde against the slug *Deroceras reticulatum* (Müller). *Journal of Molluscan Studies.* **51**, 228-230.

Kemp, N J; Newell P F (1987) Slug damage to the flag leaves of winter wheat. *Journal of Molluscan Studies.* **53**. 109-111.

Mølgaard, P (1986) Food plant preferences by slugs and snails: a simple method to evaluate the relative palatability of food plants. *Biochemical Systematics & Ecology.* **14**, 113-121.

Raut, S K; Ghose K C (1983) The rôle of non-crop plants in the protection of crop plants against the pestiferous snail *Achatina fulica*. *Malacological Review.* **16**, 95-96.

Stephenson, J W (1975) Laboratory observations on the effect of soil compaction on slug damage to winter wheat. *Plant Pathology.* **24**, 9-11.

PROTEOLYTIC ENZYMES PRESENT IN THE DIGESTIVE SYSTEM OF THE FIELD SLUG, DEROCERAS RETICULATUM, AS A TARGET FOR NOVEL METHODS OF CONTROL.

A J WALKER, D M GLEN, P R SHEWRY
IACR - Long Ashton Research Station, Department of Agricultural Sciences, University of Bristol, Long Ashton, Bristol, BS18 9AF, UK

ABSTRACT

Proteinase activities present in the crop, digestive gland and salivary glands of the field slug, *Deroceras reticulatum*, were investigated. The digestive gland was found to be responsible for approximately 80% of the total proteolytic activity. This activity, detectable by the maximal hydrolysis of the synthetic substrate Z-Phe-Arg-*p*NA at pH 6.0, was almost totally inhibited by the cysteine proteinase inhibitors E-64, CEW cystatin and leupeptin, and was activated by thiol compounds. Purification of the major cysteine proteinase was achieved using cation-exchange chromatography. Feeding juvenile slugs E-64 over a 35-day period resulted in a significant reduction in slug weight gain. This suggests that the expression of cysteine proteinase inhibitors in transgenic crop plants may have beneficial effects in reducing the growth rates of slug populations.

INTRODUCTION

The economic importance of slugs as pest species in horticultural and agricultural crops is recognised world-wide and the extent of slug damage has become increasingly apparent over the past 30 years. The development of an effective and specific method of slug control has great economic and environmental importance since currently used molluscicides often fail to protect against slug damage and are also toxic to non-target organisms (Bieri *et al.*, 1989; Kennedy, 1990; South, 1992).

An attractive alternative to molluscicide usage is the inhibition of slug digestive proteinases with plant-derived proteinase inhibitors expressed in transgenic crop plants. These inhibitors, which occur in storage organs, reproductive organs and vegetative cells of most plant families (Garcia-Olmedo *et al.*, 1987; Shewry and Lucas, in press) have been extensively researched and are considered to be involved in plant defence against pests or pathogens. Inhibition of proteolysis by the gut extracts of many phytophagous insects has been demonstrated *in vitro* and some studies have shown that these inhibitors act as anti-metabolites when presented to insects in whole plants or artificial diets (see, for example, Hilder *et al.*, 1987; Boulter, 1993; Burgess *et al.*, 1994).

As a basis for developing transgenic strategies using proteinase inhibitor genes for the control of slug species, it is necessary to understand more about the basic biology of slug digestion. Our investigations have therefore focused on the crop, digestive gland and salivary glands of the field slug, *Deroceras reticulatum*, in order to characterise the proteolytic enzymes in these organs and identify inhibitors of the major activities present.

MATERIALS AND METHODS

Materials

NAP-5 desalting and HiTrap SP cation exchange prepacked columns were purchased from Pharmacia Biotech (Uppsala, Sweden). Ribulose bisphosphatecarboxylase/oxygenase (Rubisco), and azocasein were purchased from Sigma-Aldrich Co Ltd (Poole, UK) and Carbobenzoxy-L-Phenylalanyl-L-Aginine-*p*-Nitroanalide (Z-Phe-Arg-*p*NA) from BACHEM Ltd (Saffron Waldon, UK). Trans-epoxysuccinyl-leucylamido-(4-guanidino)butane (E-64) and Leupeptin were from Peptide Institute (Osaka, Japan), ditiothreitol (DTT) was from Melford Laboratories Ltd (Ipswich, UK). Pepstatin, CEW (chicken egg white) cystatin, phenylmethanesulphonyl flouride (PMSF), soybean and lima bean trypsin inhibitors (SBTI and LBTI) and ethylenediaminetetraacetic acid (EDTA), 2-(N-morpholino)ethanesulphonic acid (MES), 1,3-bis(tris[Hydroxymethyl]methylamino)propane (Bis-Tris Propane), 3-(cyclohexylamino)-1-propanesulphonic acid (CAPS), cysteine, sodium dodecyl sulphate (SDS), trichloroacetic acid (TCA), 2,4,6-trinitrobenzsulphonic acid (TNBS), dimethyl sulphoxide (DMSO), Brij 35 and sodium tetraborate were obtained from Sigma.

Enzyme preparations

Slugs were trapped and collected from various arable sites at IACR-Long Ashton Research Station using upturned flower pot saucers baited with bran. They were then placed in ventilated plastic boxes lined with moist absorbent cotton wool and kept at 10°C. The crop, digestive glands and salivary glands were carefully dissected out on ice, frozen in liquid nitrogen and stored at -20°C until required. The thawed tissues were homogenised at 5°C in double-distilled water, centrifuged at 20,000g for 15 min (5°C) and the aqueous supernatant was frozen and stored at -20°C.

Enzyme Assay Methods

General proteinase activity was assessed using Rubisco as the substrate protein. 10 μl of crop, digestive gland or salivary gland extract was mixed with 70 μl of buffer (citrate phosphate, citrate phosphate+MES, MES, MES+Bis-Tris Propane, Bis-Tris Propane and CAPS for pH 3.0-5.0, 5.5, 6.0, 6.5, 7.0-9.5 and 10-10.5 respectively; final buffer concentration 50mM) and 10 μl Rubisco (1% (w/v) in 0.06% (w/v) SDS) and incubated at 25°C for 1 h. The reaction was then stopped by adding 30 μl of 10% (w/v) ice-cold TCA and the tubes were placed on ice for 30 min before centrifuging at 8,000g for 4 min to remove precipitated protein. The supernatant (90 μl) was then added to 300 μl of TNBS reagent (one volume of 0.3% (w/v) TNBS in water mixed with nine volumes of 4% (w/v) sodium tetraborate in 0.15M NaOH (pH 9.9)), and the mixture was incubated in the dark for 1 h at 25°C prior to the addition of 150 μl of 0.5M HCl. The absorbance of the resulting solution was then read at 340 nm against both reagent and enzyme controls and the activities calculated using a standard curve of L-leucine in MES buffer.

The effect of inhibitors (pepstatin (5μg ml^{-1}), CEW cystatin (20μM), E-64 (20μM), PMSF (5mM), SBTI (1mg ml^{-1}), LBTI (200μg ml^{-1}), EDTA (10mM), Leupeptin (100μM)) and activators (cysteine (1mM) and DTT (1mM)) on digestive gland proteolytic activity against azocasein was also determined. 10 μl of digestive gland extract was mixed with 20 μl of 100mM citrate-phosphate buffer (pH 5.0) containing the inhibitor/activator and pre-incubated for 20 mins at 25°C. 10 μl of azocasein solution (1% (w/v) in 0.06% (w/v) SDS) was then added to the tubes and the mixture was incubated at 25°C for 1 h before termination with 30 μl of ice-cold TCA (10% (w/v)). Following precipitation on ice, the tubes were centrifuged at 8,000g for 4 min after which 60 μl of the supernatant was withdrawn and added to a well of a mirotitre plate containing 40 μl of 1M NaOH. The absorbance of the resulting solution was then read at 405 nm in a Dynatech microtitre plate reader together with those for reagent and enzyme controls. Azocasein hydrolysis by column fractions was determined without activators or inhibitors for 3 h.

Digestive gland cysteine proteinase activity was measured from pH 3.0-7.5 using the synthetic substrate Z-Phe-Arg-pNA. 2.5 μl of crude extract was pre-incubated for 20 min at 25°C in 235 μl buffer (see above) containing cysteine, EDTA and Brij 35 (final concentrations 5mM, 1mM and 0.1 % (w/v) respectively). The reaction was then initiated by adding 12.5 μl Z-Phe-Arg-pNA (10mM stock in DMSO) and the change in absorbance at 405 nm followed for 30 min at 25°C in the microtitre plate reader; absorbancies were also measured for reagent and enzyme controls. Proteolytic activity of column fractions (see below) was analysed at pH 6.0 for 3 h.

Fractionation on SP Sepharose High Performance

Digestive gland extract (1ml, previously desalted on a NAP-5 column) was chromatographed on a Pharmacia HiTrap SP 1ml cation exchange column equilibrated in 20 mM citrate phosphate, 5mM cysteine, 1mM EDTA (pH 5.0) at a flow rate of 1ml min^{-1}. The column was washed for 5 mins before applying a linear gradient of 0-500mM NaCl at 1ml min^{-1} for 15 min; 500 μl fractions were collected and assayed with the substrates azocasein and Z-Phe-Arg-pNA (see above).

In vivo effects of inhibitors

The *in vivo* effects of the cysteine proteinase inhibitor E-64 on juvenile slugs (initial weight 2.5 - 4.5 mg, 30 per treatment) was determined over a 35-day feeding period at 19°C. Eggs laid by field-collected slugs were incubated at 19°C in Petri dishes lined with moist filter paper until the juvenile slugs emerged. Individual slugs were then removed and placed in 500 μl microfuge tubes (pin-holed for gaseous exchange) together with a piece of moistened filter paper (1mm^2). E-64 was administered to the juvenile slugs at 0, 0.15, 0.30 and 0.45 % (v/v) in a Ready Brek (Weetabix Ltd, Kettering, England) diet (3ml distilled water g^{-1}) by spreading the diet onto a mesh wedged into the lid of the microfuge tube. Diets were changed every three days and the amount of diet eaten was estimated using the mesh grid. Slug mortality was also recorded. Slug weight gain was determined every nine days.

RESULTS

Preliminary experiments showed that slugs which were administered antibiotics in an artificial diet (to reduce the total and proteolytic bacterial populations in the crop, digestive gland and salivary gland) had similar total proteinase activities and pH profiles to control slugs fed artificial diet without antibiotic. There did not, therefore, appear to be any major proteinases produced by the bacterial microflora of these regions (data not presented).

The proteolytic activity against Rubisco was determined for crop, digestive gland and salivary gland extracts of *D. reticulatum* (expressed as activity per slug, based on the mean weights of these organs) (Fig. 1). The pH optimum of the digestive gland (5.0) was much lower than for the crop and salivary glands, which shared a similar pH optimum (7.5). Furthermore, proteolysis occurred at pH values corresponding to the physiological pH of these organs, as determined using ion-selective microelectrodes (5.99, 6.52 and 6.94 for the crop, digestive gland and salivary gland respectively) (Walker *et al.*, 1996). Analysis of variance revealed significant differences between the proteolytic activities of the digestive gland, crop and salivary glands ($P<0.001$). Whereas the digestive gland is responsible for approximately 82% of the proteolytic activity within the slug gut, the crop (13%) and salivary glands (5%) are less important. Further studies were therefore focused on characterising activity in the digestive gland.

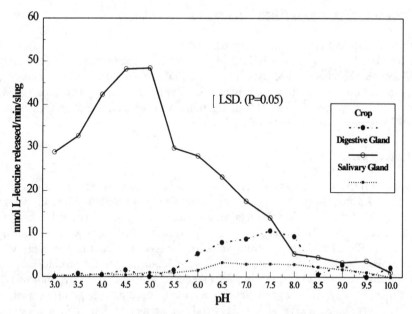

Fig. 1. The proteolytic activities of crop, digestive gland and salivary gland extracts of *D. reticulatum*. LSD = least significant difference, $P = 0.05$, 90df.; LSD = SED x t_{90df}, $P=0.05$.

Maximum hydrolysis of the synthetic substrate Z-Phe-Arg-pNA by digestive gland extract occurred at pH 5.5-6.0. Furthermore, proteolysis compared to controls (azocasein as substrate) was significantly reduced ($P<0.001$) to 3% with E-64 and 20% with CEW cystatin, which are inhibitors specific for cysteine proteinases (Fig. 2). The multicatalytic inhibitor leupeptin, which inhibits both cysteine and serine proteinases, had a similar effect on proteolysis since activity was reduced to 7% of the controls. Inhibitors specific for aspartic acid (pepstatin), metallo (EDTA) and serine proteinases (PMSF, LBTI, SBTI) had no effect on proteolysis against azocasein. Finally, the reducing agents used for activating cysteine proteinases (cysteine and DTT) significantly increased proteolysis when added to the incubation mixture ($P<0.001$, Fig. 2).

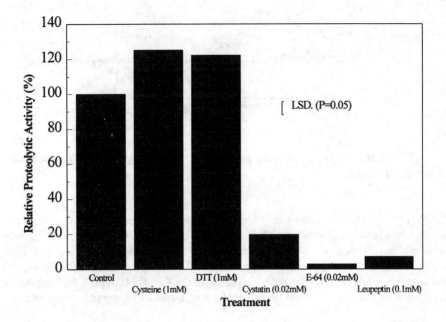

Fig.2. The effects of inhibitors and activators on digestive gland proteolytic activity.

Separation of *D. reticulatum* digestive gland extract on SP Sepharose high performance revealed a major peak of proteolytic activity against both azocasein and Z-Phe-Arg-pNA at a salt concentration of aproximately 0.13M NaCl (Fig. 3). The corresponding fractions contained a polypeptide (approximately 40kDa) when analysed by SDS-PAGE under reducing conditions and we are currently determining the catalytic properties of this protease.

The *in vivo* effects of E-64 on slug growth, feeding and mortality of juvenile slugs were assessed over a 35-day feeding period. This inhibitor, when administered at 0.15%, 0.30%

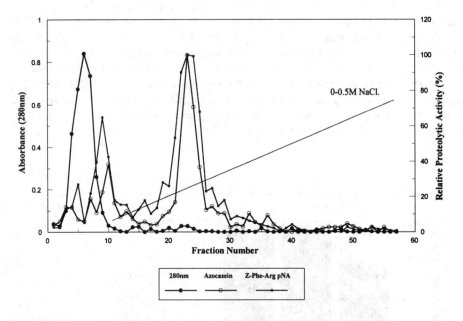

Fig. 3. SP Sepharose high performance ion-exchange chromatography of *D. reticulatum* digestive gland extract.

and 0.45% (v/v), caused a significant decrease in growth rate compared to the untreated, control group (Table 1). Moreover, slugs fed diet containing 0.30% or 0.45% (v/v) E-64 gained less weight than slugs fed on the lower concentration (0.15% (v/v)) of E-64 (P<0.05, Table 1). The weight gain of slugs fed on diet containing 0.15% and 0.30% (v/v) E-64 was 40% and 50% respectively less than that of control slugs. No significant differences were observed between treatments for slug mortality or amount of diet eaten.

Table 1. The effects of E-64 on *D. reticulatum* mean weight gain (%) over a 35 day feeding period. Each group contained 30 replicates.

	Inhibitor (E-64) concentration (v/v)			SED
0	0.15	0.30	0.45	(394df)
58.7	35.5	24.9	25.3	4.1

DISCUSSION

The decreases in proteolytic activity following pre-incubation of crude digestive gland extract with E-64, CEW cystatin and Leupeptin, the enhancement of activity by thiol reagents and the lack of inhibition by inhibitors diagnostic for other mechanastic classes of proteinase, suggest the dominance of cysteine proteinase activity in the digestive gland of *D. reticulatum*. Incubation of these extracts with the cysteine proteinase specific substrate Z-Phe-Arg-*p*NA has provided further evidence that this class of enzyme is responsible for the major proteinase activity in the gut. Separation of digestive gland extract by SP Sepharose high performance and subsequent analysis by SDS-PAGE indicates that this activity is due to one major protein.

While the slightly acid pH optima for hydrolysis of Rubisco (pH 5.0) and Z-Phe-Arg-*p*NA (pH 5.5-6.0) by *D. reticulatum* digestive gland extracts are typical for cysteine proteinases and are close to the physiological pH of these organs, they are far removed from the optimum (pH 1.7) reported by Johnston (1989) for proteolysis of haemoglobin by whole gut extracts of *D. reticulatum*. Our data are in agreement with a study carried out by Walker (1969) who reported proteolysis of gelatin by gut extract to occur between pH 4.0-6.5.

In our study where slugs were fed on artificial diet, ingestion of E-64 reduced juvenile slug weight gain but did not significantly affect slug mortality or reduce feeding. Even at the lowest concentration used (0.15% v/v), juvenile slug weight gain was reduced by an average of 40% over a 35-day feeding period. While the immediate effects of this reduction in weight gain may appear unimportant, the long term chronic effects of inhibitor ingestion may be more significant. Assuming that the observed effects would continue and death was not premature due to reduced amino acid availability, slugs would consume less food material due to their smaller size. They would also take much longer to mature and, since *D. reticulatum* is estimated to lay up to 500 eggs in a year (Carrick, 1938), this would have large implications for the rate of increase of slug populations. Furthermore, egg quality may be compromised. Finally, while there was no evidence that juvenile slugs avoided feeding on diet containing inhibitors (no significant differences in the amount of diet eaten between treatments) presentation of an alternative food material may result in differential feeding on diet containing inhibitors.

Studies with insects, using artificial diets and transgenic plants, suggest that the results of *in vitro* assays of the relative effects of inhibitors on gut enzymes are consistent with effects on insect development and survival *in vivo* (Gatehouse *et al.*, 1993). We have found a similar relationship, and can now tentatively exploit this by screening for inhibitors effective against slug gut proteolysis *in vitro*. Ultimately, however, the potential for the use of proteinase inhibitors for the control of slug species must be studied *in vivo* with transgenic plants expressing cysteine proteinase inhibitors. Any altered life-history or feeding parameters, resulting from chronic inhibitor ingestion, may then be studied in a model environment where alternative food is available to enable realistic interactions to be assessed and to evaluate the feasibility of various possible strategies for using cysteine proteinase inhibitors as the basis for novel methods of slug control.

ACKNOWLEDGEMENTS

We thank the Biotechnology and Biological Sciences Research Council (BBSRC) of the United Kingdon for a PhD studentship (to AJW). IACR receives grant-aided support from BBSRC.

REFERENCES

Bieri, M; Schweizer, H; Christensen, K; Daniel, O (1989). The effect of metaldehyde and methiocarb slug pellets on *Lumbricus terrestris* Linne. In: *Slugs and Snails in World Agriculture*. I Henderson (ed.). BCPC Monograph No.41, pp 237-244.

Boulter, D (1993). Insect pest control by copying nature using genetically engineered crops. *Phytochemistry*, **34**, 1453-1466.

Burgess, E P J; Main, C A; Stevens, P S; Christeller, J T; Gatehouse, A M R; Laing, W A (1994). Effects of protease inhibitor concentration and combinations on the survival, growth and gut enzyme activities of the black field cricket, *Teleogryllus commodus*. *Journal of Insect Physiology*, **40**, 803-811.

Carrick, R (1938). The life history and development of *Agriolimax agrestis* L. the grey field slug. *Transactions of the Royal Society of Edinburgh*, **59**, 563-579.

Gatehouse, A M R; Shi, Y; Powell, K S; Brough C; Hilder, V A; Hamilton W D O; Newall, C A; Merryweather, A; Boulter, D; Gatehouse, J A (1993). Approaches to insect resistance using transgenic plants. *Philosophical Transactions of the Royal Society of London B*, **342**, 279-286.

Garcia-Olmedo, F; Salcedo, G; Sanchez-Monge, R; Gomel, L; Royo, J; Carbonero, P (1987). Plant proteinacious inhibitors of proteinases and α-amylases. In: *Oxford Surveys of Plant Molecular and Cell Biology*. B Miflin (ed.), Oxford University Press, Oxford, pp 275-334.

Hilder, V A; Gatehouse, A M R; Sheerman, S E; Barker, R F; Boulter D (1987). A novel mechanism of insect resistance engineered into tobacco. *Nature*, **300**, 160-163.

Johnston, K A (1989). *Biochemical mechanisms of resistance of potato cultivars to slug attack*. PhD Thesis, University of Newcastle Upon Tyne.

Kennedy, P J (1990). *The effects of molluscicides on the abundance and distribution of ground beetles (Coleoptera, Carabidae) and other invertebrates*. PhD Thesis, University of Bristol.

Shewry, P R; Lucas, J A (In Press). Plant proteins that confer resistance to pests and pathogens. *Advances in Botanical Research*.

South, A (1992). *Terrestrial Slugs. Biology Ecology and Control*. Chapman and Hall, London, 428pp.

Walker, A J; Miller, A J; Glen, D M; Shewry, P R (1996). Determination of pH in the digestive system of the slug *Deroceras reticulatum* (Müller) using ion-selective microelectrodes. *Journal of Molluscan Studies*, **62**, 387-390.

Walker, G (1969). *Studies on digestion of the slug Agriolimax reticulatus* (Müller) (Mollusca, Pulmonata, Limacidae). PhD Thesis, University of Wales.

Poster Session

Session Organiser   Dr K A Evans

ELECTROPHORETIC COMPARISON OF *DEROCERAS RODNAE* AND *D. JURANUM* (PULMONATA, AGRIOLIMACIDAE)

H REISE

Museum of Natural History Görlitz, PF 300154, D-02806 Görlitz, Germany

ABSTRACT

The terrestrial slug *Deroceras juranum* Wüthrich, 1993 (Pulmonata, Agriolimacidae) was described as an endemic species of the Swiss Jura Mountains. The description was mainly based on its blackish-violet body colour. Previous breeding experiments and comparative studies of the genital anatomy and mating behaviour indicated that *D. juranum* is a colour morph of *D. rodnae* Grossu & Lupu, 1965. Twelve specimens of *D. juranum*, eighteen *D. rodnae*, three *D. reticulatum* (O.F. Müller, 1774), one *D. laeve* (O.F. Müller, 1774) and one *D. panormitanum* (Lessona & Pollonera, 1882) were electrophoretically (PAGE) assayed for twelve enzymes, yielding information about sixteen putative loci. Whereas *D. reticulatum*, *D. laeve* and *D. panormitanum* are clearly separated from *D. rodnae* and *D. juranum* the latter two are genetically extremely similar and cluster together. These results support the previous comparative studies. It is concluded that *D. juranum* and *D. rodnae* are most probably conspecific.

INTRODUCTION

The terrestrial slug *Deroceras juranum* was described by Wüthrich (1993) as an endemic species living in a restricted area of the Swiss Jura (NW Switzerland) at altitudes above 900 m. The species description is based on its striking blackish-violet body colour which differs from most other whitish, cream or brownish coloured agriolimacids in Europe. Before it was recognized as a distinct species, *D. juranum* was considered to be a dark form of *D. agreste, D. rodnae* or *D. reticulatum* (Mermod, 1930, Dufour-Humblet, 1982, Forcard see Wüthrich, 1993). The most similar species is the cream-coloured *D. rodnae* Grossu & Lupu, 1965, which covers a much larger area than *D. juranum* and occurs syntopically. Intermediate colour morphs do not occur.

A morphological comparison of cream-coloured *rodnae*-like and violet slugs from the Swiss Jura and *D. rodnae* from different other areas revealed no differences. Alleged slight differences in the penis morphology mentioned in the species description range within the intrapopulational variability; the penis of violet specimens is merely more or less pigmented. In addition, there seems to be no difference in the mating behaviour (Reise, in press). The latter is remarkable as mating behaviour can allow us to distinguish between even very similar *Deroceras* species (Reise, 1995). Breeding experiments have shown that *D. juranum* can also be cream-coloured like typical *D. rodnae* and suggest strongly that it is a colour morph of the latter species. The body colour seems to be genetically determined by a single locus with two alleles of which violet is dominant over cream (Reise, in press).

In order to further investigate the species status of *D. juranum* a comparative allozyme study of *D. rodnae* and *D. juranum* was conducted. The results of the study are presented in this paper.

MATERIAL AND METHODS

Eighteen *D. rodnae* from four localities in Poland, Czech Republic and Germany and twelve *D. juranum* from four localities in the Swiss Jura mountains (Switzerland) were studied. Moreover, three *D. reticulatum*, one *D. laeve* and one *D. panormitanum* were included as outgroups. The slugs were collected at the localities listed in table 1. However, some specimens are offspring of slugs collected at the given localities. The *D. juranum* specimens from site no. 2 (table 1) included one cream-coloured and one violet individual, both from the same egg clutch. All other *D. juranum* used for the study were violet.

Table 1: Tested specimens. N: number of specimens. For further descriptions of sampling localities of *D. juranum* see Reise (in press).

sample group	N	species/colour form	locality
RET	3	*D. reticulatum*	Görlitz, Saxonia (Germany)
LAE	1	*D. laeve*	near Neuquitzenow, Mecklenburg (Germany)
PAN	1	*D. panormitanum*	Görlitz, Saxonia (Germany)
D. rodnae			
P	4	*D. rodnae* (cream)	Babia Góra (S Poland)
C	8	*D. rodnae* (cream)	Lusicky hory (Czech Republic)
G	3	*D. rodnae* (cream)	near Tübingen, Baden-Württemberg (Germany)
G	3	*D. rodnae* (cream)	near Auma, Thüringen (Germany)
D. juranum			Swiss Jura mountains (Switzerland), May 1994
J2	1	*D. juranum* (cream)	site no. 2 (Hasenmatt)
J2	1	*D. juranum* (violet)	site no. 2 (Hasenmatt)
J3	4	*D. juranum* (violet)	site no. 3 (Oberschwang, North of Hasenmatt)
J4	6	*D. juranum* (violet)	site no. 4 (Chasseral, Combe Grède)
J5	4	*D. juranum* (violet)	site no. 5 (Balmberg, near Röti)

Homogenates of the digestive gland of the specimens were assayed for 12 enzymes by polyacrylamide gel electrophoresis (PAGE). The enzymes are listed in table 2. Sample preparation and electrophoretic procedures of PAGE followed in general Backeljau (1987, 1989). The electrophoretic buffer systems were TG (Tris/HCl, pH 9.0 for the gel and Tris/Glycine, pH 9.0 for the electrode), TC (Tris/Citric acid, pH 8.0 for both the gel and the electrode) or TBE (Tris/EDTA/borate, pH 8.9 for both the gel and the electrode), respectively. Staining procedures were adapted from Richardson et al. (1986). For Peptidases, the substrates Gly-Leu, Leu-Leu-Leu and Leu-Ala were used. Four peptidase loci were distinguished: locus 1 appeared with the substrate Leu-Ala only, locus 2 with Leu-Ala and with Leu-Leu-Leu, loci 3 and 4 with Gly-Leu. Only two esterase loci were interpretable.

Table 2: Enzymes studied. Abbr.: abbreviated enzyme name, EC-code: Enzyme Commission number, Buffer: employed electrophoretic buffer system, N: number of putative loci.

Name	Abbr.	EC-code	Buffer	N
Aspartate aminotransferase	AAT	2.6.1.1.	TG	1
Dihydrolipoamide dehydrogenase	DIA	1.8.1.4.	TC	1
Esterases	EST	3.1.1.1.	TC	2
Glucose-6-phosphate isomerase	GPI	5.3.1.9.	TC	1
Glycerol-3-phosphate dehydrogenase	G3PDH	1.1.1.8.	TBE	1
3-Hydroxybutyrate dehydrogenase	HBDH	1.1.1.30.	TC	1
Isocitrate dehydrogenase	IDH	1.1.1.42.	TC	1
Malate dehydrogenase	MDH	1.1.1.37.	TBE	1
Peptidases	PEP	3.4.11/13.	TG	4
Phosphogluconate dehydrogenase	PGDH	1.1.1.44.	TC	1
Phosphoglucomutase	PGM	5.4.2.2.	TBE	1
Superoxide dismutase	SOD	1.15.1.1.	TBE	1

The electromorphs at each locus were named in alphabetical order with A for the slowest (most cathodal) one. The data were analysed using the BIOSYS-1 program of Swofford & Selander (1981). However, for some specimens there were insufficient data, therefore some analyses were run on subgroups only. The specimens of *D. rodnae* from the two German populations were pooled. Nei's (1978) distances were used to construct a UPGMA dendrogram.

RESULTS

The 12 enzymes studied yielded information about 16 putative loci. The results are given in table 3. There seems to be considerable polymorphism within the genus *Deroceras*. All loci were polymorphic among and/or within species, except PGDH which was fixed for the same electromorphs in all populations.

A number of enzymes separate the groups (RET, LAE, PAN, *D. rodnae/D. juranum*) from one another. Among the 8 loci available for *D. laeve* and *D. panormitanum*, 4 showed alleles unique for one or both species (AAT-C or -D respectively, IDH-C, PEP-2-B, PEP-3-D). For *D.reticulatum*, EST-1-A, IDH-B, MDH-A, PEP-1-A, DIA-A, HBDH-A, PEP-3-A and PGM-C separated the species from *D. rodnae*. However, the last four alleles were only partially fixed.

Within the group *D. rodnae/D. juranum*, there are a number of unique alleles: EST-1-B and -C, IDH-A, MDH-B, PEP-1-B, AAT-A, DIA-C, EST-2-B and -C, GPI-A, HBDH-A, PEP-3-A, PEP-4 and PGM-C, of which the first five alleles were fixed. EST-1, EST-2, AAT, G3PDH, MDH, PEP-3, PEP-4, DIA, PGM and GPI were polymorphic within the group. Nevertheless, not more than four enzymes (AAT, EST-1, GPI and PEP-4) showed electromorphs unique to *D. juranum*. However, only allele B of PEP-4 occured at all localities where *juranum* had been found, but this was not fixed for *D. juranum* either. J5 is a remarkable population fixed for EST-1-C whereas all other populations of *D. juranum* and *D. rodnae* are fixed for EST-1-B.

Table 3. Allele frequencies. N: number of specimens for which data are available.

Locus		RET	LAE	PAN	P	C	G	J3	J2	J4	J5
					Population						
AAT	(N)	3	1	1	2	4	4	3	2	4	4
	A	-	-	-	-	-	-	0.667	-	1.000	0.083
	B	1.000	-	-	1.000	1.000	1.000	0.333	1.000	-	0.917
	C	-	1.000	-	-	-	-	-	-	-	-
	D	-	-	1.000	-	-	-	-	-	-	-
DIA	(N)	3			2	5	4	4	2	5	3
	A	0.167	-	-	-	-	-	-	-	-	-
	B	0.333	-	-	-	-	0.250	-	-	-	-
	C	-	-	-	1.000	0.900	0.750	1.000	1.000	1.000	1.000
	D	0.500	-	-	-	0.100	-	-	-	-	-
EST-1	(N)	1			1	2	2	3	2	4	4
	A	1.000	-	-	-	-	-	-	-	-	-
	B	-	-	-	1.000	1.000	1.000	1.000	1.000	1.000	-
	C	-	-	-	-	-	-	-	-	-	1.000
EST-2	(N)	1			1	2	2	3	2	4	4
	A	1.000	-	-	1.000	-	-	1.000	1.000	0.500	1.000
	B	-	-	-	-	-	0.500	-	-	0.500	-
	C	-	-	-	-	1.000	0.500	-	-	-	-
GPI	(N)	1			1	2	2	2	2	4	3
	A	-	-	-	0.500	1.000	0.500	0.500	0.500	0.125	0.500
	B	1.000	-	-	0.500	-	0.500	0.500	0.500	0.625	0.500
	C	-	-	-	-	-	-	-	-	0.250	-

Locus		RET	LAE	PAN	P	C	G	J3	J2	J4	J5
					Population						
G3PDH	(N)	1			1	2	2		2	3	3
	A	1.000	-	-	1.000	-	-	-	-	-	-
	B	-	-	-	-	1.000	1.000	-	1.000	1.000	1.000
HBDH	(N)	1	1	1	3	3	4	3	2	4	3
	A	0.500	-	-	-	-	-	-	-	-	-
	B	-	-	-	1.000	1.000	1.000	1.000	1.000	1.000	1.000
	C	0.500	1.000	1.000	-	-	-	-	-	-	-
IDH	(N)	3		1	3	6	4	3	2	5	3
	A	-	-	-	1.000	1.000	1.000	1.000	1.000	1.000	1.000
	B	1.000	-	-	-	-	-	-	-	-	-
	C	-	-	1.000	-	-	-	-	-	-	-
MDH	(N)	1			1	2	2	1	1	3	3
	A	1.000	-	-	-	-	-	-	-	-	-
	B	-	-	-	1.000	0.500	-	1.000	1.000	1.000	1.000
	C	-	-	-	-	0.500	1.000	-	-	-	-
PEP-1	(N)	2	1		4	3	3	4	2	6	4
	A	1.000	-	-	-	-	-	-	-	-	-
	B	-	-	1.000	1.000	1.000	1.000	1.000	1.000	1.000	1.000
PEP-2	(N)	1	1	1	3	6	4	4	2	6	4
	A	1.000	-	-	1.000	1.000	1.000	1.000	1.000	1.000	1.000
	B	-	1.000	1.000	-	-	-	-	-	-	-
PEP-3	(N)	1	1	1	2	5	3	4	2	5	4
	A	0.500	-	-	-	-	-	-	-	-	-
	B	-	-	-	-	0.300	-	0.875	-	0.500	0.625
	C	0.500	-	1.000	1.000	0.700	1.000	0.125	1.000	0.300	0.375
	D	-	1.000	-	-	-	-	-	-	0.200	-
PEP-4	(N)	1			1		1	3	2	5	4
	A	1.000	-	-	1.000	-	1.000	0.333	0.500	0.800	0.250
	B	-	-	-	-	-	-	0.667	0.500	0.200	0.750
PGDH	(N)	1	1	1	2	2	4	3	2	5	3
	A	1.000	1.000	1.000	1.000	1.000	1.000	1.000	1.000	1.000	1.000
PGM	(N)	1			1	2	2		2	4	3
	A	-	-	-	1.000	-	-	-	-	0.500	-
	B	0.500	-	-	-	1.000	1.000	-	1.000	0.500	1.000
	C	0.500	-	-	-	-	-	-	-	-	-
SOD	(N)	3	1	1	3	7	4	4	2	5	4
	A	1.000	-	1.000	-	-	-	-	-	-	-
	B	-	1.000	-	1.000	1.000	1.000	1.000	1.000	1.000	1.000

Table 4: Nei's (1978) unbiased genetic identities (above diagonal) and distances (below diagonal).

Population	RET	LAE	PAN	P	C	G	J3	J4	J2+J5
RET	---	0.283	0.567	0.661	0.659	0.661	0.486	0.432	0.645
LAE	1.261	---	0.500	0.333	0.347	0.333	0.357	0.390	0.355
PAN	0.567	0.693	---	0.333	0.295	0.333	0.201	0.230	0.281
P	0.413	1.099	1.099	---	0.989	1.000	0.797	0.762	0.975
C	0.417	1.058	1.221	0.011	---	0.989	0.872	0.804	1.000
G	0.413	1.099	1.099	0.000	0.011	---	0.797	0.762	0.975
J3	0.722	1.029	1.604	0.227	0.137	0.227	---	0.977	0.912
J4	0.840	0.942	1.468	0.272	0.218	0.272	0.023	---	0.842
J2+J5	0.439	1.037	1.270	0.025	0.000	0.025	0.092	0.172	---

A statistical analysis was run for the 6 loci (AAT, HBDH, PEP-2, PEP-3, PGDH and SOD) on which there were data available for all groups. Nei's (1978) unbiased identities and distances are given in table 4. The UPGMA dendrogram (fig. 1) shows that *D. laeve* and *D. panormitanum* are clearly separated from *D. rodnae/D. juranum*. *D. reticulatum* is also well separated from this group. *D. juranum* and *D. rodnae* from the different localities cluster together. They do not form two separate groups.

Because the data set was incomplete, the program was run several times with subgroups. The results are consistent with those presented here, i.e. *D. juranum* and *D. rodnae* always cluster together and differ strongly from the other groups.

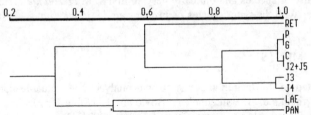

Fig. 1: UPGMA dendrogram of *D. reticulatum* (RET), *D. laeve* (LAE), *D. panormitanum* (PAN), *D. rodnae* (P, C, G) and *D. juranum* (J) based on Nei's (1978) unbiased genetic identities. Cophenetic correlation = 0.942.

DISCUSSION

The observed differences between *D. reticulatum*, *D. laeve* and *D. panormitanum* suggest that there is sufficient enzyme polymorphism within the genus to allow species separation.

D. laeve and *D. panormitanum* form a group very different from the other groups. They belong to the subgenus *Deroceras* s. str. Whereas the external appearance of most species of the genus *Deroceras* is extremely similar, *Deroceras* s. str. can be recognized rather easily: relatively small, large mantle, thin and relatively transparent skin (Wiktor, 1973). Although the results reported here must be interpreted cautiously, as only one specimen of each was included into the study, they do seem to correspond with morphological and anatomical differences between this subgenus and other subgenera, at least for *D. reticulatum* (subgenus *Agriolimax*) and *D. rodnae* or *D. juranum* (subgenus *Plathystimulus)*.

Similarly, Wolf (1990) found that *D. laeve* and *D. panormitanum* together with *D. sturanyi* form a distinct species group of the subgenus *Deroceras* s. str. clearly different from *D. reticulatum*. Like the results presented here, the *D. reticulatum* of Wolf's study together with *D. agreste* is more similar to *D. rodnae* but nevertheless clearly separated (fig.2).

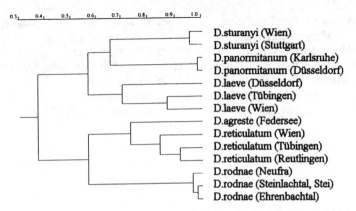

Fig. 2: UPGMA-dendrogram with NEI's genetic identities, after Wolf (1990), slightly changed

However, the high similarity values within the *D. rodnae/D. juranum*-group do not give the slightest hint to the species status of *D. juranum*. The data confirm the results of the morphological and behavioural studies and the breeding experiments. There seems to be no reason to consider *D. rodnae* and *D. juranum* as two different species. *D. juranum* is a colour morph of *D. rodnae*. Nevertheless, more data about enzyme polymorphism within the genus *Deroceras* especially of species anatomically very similar with *D. rodnae* are needed. Further comparative studies with the sister species *D. praecox* are in preparation.

ACKNOWLEDGEMENTS

I am much indebted to T. Backeljau for statistical analysis and valuable comments on the manuscript. Many thanks to B. Balkenhol, H. Boyle and L. Noble for reading the manuskript and linguistic help. This work was supported by DFG (Du 202/2-2).

REFERENCES

Backeljau, T (1987) Electrophoretic distinction between *Arion hortensis, A. distinctus* and *A. owenii* (Mollusca: Pulmonata). *Zoologischer Anzeiger.* **219**, 33-39.

Backeljau, T (1989) Electrophoresis of albumen gland proteins as a tool to elucidate taxonomic problems in the genus *Arion* (Gastropoda, Pulmonata). *Journal of Medical and Applied Malacology.* **1**, 29-41.

Dufour-Humblet, C (1982) Introduction a la systematique et à l'écologie des Limaces dans des forêts de la region Neuchateloise. Masters degree thesis, Univ. Neuchâtel, Switzerland.

Mermod, G (1930) Catalogue des Invertébrés de la Suisse, 18. Gastéropodes, *Agriolimax agreste.* Genève, 38-41.

Nei, M (1978) Estimation of average heterozygosity and genetic distance from a small number of individuals. *Genetics.* **89**, 583-590.

Reise, H (1995) Mating behaviour of *Deroceras rodnae* Grossu & Lupu, 1965 and *D. praecox* Wiktor, 1966 (Pulmonata: Agriolimacidae). *Journal of Molluscan Studies.* **61**, 325-330.

Reise, H *Deroceras juranum* - a Mendelian colour morph of *D. rodnae* (Gastropoda: Agriolimacidae). - *Journal of Zoology,* in press.

Swofford, D L; Selander, R B (1981) BIOSYS-1: a Fortran program for the comprehensive analysis of electrophoretic data in population genetics and systematics. *The Journal of Heredity.* **72**, 281-283.

Wiktor, A (1973) Die Nacktschnecken Polens. Arionidae, Milacidae, Limacidae (Gastropoda, Stylommatophora). Monographie Fauny Polsky, **1**, Warsaw, Cracow, pp .

Wolf, M (1990) Verwandtschaftsverhältnisse und Fortpflanzungsmechanismen von sechs westeuropäischen Arten der Gattung *Deroceras* Rafinesque 1820 (Mollusca, Gastropoda, Agriolimacidae). Masters degree thesis, Univ. Tübingen, Germany.

Wüthrich, M (1993) *Deroceras (Plathystimulus) juranum* n. sp., eine endemische Nacktschnecke aus dem Schweizer Jura. *Archiv für Molluskenkunde.* **122**, 123-131.

SLUG DAMAGE TO SUNFLOWER CROPS IN THE SOUTH-WEST OF FRANCE

Y BALLANGER, L CHAMPOLIVIER

CETIOM, C.B.A., Rue de Lagny - 77178 St Pathus (France)

ABSTRACT

A 4 year field study on slugs in sunflower crops was performed in the area of Toulouse. Difficulties in crop establishment were numerous but not especially due to soil pests - Slugs are usually present in ploughed fields in winter even if a summer drought occurs every year. At the end of winter, the mature grey field slugs disappear progressively and become scarce by the sowing time - plant emergence period. After that, as temperatures increase and soil surface becomes dry (no canopy), the appearance of new generations is linked to repeated rainfall. This typical alternation is not so obvious with black field slugs or other slugs species but these other species are not so damaging. Out of numerous cases of low damage due to slugs and/or wireworms, we only had the opportunity to characterize 7 severe attacks of slugs. Problems arise from emerged plants cut off by developed animals, related to a short period of activity, and/or progressive consumption of hypocotyls by newly hatched animals. In the first case, only a preventive treatment will solve the problem ; in the second case, molluscicides are inefficacious.

INTRODUCTION

In the South-west of France (area of Toulouse), we observed the dynamics of slug populations in relation to sunflower crops. From observations repeated all year long, we concluded that - in ploughed lands - slugs populations decline at the end of winter and become scarce at the time of sunflower emergence. As it is several weeks before the canopy becomes sufficiently thick to protect the soil surface from drying, this spring crop does not allow new generations to prosper. After first investigations in sunflower, it was difficult to make further observations upon slugs in the same fields (BALLANGER, CHAMPOLIVIER - 1990). At the same time, a survey was conducted to define difficulties encountered in sunflower farming. Sunflower crop establishment is exposed to numerous hazards. So out of 35 fields studied : in 32 cases, we could not observe 90 productive plants for 100 seeds drilled in to the soil ; in 21 cases, not 80 healthy plants (BALLANGER, CHAMPOLIVIER - 1992). In this context, what were the parts of the slugs ?

MATERIALS AND METHODS

Observations were conducted in 35 fields, 6 fields (601 to 606) in 1986, 8 (801 to 808) in 1987, 12 (901 to 912) in 1988 and 9 (901 to 909) in 1989. Fields 601, 701 and 801 also supported studies on the dynamics of slug populations (based on soil washing).

During 4-5 months, from January-February to May-June, traps and treated small plots (m²) were deployed, from time to time, in the same areas of the different fields. "Traps" and "m²" were put in situation (4 traps and 4 m² by field), after rain, on wet soil in late afternoon. Subsequent observations were performed early morning, the day after. "ACTA-traps" (Cardboard - 50x50 cm - covered by a black plastic sheet - 75x75 cm) were used in 1986, "INRAtraps'' (50x50 cm) the others years. Traps were baited with molluscicide (No bait in 1989). ''m²'' (1x1 m) represent the central parts of treated small square plots (2x2 m) treated (20 kg/ha) with molluscicide (unused in 1989). MESUROL (mercaptodimethur) was used.

In each field, four small plots - the four corners of a virtual square of 25-30 meter sides - consisting in 4 rows of 4 meters each were marked out before the first plants appeared (row space varying between fields). Then, each sowing site was characterized. In case of crop failure, we tried to determine the cause of the lack of plant by hand sorting from soil and/or complementary observations. Each field investigation concerned more than 200 sowing sites (30 fields).

At first (1986, 1987), the four plot design was doubled - The plots of the twin design were covered with nets, to prevent bird damages, and treated (as ''m²'') 2, 3 or 4 times from drilling, to prevent slug injuries.

RESULTS

Each year, the first investigations were conducted on ploughed land, the last ones in crops with sufficiently developed plants (2.6 - 2.8 : 3-4 twin leaves). We observed grey field slugs (GFS), *Deroceras reticulatum*, black field slugs (BFS), *Arion hortensis*, and Other slug species (OSS), mainly, *Arion subfuscus*. All the fields were treated with microgranular insecticides put in the furrow to control wireworms - Field 601 and field 811 (second drilling) excepted, possible molluscicide treatments were only applied to the rest of the fields.

Climatic context (Figure 1).

The area of Toulouse enjoys an agreeable climate, rarely cold in winter, often too hot in summer. But, rainfall is rather important (680 mm) and well partitioned all year long, between 46 mm, for July, to 65 mm, for December. During the time of the survey some ''abnormal'' sequences occurred with - for example - a severe drought, from March 1986 to February 1987, and an extensive period of rainfall (leading to risings of water and diverse local engineering catastrophies) at the end of the winter in 1988. For sunflower crops, drilling usually take place in the second half of April, when the soils are warmed up and when the soil surface humidity has become fully dependant on rainfall.

Figure 1 : Rainfall per 10 days (1-3) for 3 months (March-May) over 4 years (1986-1989).

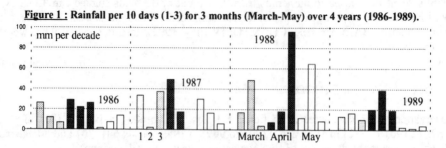

Damages to sunflower by slugs (and/or wireworms)

In many cases, when a predominant pest is not clearly recognized in the field or on the plant, it is quite impossible to distinguish between soil pests. During the study, we were often confronted by different sorts of minor damage coming from the different slugs species, acting as developed animals or as newly hatched stages, and from wireworms.

When foliar damage occurs after emergence, the life and even the good growth of the plant is not really handicapped - The problems come from plants killed, as seed (germinated or not), as young plants attacked in the soil before emergence or after emergence, and as young emerged plants cut off, without being otherwise damaged, by slugs crawling on the soil surface.

Survey 1986 : Field 601 to 606 (Figure 2-A, Table 2)

Slug trapping was only performing in fields 601 and 602 (The main part of the BFS observed in 605 corresponded to a late hatching - between 28 May and 6 June - and no damage occurred) - Presuming a high risk to the crop, field 601 was treated at drilling (but in a way considered inefficacious) however slug population seemed to decline by itself. Damage were not important. After first preparation of the soil, traps and "m²" did not detect any slugs in field 602 and the emergence of the sunflower seemed to go on in the best conditions. But, in a few days after a rainfall, many plants were cut off by slugs which had resumed activety.

Survey 1987 : Field 701 to 708 (Figure 2-B, Table 2)

After the long period of drought, it was difficult to detect slugs in the fields, even in winter. Meanwhile, newly hatched BFS were observed (705) but only in the first part of June. All the injuries due to soil pests were performed underground (119 among 120 per 8 fields). Thus, important losses of plants were detected, for example in field 708, where it was never possible to find slugs, only wireworms. But, comparing molluscicide treated plots and untreated plots (Fields 701 to 704), a trend favouring slug protection appeared.

Figure 2 - Trapping slugs in sunflower crops from the area or Toulouse (GFS per 1 trap) - A : Field 601 - 12 traps (E : emergence of the crop) - B : Fields 701 to 708 - 8 x 4 traps.

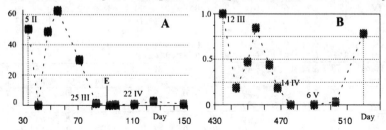

Survey 1988 : Field 801 to 812 (Figure 3-A, Table 1, Table 2)

Winter catches were unequal but significant everywhere, for GFS, BFS and also OSS. After that, the particular moisture regime was associated with unusual hatchings of eggs and sustained activity by new generations of slugs.

- Field 811 - Slugs were numerous and diverse, GFS, BFS and OSS, *A. subfuscus* (young animals, 1-2 cm long, with a variable coloration ?), .. and *Milax* species. Damage was important (so, hand sorting to characterize empty sowing sites was not realized). Firstly, emergence of a plant was observed for only a half (49 %) of the sowing sites. Secondly, half of the young plants of the limited population established disappeared (27 %). A second, direct drilling had been attempted (1 June), with a molluscicide spread on the soil. 18 % of the empty sowing sites were associated with soil pests injuries, damaging seeds or newly germinated seedlings, without foliar attacks to emerged plants.

- Fields 807 and 808 - The two fields were only separated by a road and had received the same farming practices for several years. Slugs, GFS and BFS, were numerous in the two parts, before drilling. After a short period of inactivity, traps and "m²" worked again and damage occurred. In 807, after initially limited injuries and losses of seeds and young plants, numerous GFS hatched and gathered on the hypocotyles of the emerged young plants. Constant humidity of the superficial layers of the soil allowed the pests to eat regularly day after day and to gnaw irreversibly until the plant axis was cut off. In 808, the damage was less. Lying lower than the other field, after the first attack, this part became flooded. We could observe dead (drowned) slugs on the soil, hatching was limited and attacks ceased.

- Field 801 - The first period favourable for drilling was missed (as in 803) and sowing was delayed until the end of the rainy period. Then, GFS seemed to disappear without giving a new generation, BFS persisted and gave a new generation. It was often possible to find a young BFS when examining the basis of stained plants, in June (Symptoms similar to wireworm injury seen elsewhere, field 804, for example).

Table 1 - 1988 (Field 801 to 812) - GFS per 4 traps per field (detail), GFS, BFS and OSS per 4 traps or "m²" (total all dates)

DD	MM	Day*	801	802	803	804	805	806	807	808	809	810	811	812
22	I	752	41	32	3	11	0	30	7	23	0			
26	I	756	59	23	5	9	0	17	4	11	2	11	11	4
29	I	759	89	57						20				
4	II	765	21	53		15		8	21	17		21	35	9
8	III	798	23	9	0	1	0	3	4	7	0	5	14	0
22	III	813	30	15		6		0	3	2		11	37	
1	IV	822	2	6		1			15	0		5	4	
6	IV	827	8	5	0	3	0		1	1		2	28	
17	IV	838				0	0		2	0	0			
20	IV	841	3	0	0	0	0	0	4	1	4	1	17	0
24	IV	845	2	6	0	0	0	0	11	1	2	0	10	0
3	V	854	0	11		0	0	1	16	0	5	1	33	0
17	V	868	0	5	0	0	0	0	30	2	2	6	23	0
27	V	878	0	0	0	0	0	0	55	0	18	4	17	0
2	VI	884	0	0	0	0	0	0	8	1	0	5	7	0
GFS	Traps		278	222	8	46	0	59	181	86	33	72	236	13
	m²		65	74	6	19	2	9	159	35	14	54	87	1
BFS	Traps		56	27	5	49	0	0	28	74	0	20	32	7
	m²		44	14	9	29	0	1	14	30	0	8	4	3
OSS	Traps		12	4	2	0	0	2	2	3	4	8	63	3
	m²		10	3	0	0	0	0	2	3	3	4	12	0

DD : day, MM : month - *Day 1 = 1 January 1986.

Figure 3 - Plant population development per 4 plots x 4 rows x 4 meters (Percentage)
A : Field 807 (195 sowing points) - B : Field 901 (208 sowing points)

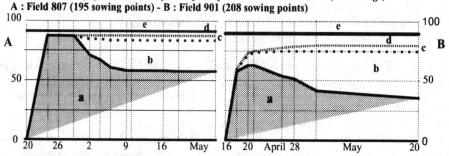

a : plants, b+c : plants killed after emergence, d : seeds and plants killed before emergence,
e : others causes of lack of plant - b+c+d : soil pests.

Slugs were detected in 8 of the 9 fields in the survey, catches were not so numerous (unbaited traps) despite conditions often favorable to slug activity, wether slug numbers were low in the fields or in the presence of a serious infestation of GFS, as in field 901.

Table 2 - 1986 - Seed-plants damaged by soil pests (per 4 plots x 4 lines x 4 meters).

-	601		602		603		604		605		606	
DRILLING -	29 III		29 IV		5 IV		2 IV		15 IV		17 IV	
Sowing sites -	208		237		204		X		184		276	
Soil pests -	14	4	62	15	1	2	X	X	2	2	5	8
- seed -	7	3	10	9	0	0			1	0	0	2
- plant -	4	1	14	2	1	2			0	2	0	6
- cut off -	3	0	38	4	0	0			0	0	0	0
-	uT	T	uT	T	uT	T	uT	T	uT	T	uT	T

	701		702		703		704		705	706	707	708
DRILLING	23 IV		24 IV		17 IV		21 IV		25 IV	23 IV	28 IV	28 IV
Sowing sites	245		220		220		204		238	209	221	328
Soil pests	14	4	18	9	10	7	17	5	19	3	8	31
- seed	3	4	8	2	6	4	8	3	11	1	3	1
- plant	10	2	10	7	4	3	9	2	8	2	5	30
- cut off	1	0	0	0	0	0	0	0	0	0	0	0
	uT	T	uT	T	uT	T	uT	T	uT	uT	uT	uT

-	801	802	803	804	805	806	807	808	809	810	811	811	812
DRILLING -	25 V	18IV	15V	11IV	15IV	15IV	12IV	13IV	13IV	18IV	18IV	30V	18IV
Sowing sites -	270	X	212	208	180	208	195	192	185	248	248	212	240
Soil pests -	55	X	0	15	1	4	66	26	11	5	189	40	15
- seed -	19	0	3	1	0	6	4	4	0		122	27	3
- plant aE -	11	0	12	0	1	0	7	1	5			1	4
- plant pE -	7	0	0	0	3	52	8	6	0		29	8	4
- cut off -	18	0	0	0	0	8	7	0	0		38	2	0

	901	902	903	904	905	906	907	908	909
DRILLING	30 III	6 IV	30 III	30 III	7 IV	7 IV	23 IV	24 III	6 IV
Sowing sites	208	221	204	207	(141)	238	223	240	209
Soil pests	120	1	8	12	8	5	1	12	9
- seed	0	0	2	1	0	1	1	2	1
- plant	26	1	6	11	7	4	0	7	8
- cut off	94	0	0	0	1	0	0	3	0

seed (killed, germinated or not), plant (killed by injuries in the soil), (hypocotyles) cut off (at the soil level) - aE : before emergence, pE : after emergence - uT : untreated, T : protection against slugs and birds

N.B. : A molluscicide trial (7 actives) was performed in a field (908) poorly infested by slugs but provided with wireworms - After drilling, emergence in small plots was perfect (97 plants for 100 sowing sites). Later, progressively, plants disappeared (14%), fully dried under the sun, following hypocotyl injury in the soil. As mean losses are not significantly different, the tendencies observed in 1986 and 1987 reappeared (untreated : 45 plants killed, treated : 15 to 27 / per 24 m² : 4 replicates x 2 rows x 6 meters).

DISCUSSION, CONCLUSION

In the area of Toulouse, we found slugs widely distributed - and sometimes at high population levels -, in winter, in ploughed lands destined for sunflower. Later, as populations of active slugs declined and environmental conditions become adverse, the risk posed to crops - often - resolved itself.

- In theory, a simple presence of phytophagous animals in a sunflower field can lead to consequent damage, firstly, because we attempt to establish a crop from only 6-7 seeds per m², secondly, because the young plant poorly overcome injuries to hypocotyl and dehydrate, and thirdly, because active slugs tend to kill many plants, cutting off hypocotyls without eating the damaged parts - So advisers propose a prophylactic treatment, if 1-2 slugs are detected per m². But, at sowing time, it also become difficult to detect slugs in the fields.

- In practice, in a region favourable to slugs (difficulties occur every year in winter rapeseed), the situation of sunflower is not so critical. We found only 7 severe attacks due to slugs by surveying 35 fields, chosen as risky slugs situations.

1987 and 1989 were not favouble to slug populations and no damage occurred except in a particular situation (Field 901 : sunflower coming after lucern). 1986 (12 fields) allowed slug attacks in infested fields (602 : cattle and forage production, 601 : technical mistake with the straw of the previous crop). Field 601 1988 was the year of all damages, involving diverse species and different stages (Fields 801, (804), 807, (808), 811 and 812), but 6-8 crops were not attacked . At all, only one crop was destroyed by slugs and necessitated a second drilling (811). In this case, numerous slugs (and wireworms?) in a favorable situation destroyed 67% of the sowing points, representing 4.3 seeds or young plants per m² (43.010 seeds or plants per ha).

The finding of high numbers of slugs in winter is not sufficient to decide a strategy. It probably only means that eggs are also numerous in the soil.
- Active slugs can disappear or be inactive at emergence time of the crop. A short but damaging period of activity is always possible. In that context a molluscicide treatment applied at drilling or before the first rainfalls may be useful.
- Spring time is also the period of appearance and development of new slug generations. Newly hatched slugs were just detected, often lately (June) in all years for BFS, only if the moisture regime was really favorable (1988) for GFS. So BFS damage may not be so unusual (Fields 801, 804, 809, 905), but these species are not as important and widespread as GFS. Paradoxically, we have learnt that excessive rainfalls may be a factor of limited risk in damp places that can be easily flooded.

It is probable that restricting our attention to regular sunflower growers would have revealed a rather less uncomfortable situation. Spring crops, and especially sunflower, constitute a good option for limited problems with slugs.

This study also revealed that slugs are only part of the many minor difficulties that add up at sowing time - emergence period of the crop. It seems that wireworms are not underestimated (all crops are insecticide treated at sowing) but not really controlled.

REFERENCES

BALLANGER, Y; CHAMPOLIVIER, L (1990) Dynamique des populations de Limace grise (Deroceras reticulatum Müller) en relation avec la culture de Tournesol. Proc. Conférence Internationale sur les Ravageurs en Agriculture, Versailles, *Annales ANPP*, **3**, I : 143-150.

BALLANGER, Y; CHAMPOLIVIER, L (1992) Sunflower and slugs. *Proc. 13th International Sunflower Conference. Pisa*, I : 874-879.

SLUG AND SNAIL PESTS IN SPANISH CROPS AND THEIR ECONOMICAL IMPORTANCE

J CASTILLEJO, I SEIJAS, F VILLOCH

Department of Animal Biology. Faculty of Biology. University of Santiago. E-15706 Santiago de Compostela. Spain

ABSTRACT

In Spain as in many other countries there is very little information on the damage that slugs and snails cause on vegetables crops and fruit trees. The species most frequently found in Spanish crops are *Deroceras reticulatum, Deroceras panormitanum, Milax gagates, Limax flavus, Arion ater, Theba pisana, Ceapea nemoralis* and *Helix aspersa*. For control of these pests the Spanish farmers apply 2500 tonnes of molluscicides/year, at a cost of 1000 millions pesetas - £5 millions. To resolve this problem we are developing a strategy to kill slug and snail eggs in the soil with light chemical compounds. The possibility of denaturing the outer layer of slug eggs with tannins from powdered bark of various trees (especially oak, holm oak, eucalyptus or chestnut trees) which have high quantities of tannin in their bark, will be explored in laboratory and field experiments.

SNAIL AND SLUG PESTS

Terrestrial molluscs as a group have fairly generalized feeding habits, but some species are predominantly phytopatogenus while others feed mainly on decaying organic matter. They cause considerable damage on many Spanish field crops, horticultural crops and fruit crops. As with many other agricultural pests, there is very little information on the extent or the value of damage they do, and it is very difficult to estimate where they stand in order of importance as farm and garden pests. Very little attention has been given to the pest status of slugs mainly because they are so difficult to control and because their damage tends to be localised and unpredictable.

In our country research on slugs and snails was formerly directed mostly at the intermediate host of trematode parasites of man and domestic animals and it is only recently the attention has been directed to terrestrial gastropods on cultivated land. As long ago as the past century, (Seoane,1866), already alluded to *H. aspersa* and *D. reticulatum* as pests in Galician gardens and allotments. Occasionally the Spanish Ministry of Agriculture publish leaflets or include chapters in official bulletins on their control (Pérez *et al.* 1973; Canales, 1990; Rivero,1990)

In horticultural crops -beans, cauliflower, cabbages, lettuces, tomatoes, etc.- the most destructive species are *D. reticulatum, D. panormitanum, M. gagates, L. flavus, A. ater, C. nemoralis* and *H. aspersa*.

In the Spanish Mediterranean area the *Citrus* and *Pomatoceus* trees are attacked by the snails *Otala lactea*, *H. aspersa* and *T. pisana*. The damage appear on young shoots, leaves and fruits. These losses are more intense in autumn, warm winters and early springs.

In the North of Spain the fruit trees are attacked more by slugs than snails. *D. reticulatum*, *L. marginata*, *L. flavus*, *Arion intermedius* and *Arion flagellus* are the slugs more frequently found. *H. aspersa* is the most damaging snail.

One curious case of a possible slug pest on trees is *Geomalacus maculosus*. It is a Lusitanic endemic species. In Spain and Portugal it is easy to find *Geomalacus* feeding on old chestnuts. It hides and rests behind the bark covered with moss and lichens. Possibly it can destroy green-shoots, leaves and spread the fungi and blight cankers into trunk cracks. In one wet and warm night it is possible to find more than 40 *Geomalacus* on one chestnut trunk.

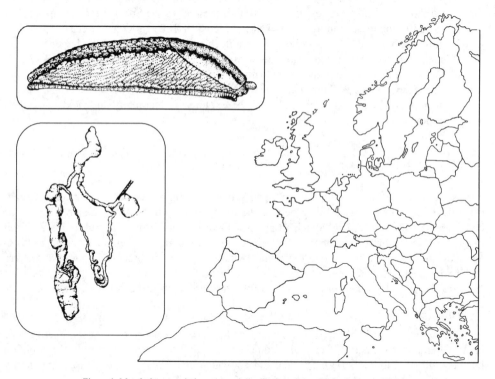

Figure 1. Morphology, genital system and distribution of slug *Geomalacus maculosus*

The largest adult specimens of *G. maculosus* measure 70 mm long. The dorsum of live individuals can be either greyish sprinkled with white spots, or greenish sprinkled with yellow spots. Body mucus is yellow. The epiphallus is a helicoidally wound tube, its internal surface is profusely lined with papillae which make up longitudinal folds (Castillejo, 1981).

G. maculosus exhibits a typical Lusitanian or Atlantic range distribution, inhabiting the Southwest of Ireland and the Northwestern coastal stripe of the Iberian peninsula, with records from the Serra da Estrela (central Portugal), as its known southernmost limit. This species appears upon the bark of oaks and beeches, and on the surface of stony walls, where they were browsing lichens. In some places is an anthropophilous specie, however it shows a preference for montane forests and chestnut and oak-tree groves. (Castillejo *et al.*, 1994)

MOLLUSCICIDES AND SPANISH GOVERNMENT POLICY

In the early fifties the Spanish Jefatura Agronómica recommended the use of lead Arsenite or calcium Arsenite against slug and snail pest on vegetable crops. In 1967 the Central Service of Farming Policy noticed about damage of slugs and snails on citrus trees, almond trees, olive trees and also in viticulture. In this leaflet recommended that Metaldehyde as the specific remedy for control these pests. Five years later, in Alicante -SE of Spain- the Service of Defence against Crop Pests spread granulate Mesurol (Methiocarb) by helicopter over bean crops. The quantity applied was 4 kg/ha. All snails that ate Mesurol died, but many of them remained alive stuck on the stem and leaves. A side effect observed was ant, myriapod and woodlice mortality (Pérez *et al.*, 1973).

In The Valencian Country on the Mediterranean Canales (1990) found *H. aspersa* and *T. pisana* damaging citrus crops. The degree of damage in some areas could reach nearly 90% of citrus trees. The damage was observed on leaves, flowers and on young or ripe fruits. This author recommend the use of Metaldheyde (5%) or Methiocarb (1%) at the end of autumn for control of damage. Also in Valencia, in the same year, Rivero (1990) implement a screening test of herbicides as molluscicides against the snails *H. aspersa* and *T. pisana*. These experiences was carried out in the laboratory and greenhouses, and compared the results with Metaldehyde and Metiocarb. The results were than Ioxynil and Diquat were promising against both snails, Paraquat was below Diquat and irregular, MSMA was promising against *H. aspersa* and Cacodylic acid plus Sodium Cacodylate was promising against *T. pisana*. Rivero did not continue this experiments for lack of government support and interest.

USAGE OF MOLLUSCICIDES

Since 1990 the amount of molluscicides applied to Spanish crops has stabilized at about 2500 tonnes/year, costing 1000 millions of pesetas - £5 millions - per year. All this money being spent by farmers on molluscicides according to AEPLA -Spanish Association for Plant Protection-. Among Spanish Autonomic Communities Valencia and Madrid lead the ranking with 500 tonnes/year. In the other end are La Rioja and Extremadura with only 10 tonnes/year.

	1990	1991	1992	1993	1994	1995
Andalucia	321 (£2m)	405 (£2m)	224 (£1m)	466 (£2.4m)	166 (£0.9m)	139 (£0.7m)
Cataluña	297 (£1.5m)	397 (£2m)	265 (£1.5m)	291 (£1.5m)	196 (£1m)	178 (£1m)
Galicia	119 (£0.6m)	144 (£0.7m)	149 (£0.7m)	215 (£1m)	247 (£1m)	208 (£1m)
Madrid	28 (£0.2m)	32 (£0.2m)	492 (£2.5m)	348 (£2m)	624 (£3m)	594 (£3m)
Murcia	526 (£3m)	108 (£0.6m)	129(£0.7m)	205 (£1m)	75 (£0.4m)	51 (£0.3m)
Valencia	1541 (£8m)	951 (£5m)	636 (£3m)	596 (£3m)	394 (£2m)	495 (£2.5m)
Total in Spain	3303 (£17m)	2493 (£13m)	2367 (£12m)	2925 (£15m)	2325 (£12m)	2079 (£11m)

Table 1. Tonnes of molluscicides applied over some Spanish Autonomic Community crops in the last six years and their cost in millions of pound sterling (after AEPLA, 1996).

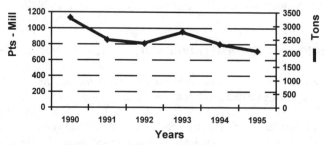

Figure 2. Tonnes of molluscicides applied over all Spanish crops during last six yearsand their cost in millions of pesetas. Usage reduced, but cost remained stable (after AEPLA, 1996).

OUR PARTICULAR BATTLE

It is almost impossible to destroy any significant number of terrestrial gastropods by use of bait; probably no more than 10% of a population can be destroyed in this way (Frömming & Plate, 1952). For a control of pest Gastropods it is essential to have precise knowledge, not only of population density, but also of the causal factors affecting it. It is also necessary to know the mortality rate, the development periods of eggs and juveniles, and the longevity of adults and senile, to establish the life span an number of generations per year for any particular species. In Galicia *H. aspersa* is one of the most dangerous species in vegetable crops and orchards.

In the last five year the Faculty of Biology of University of Santiago has been studying the biological cycle of *H. aspersa* in the Galicia Atlantic and Mediterranean. The Atlantic Galicia has precipitation and mild temperatures all year. In this area *H. aspersa* hibernate from December to March, and do not aestivate. The largest number of juveniles appear in August and September. The mating occur from May to September. *H. aspersa* has a similar cycle in the North of Gales (Bailey, 1981). The Mediterranean Galicia has a large deficit of rain in summer, with high temperature in summer and low in winter. In inland Galicia *H. aspersa* hibernate from November to March, and aestivate from June to September. The highest number of juveniles appear in April and May, and the adults in September and October. Mating occurs in Spring and Autumn. This is a typical semiarid Mediterranean cycle (Madec, 1989). In both areas the mean life span of *H. aspersa* is four years. (Iglesias, 1995).

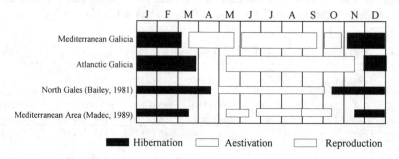

Figure 3. Circannual cycle of *Helix aspersa* in Galicia (Mediterranean and Atlantic area), North of Gales and Mediterranean area.

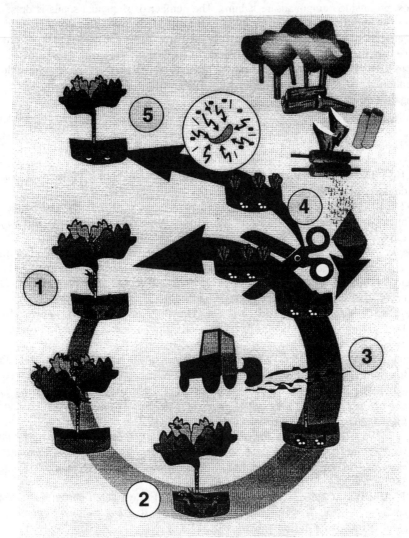

Figure 4. Schematic cycle of snails and slugs in farmer crops and the hypothetical effect of tannins on slug and snail eggs.

With the investigation on biological cycle of *H. aspersa* we drew the conclusion that the weak point in the biological cycle of *H. aspersa* is the egg phase. It could be more feasible to destroy slug eggs than kill slugs, but there are currently no effective methods of killings slug eggs. Our present investigation is focused on trying to eliminate the slug and snail eggs in the soil before they hatch. We are working on *H. aspersa, D. reticulatum* and *A. ater* eggs. In this trial we are assaying light chemical compounds for denaturing the outer albumen layer of the eggs. The compound under examination are tannin, potash alum, quick lime and slaked lime and formalin.

The most promising compoundsare tannins. The number of slugs and snails is drastically reduced in Galician forest soil with pine, eucalyptus and oak trees. This diminution in number and species could be caused by tannins from their tree barks and leaves. The use of tannins from tree bark may be an environmentally friendly method of low-chemical control, as tannin is known to form complexes with and denature proteins and slug eggs may be vulnerable to such denaturation. This method would address a current weakness in slug control strategies, especially against the fiel slug.

REFERENCES

Aepla, 1996. Dosier informativo de la Asociación Empresarial para la Protección de la Plantas desde 1990 a 1996. pp. 1-98. Madrid.

Bailey, S E R; 1981 Circannual and circadian rhythms in the snail *Helix aspersa* Müller and the photoperiodic control of annual activity and reproduction. *Journal of Comparative Physiology,* **142 A,** pp. 89-94.

Canales, R (1990) Los caracoles: una plaga a tener en cuenta en los cítricos de la Comunidad Valenciana. *Agrícola Vergel. Abril..* pp. 279- 280.

Castillejo, J (1981). *Los Moluscos terrestres de Galicia (Subclase Pulmonata).* Tesis Doctoral. Facultad de Biología. Universidade de Santiago.

Castillejo, J; Garrido, C; & Iglesias, J (1994) The slugs of the genus *Geomalacus* Allman, 1843, from the Iberian Peninsula (Gastropoda: Pulmonata: Arionidae). *Basteria.* **58:** 15-26

Frömming, E; Plate, H P (1952) Sind die Metaldehydeköder zuverlässige cheneckenbekämpfungsmittel? *Schädlingsbkämpfung.* **44,** 130-131.

Iglesias, J (1995) *Biología del caracol común Helix (Cornu) aspersa (O.F. Müller, 1774), en poblaciones naturales de Galicia. Implicaciones en su conservación y cría zootécnica (Helicicultura).* Ph D. Universidad de Santiago, pp. 274.

Madec, L (1989) *Etude de la différenciation de quelques populations géographiquement séparées de l'espêce Helix aspersa Müller (Mollusque, Gastropode, Pulmoné): aspects morphologiques, ecophysiogiques et biochimique.* Ph D. Universidad de Rennes I. pp. 380.

Pérez, T; Alberti, J; Calderón, E; Vinaches, P; (1973) Ensayo aéreo del producto "Mesurol" contra caracoles. *Servicio de Defensa contra Plgas e Inspección Fitopatológica, Alicante. Estación de Avisos Agrícolas.* pp. 81 a 86.

Rivero, del J M; (1990) Investigación sobre otros efectos plaguicidas potenciales diferentes de varios herbicidas; especialmente bipiridilos y organo arsenicales. *Boletín de Sanidad Vegetal. Plagas.* **16**: 605-611.

Seoane, V L; 1866 *Reseña de la Historia Natural de Galicia.* 66 pp. Lugo, Spain

NATURALLY OCCURRING AND INTRODUCED SLUG AND SNAIL PESTS IN NORWAY

A ANDERSEN

The Norwegian Crop Research Institute, Plant Protection Centre, Department of Entomology and Nematology, Fellesbygget, N-1432 Ås, Norway

ABSTRACT

The most important pest species are commented upon: *Deroceras reticulatum* (Müller) is the only species appearing commonly in agricultural fields. In Central Norway it may totally defoliate spring barley plants when following oilseed rape. *Arion lusitanicus* Mabille was found in Norway for the first time in 1988. It has increased its range dramatically in the 1990s. *Limax maximus* Linnè has also increased its range and become more common in the 1990s. *Helix aspersa* Müller is established locally in a few areas in south-eastern Norway after escaping from snail farms.

THE PEST SPECIES

Deroceras reticulatum (Müller) appears in fields and gardens all over the country (Hofsvang, 1995). This is the only slug species appearing as a regular pest in agricultural fields. However, in most areas little or no economic damage is done. Economic damage occurs mainly in coastal areas, especially along the Trondheimsfjord in Central Norway. In this area, if spring barley is sown the year following oilseed rape, the snails may almost totally defoliate the plants during the summer. In recent years the species have been shown to be more common in fields with reduced tillage, especially direct drilling. This is probably due to less disturbance and increased humidity.

Arion lusitanicus Mabille was found in Norway for the first time in 1988 (Proschwitz & Winge, 1994). The distribution has exploded in the 1990s, probably due to several mild winters and cool summers. Today the species is distributed along the coast north to the Trondheimsfjord (63° 30' N) in Central Norway. The true identity of the species has not been stated with certainty in all cases, and possibly it is mixture of two species (Dag Dolmen pers.comm.). It is mainly a problem in gardens, but has also been reported to destroy strawberries and oilseed rape. Several times the species has been shown to spread into new areas with plant material and the soil in flower-pots. It is often found locally in large numbers, especially along the West Coast.

Limax maximus Linné is distributed along the coast north to the Trondheimsfjord. Further north along the coast there are scattered finds in Fauske (67° 15' N) and Tromsø (69° 45' N) (Winge, 1993, Winge & Vader, 1995). The species has expanded its range and become much more common in the 1990s. Several times the species has been shown to spread into new areas with plante material and the soil in flower-pots. It is mainly causing damage to flowers and vegetables in gardens, but rarely in large numbers.

Helix aspersa Müller is only locally established in a few areas in south-eastern Norway after escaping from snail farms or illegal imports. Populations shown to overwinter for several years out of doors have probably taken advantage of compost areas.

PEST CONTROL

Snails and slugs can normally be controlled in garden flowers and vegetables by hand-picking or chemical control. However, for the time being no satisfactory control is available in fields. This is due to high cost of the chemicals when used in the doses necessary for sufficient effect.

REFERENCES

Hofsvang, T (1995) Snegl som skadedyr i jord- og hagebruk (in Norwegian). *Faginfo* no. 3 1995, 120-123.

Proschwitz, T von; Winge, K (1994) *Arion lusitanicus* Mabille - an anthropochorous slug species spreading in Norway. *Fauna* **47**, 195-203.

Winge, K (1993) *Limax maximus* in Sør-Trøndelag and Møre and Romsdal. *Fauna* **46**, 106-109.

Winge, K; Vader, W (1995) Northerly records of the slug *Limax maximus* in Norway. *Fauna* **48**, 34-35.

THE CONTROL OF SLUG DAMAGE USING PLANT-DERIVED REPELLENTS AND ANTIFEEDANTS

C J DODDS

IACR-Rothamsted, Harpenden, Herts., AL5 2JQ, UK

ABSTRACT

Exposure of the sensory pad on the large posterior tentacle of the slug *Deroceras reticulatum* to volatile components of certain umbellifer plant extracts induced intense electrical activity in the nerve connecting it to the metacerebrum. Extracts inducing activity included hemlock (*Conium maculatum*), and curled chervil (*Anthriscus cerefolium*). A complementary bioassay assessed the effect of plant extracts on the feeding behaviour of intact slugs. The most marked reduction in feeding was caused by extracts of hemlock, rock samphire (*Crithmum maritimum*) and curled chervil, indicating a possible link between intensity of nerve response and strength of antifeedant properties. Two active antifeedants were found in curled chervil, (+)fenchone and 4-allylanisole. Most of the antifeedant activity of hemlock was caused by a single new slug antifeedant compound.

INTRODUCTION

Molluscs can be highly selective feeders, using receptors located in the tentacles. The posterior tentacles detect airborne (volatile) cues, while the anterior pair act as contact (gustatory) chemoreceptors (Bailey & McCrohan, 1989; Cook, 1985; Garraway, 1992; Getz, 1959). The cues used by slugs in discriminating between plants are thought to be secondary metabolites which act as attractants or deterrents and toxins.

The ability of some plants to influence slug feeding could perhaps be used to reduce damage to crop plants which do not normally produce such chemical defences. A prerequisite of such a strategy would be to screen plants for antifeedant activity and determine which chemicals are responsible. This study is of a potential rapid screen for suitable activity. Although behavioural tests were also used, initial screening for activity and identification of active components was done by electrophysiological recording from chemoreceptors of the field slug *Deroceras reticulatum*.

The Umbelliferae, or Apiodeae (carrot family), are annual or perennial herbs, rarely shrubs, and many are strong smelling. The family contains some 42 different genera (Tutin, 1980), any of which are potential sources of behaviour modifying chemicals (semiochemicals). In the work described, extracts of 33 representative species of Umbelliferae were tested for activity by electrophysiological recording and by a feeding bioassay.

MATERIALS AND METHODS

Plant extraction

Plants were extracted using Microwave-Assisted Distillation. This technique was based on the method of Craveiro et al (1989), in which fresh plant material was heated in a microwave oven until the rise in pressure inside the cells caused them to burst, releasing volatile components. These volatiles were then picked up in a stream of nitrogen gas and trapped by bubbling through a flask containing cooled solvent.

Electrophysiological assays

The main olfactory organ, the posterior tentacle, of *D. reticulatum* contains two major nerves, the optic and olfactory, which both run to the metacerebrum on the cerebral ganglion. The olfactory nerve terminates peripherally in the digitate ganglion, a thickened epithelial pad positioned just below the eye. The digitate ganglion contains the main olfactory receptors which are sensory nerve endings and are used to register any volatile olfactory cues from the surrounding environment.

A posterior tentacle was dissected out of an anaesthetized slug and placed on a Sylgard dish containing a specially formulated Ringer solution (Garraway, 1992). The tentacle was then dissected to expose further the main olfactory nerve and the sensory pad. The preparation was positioned so that the nerve remained in the Ringer solution but the sensory pad was exposed to air so that airborne chemical stimuli could be applied. Two suction electrodes containing silver/silver chloride wire were placed into the Ringer solution, one acting as the 'indifferent' electrode while the other was applied to the nerve as the 'recording' electrode. Signals produced by the preparation were simultaneously displayed on a digital storage oscilloscope and recorded on magnetic tape.
An airstream purified by passage through activated charcoal and molecular sieves was passed continuously via a 3mm diameter tube over the sensory pad at a rate of 60ml/min. Test extracts were introduced into this airstream for delivery to the preparation and the amplified electrical responses were recorded on tape for computer analysis using packages such as Autospike (Syntech, Hilversum).

Behavioural assays

Field collected slugs were fed on chinese cabbage (*Brassica chinensis*) for 24 hours, then starved for 24 hours. They were then placed in separate dishes, each of which contained a pre-weighed wheat flour pellet which had been dosed with either plant extract or solvent control. After 24 hours the pellets were dried and re-weighed, and the amount consumed by the slug calculated.

RESULTS AND DISCUSSION

Recordings of electrical activity in the olfactory nerve showed marked differences in response to closely related species of Umbelliferae using only the volatile chemicals present within the plant foliage (Fig 1). Of the 33 different species screened for activity, 21 triggered nervous activity in the slug tentacle preparation, with intense responses induced by certain extracts including hemlock and curled chervil. Subsequent behavioural bioassays demonstrated that extracts from both these species were strongly antifeedant. Conversely, extracts from species such as ground elder and rough chervil caused a weak nerve response and had little deterrent effect on slug feeding (Table 1).

Analysis of the extract of curled chervil, using gas chromatograph linked to a mass spectrometer and both electrophysiological and feeding bioassays, revealed the presence of (+)fenchone and 4-allylanisole, both of which are known slug antifeedants previously found in the herb fennel (Garraway, 1992). The activity of the most potent extract, hemlock, has been shown to be due to a new slug antifeedant compound. When the feeding tests were repeated using pure synthetic chemical (at the concentration found in the plant extract), the same results were obtained as with the hemlock extract (Table 2). Tests are currently under way to assess how effective these chemicals are when applied as topical sprays to various crops.

Fig 1. Electrical activity produced by the olfactory nerve of *Deroceras reticulatum* when exposed to odours from the plant extracts of (A) rough chervil; (B) ground elder; (C) coriander and (D) hemlock.

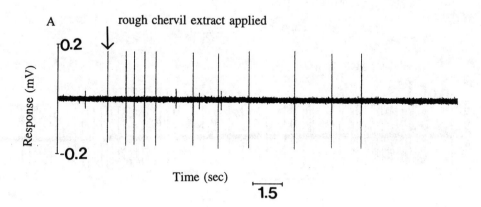

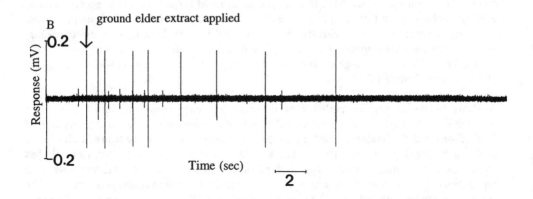

B

Response (mV)

ground elder extract applied

0.2

-0.2

Time (sec)

2

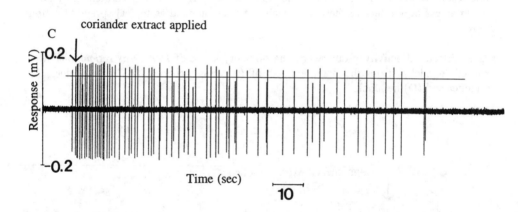

C

Response (mV)

coriander extract applied

0.2

-0.2

Time (sec)

10

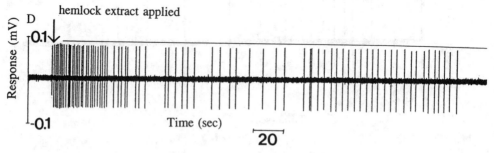

D

Response (mV)

hemlock extract applied

0.1

-0.1

Time (sec)

20

Table 1. Effects of extracts from four species of Umbelliferae on the electrical activity of the olfactory nerve of *Deroceras reticulatum* and on the weight of treated food ingested.

Plant species	Electrophysiological bioassay (total spikes/stimulus)	Feeding bioassay (percent change in feeding + S.E.)
hemlock (*Conium maculatum*)	80	-71.8±9.6
curled chervil (*Anthriscus cerefolium*)	74	-55.1±6.2
rough chervil (*Chaerophyllum temulentum*)	14	-16.5±11.9
ground elder (*Aegopodium podagraria*)	15	-11.2±8.

Table 2. Effects of pure compounds on ingestion of wheat flour pellets over 24 hours.

Plant species	Active chemical	Bioassay result (percent change in feeding + S.E. *)
hemlock	identified compound	-68.6±5.4
curled chervil	(+) fenchone	-42.9±8.7
	4-allylanisole	-26.6±9.1
	mix of fenchone & anisole	-60.2±5.6

* tested at the concentration found within the plant.

ACKNOWLEDGEMENTS

This work was funded by the BBSRC, MAFF, HGCA and the Perry Foundation. I wish to thank Mervyn Southam for the supply of many Umbelliferae and Ian F Henderson and Andrew P Martin for technical advice and assistance. IACR-Rothamsted receives grant aided support from the Biotechnology and Biological Sciences Research Council, the Home Grown Cereals Authority, the Perry Foundation and the Ministry of Agriculture Fisheries and Food.

REFERENCES

Bailey, S E R; McCrohan, C R (1989)   Foraging behaviour of terrestrial gastropods: Integrating field and laboratory studies. *Journal of Molluscan Studies.* **55**, 263-273.

Cook, A (1985) Tentacular function in trail following by the pulmonate slug *Limax pseudoflavus* Evans. *Journal of Molluscan Studies.* **51**, 240-247.

Craveiro, A A; Matos, F J A; Alencar, J W; Plumel, M M (1989) Microwave oven extraction of an essential oil. *Flavour and Fragrance Journal.* **4**, 43-44.

Garraway, R (1992) The action of semiochemicals on olfactory nerve activity and behaviour of *Deroceras reticulatum* (Müll). *PhD thesis, University of Portsmouth.*

Getz, L L (1959) Notes on the ecology of slugs: *Arion circumscriptus, Deroceras reticulatum* and *D. laeve.* *The American Midland Naturalist.* **61**, 485-498

Tutin, T G (1980) Umbelliferae of the British Isles. *Botanical Society British Isles Handbook* 2. Publ. London.

A PHARMACOLOGICAL STUDY OF THE SMOOTH MUSCLE IN THE BUCCAL MASS OF *DEROCERAS RETICULATUM*

T J WRIGHT, H HUDDART

Division of Biological Sciences, Institute of Environmental and Biological Sciences, Lancaster University, Lancaster, LA1 4YQ, UK

J P EDWARDS

Central Science Laboratory, MAFF, London Road, Slough, SL3 7HJ, UK

ABSTRACT

Conventional organ bath techniques were used to investigate the pharmacology of the smooth muscle which makes up the bulk of the buccal mass in this pest slug. Muscular contraction of the buccal mass was stimulated by K^+ depolarisation, acetylcholine, GTP and 5-hydroxytryptamine, but no specific downregulator was identified. These responses were all at least partly dependent upon external Ca^{2+}. The molluscan neuropeptide FMRFamide was without any effect. The three mammalian calcium antagonists tested gave a variety of responses, indicating differences between the mammalian and molluscan calcium channels which supply activating Ca^{2+} to the contractile apparatus.

INTRODUCTION

Slugs are major agricultural pests in many parts of the world. *Deroceras reticulatum* is generally considered to be the most important crop pest slug in the UK, particularly of winter wheat, potatoes and oilseed rape. There is a constant search for novel slug control methods, preferably ones specific to the target pests. One such possibility is interference with feeding, as molluscs have a unique feeding system. Study of the structure and physiology of the buccal mass (feeding organ) would improve the understanding of slug feeding.

The pulmonate buccal mass has an extremely complex musculature, involving 28 different muscles (Carriker, 1946, Smith, 1990). These muscles act together to carry out the rhythmic feeding movements of the buccal mass and the radula and odontophore within it. Unpublished light and electron microscopy studies of the buccal mass of *D. reticulatum* reveal the large amount of muscle present. This smooth muscle is typically molluscan in nature, with characteristic blocks each containing actin and myosin filaments surrounding a central cluster of small mitochondria. There appears to be a large amount of sarcoplasmic reticulum present throughout these fibres - a common feature of molluscan smooth muscle, but quite unlike vertebrate smooth muscle.

In this paper, the pharmacology of the feeding apparatus of *D. reticulatum* is briefly investigated. This work is being performed in conjunction with ongoing light and electron microscopy ultrastructural studies, x-ray microanalysis, and a chitin synthesis radioassay.

MATERIALS AND METHODS

A breeding population of *D. reticulatum* was maintained in plastic containers, and fed on sliced carrot, lettuce and wheat seeds. Slugs were sliced open down the dorsal midline and pinned apart. The buccal masses were removed whole by severing the overlying nerve ganglion, and cutting through the oesophagus and around the mouth opening. They were then ligated with fine cotton and suspended in a conventional organ bath setup. Tension was detected by a Grass FT.03 force-displacement transducer, amplified, and recorded as a curvilinear ink trace on a Grass model 7 polygraph. The buccal masses were maintained in an aerated artificial saline based on one used for *Helix aspersa* (Gibson & Logan, 1992). For calcium dependency experiments, a nominally calcium-free saline was made up by substituting $CaCl_2$ with choline chloride. Drugs were tested by adding known amounts of stock solutions to the organ bath. After each test, the saline was drained and replaced with fresh saline from an overhead reservoir. Tests were performed at room temperature.

RESULTS

On numerous occasions the buccal mass showed spontaneous rhythmic activity without any artificial stimulus, and in some a regular pattern could be seen (Figure 1). Even prior to placing in the organ baths, some buccal masses could be seen to visibly contract.

Addition of KCl caused significant contraction of the buccal mass (Figure 2). Concentrations of 5mM or above caused a significant increase in tonic force, up to a maximum at around 20mM. There was no phasic activity.

Acetylcholine (ACh) is a major molluscan neurotransmitter, as it is in vertebrates. It caused contraction in the form of smooth tonic force, with some phasic activity seen occasionally. Figure 3 shows a typical response to a dose of $5x10^{-5}M$.

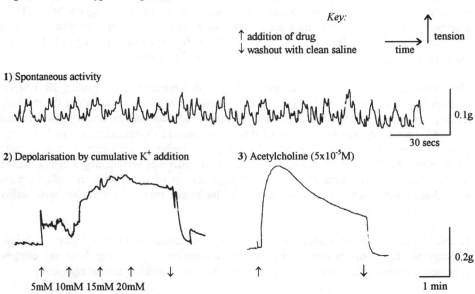

Key:

↑ addition of drug   ⟶ ↑ tension
↓ washout with clean saline   time

1) Spontaneous activity

0.1g

30 secs

2) Depolarisation by cumulative K^+ addition

↑   ↑   ↑   ↑   ↓
5mM 10mM 15mM 20mM

3) Acetylcholine ($5x10^{-5}M$)

↑   ↓

0.2g

1 min

5-hydroxytryptamine (5-HT), or serotonin, was found to be a potent upregulator of muscular activity. The preparation responded in a dose-dependent manner (Figure 4) with a threshold concentration of around 5×10^{-6}M. A concentration of 5×10^{-5}M clearly shows considerable phasic muscular twitching superimposed upon a small increase in tonic force (see also Figure 8, bottom trace). The same dose added to 5×10^{-5}M ACh gave an additive effect - increased contraction and the phasic response typical of 5-HT (Figure 5).

Guanosine triphosphate (GTP) was excitatory in the buccal mass, inducing an increase in tonic force along with some phasic activity (Figure 6). The response was dose-dependent with a very low threshold concentration around 10^{-7}M (Figure 7).

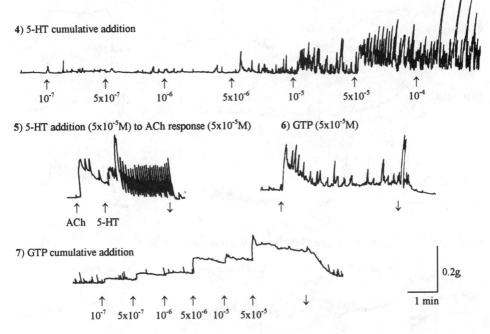

4) 5-HT cumulative addition

10^{-7} 5×10^{-7} 10^{-6} 5×10^{-6} 10^{-5} 5×10^{-5} 10^{-4}

5) 5-HT addition (5×10^{-5}M) to ACh response (5×10^{-5}M)   6) GTP (5×10^{-5}M)

ACh   5-HT

7) GTP cumulative addition

0.2g

10^{-7} 5×10^{-7} 10^{-6} 5×10^{-6} 10^{-5} 5×10^{-5}

1 min

Excitation-contraction (E-C) coupling of muscle relies on supply of Ca^{2+} to the contractile apparatus. In smooth muscle much of this comes from outside the cell. In order to assess this dependency, the effects of the four agents previously described (K^+, ACh, 5-HT and GTP) were tested on buccal masses equilibrated in a Ca^{2+} free saline (Figure 8). There was complete inhibition of ACh, K^+ and GTP responses, compared to the controls. In the case of 5-HT however, only the tonic component of the response was inhibited.

These results show the complete dependency of ACh on Ca^{2+} for E-C coupling. This was used to investigate the effects of three common mammalian organic calcium antagonists. In mammalian smooth muscle, these all block slow L-type Ca^{2+} channels. Their effects on this preparation were tested by adding them 8 minutes prior to a standard ACh dose, with controls before and after (Figure 9). As can be seen, they all inhibited contraction to varying degrees. Diltiazem and nifedipine were the most potent, though the inhibitory effect of nifedipine was longer lasting. Verapamil caused only a small inhibition of response, but curiously caused a small phasic response itself. This was also seen with nifedipine.

343

The tetrapeptide FMRFamide has been shown to have a variety of effects on molluscan smooth muscle (Painter *et al.*, 1982). In this preparation, it gave no response between 10^{-7}M and the relatively high concentration of 10^{-5}M (not illustrated).

8) Effect of Ca^{2+} free saline on various responses

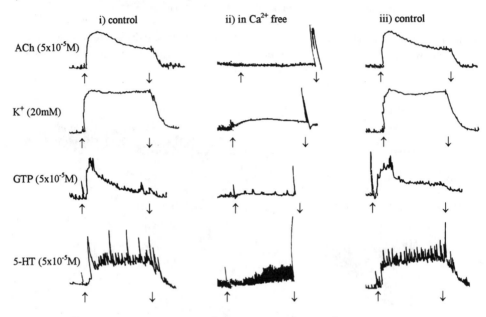

9) Effect of Ca^{2+} antagonist pretreatment (5×10^{-5}M) on ACh response (5×10^{-5}M)

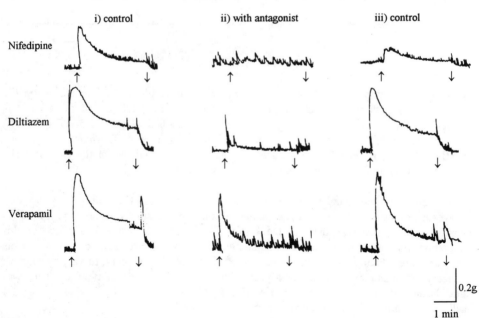

DISCUSSION

The spontaneous activity exhibited by the isolated buccal mass generally had a regular pattern. Similar activity has been shown by the buccal mass of the large slug *Arion ater* (Roach, 1966). This may well be related to the rhythmic process of feeding, involving precise movements of the buccal mass, odontophore and radula (see Huddart *et al.*, 1990). It is the complexity of this process which explains the large number of tiny muscles comprising or associated with the buccal mass. Such an intricate muscular system requires close nervous control. This preliminary pharmacological study may help to improve knowledge of this aspect of feeding control.

The buccal mass was very sensitive to KCl-induced depolarisation, opening voltage-sensitive Ca^{2+} channels and allowing a Ca^{2+} influx, thereby causing muscle contraction. The mechanical response was completely dependent on external Ca^{2+}. In some other molluscan muscles (Huddart *et al.*, 1990) and certain mammalian muscles (Langton & Huddart, 1987) KCl resulted in phasic activity at low concentrations and tonic force at higher concentrations, suggesting the presence of two different voltage-activated Ca^{2+} channels. The buccal mass of this species however, does not seem to display this KCl-induced phasic activity.

ACh appears to be the main neurotransmitter in this organ. Unpublished preliminary results with nicotinic and muscarinic agonists and antagonists show that the ACh receptor in these muscles does not consistently correspond to either of these vertebrate ACh receptor subtypes. The classic division of ACh receptors into nicotinic or muscarinic subtypes cannot be sensibly applied to the evolutionarily older invertebrate receptors. The effects of the three organic mammalian calcium antagonists on ACh responses show a similar difference between mammalian and invertebrate pharmacology. The results illustrate that the calcium channel activity of the *Deroceras* buccal mass differs from that of mammalian smooth muscle.

5-HT is well documented as a pharmacologically active substance in molluscs, with a variety of effects across the phylum, both excitatory and inhibitory, and both in cardiac and non-cardiac smooth muscle. In this preparation, it induced rapid twitches with an increase in tonic force. The tonic component was reliant on external Ca^{2+}, whilst the phasic component appeared to involve the release of intracellular Ca^{2+} stores. Consequently it is proposed that the two types of 5-HT response may act via different Ca^{2+} channels - a slow channel in the cell membrane, and a fast channel additionally in the membrane of internal Ca^{2+} stores, probably the sarcoplasmic reticulum mentioned earlier. 5-HT had an additive effect on the response to ACh, with the phasic contractions typical of 5-HT and an increase in tonic force. This suggests that 5-HT is causing an additional influx of Ca^{2+}, brought about by a different pathway.

Purines, including GTP, have a variety of effects on cardiac and non-cardiac molluscan smooth muscle. Here GTP was clearly excitatory, and dependent on external Ca^{2+}. The characteristic increase in tonic force with occasional twitching has also been recently reported in other molluscan muscles (Hill & Huddart, 1995).

Molluscan muscle is known to have a very complex innervation. A study of the radular retractor muscle of the whelk *Buccinum* (Nelson, 1994) found several types of nerve ending and four different types of neurotransmitter vesicles, with evidence for co-transmission. A similar study is currently being undertaken by transmission electron microscopy on the buccal muscles of *Deroceras*. These preliminary pharmacological studies illustrate the effects of a variety of pharmacologically active agents. Further work will hopefully aid understanding of this aspect of slug feeding control.

ACKNOWLEDGEMENTS

This work was carried out while TJW was in receipt of a Postgraduate Studentship from the States of Jersey, and a CASE Studentship from the Central Science Laboratory of MAFF.

REFERENCES

Carriker, M R (1946) Morphology of the alimentary system of the snail, *Lymnaea stagnalis appressa* Say. *Transactions of the Wisconsin Academy of Sciences, Arts and Letters.* **38**, 1-88.

Gibson, I C; Logan S D (1992) The actions of phorbol esters upon isolated calcium currents of *Helix aspersa* neurons. *Comparative Biochemistry and Physiology.* **102C**, 297-303.

Hill, R B; Huddart H (1995) Actions of GTP on molluscan buccal, cardiac and visceral smooth muscle. *Comparative Biochemistry and Physiology.* **111C**, 389-396.

Huddart, H; Brooks D D; Hill R B; Lennard R (1990) Diversity of mechanical responses and their possible underlying mechanisms in the proboscis muscles of *Busycon canaliculatum*. *Journal of Comparative Physiology B.* **159**, 717-725.

Langton, P D; Huddart H (1987) The involvement of fast calcium channel activity in the selective activation of phasic contractions by partial depolarisation in rat vas deferens smooth muscle. *General Pharmacology.* **18**, 47-55.

Nelson, I D (1994) The relation between excitation-contraction coupling and fine structure of a molluscan muscle, the radular retractor of the whelk, *Buccinum undatum*. *Journal of Comparative Physiology B.* **164**, 229-237.

Painter, S D; Morley J S; Price D A (1982) Structure-activity relations of the molluscan neuropeptide FMRFamide on some molluscan muscles. *Life Sciences.* **31**, 2471-2478.

Roach, D K (1966) Studies of some aspects of the physiology of *Arion ater*. PhD Thesis, University of Wales.

Smith, D A (1990) Comparative buccal anatomy in *Helisoma* (Mollusca, Pulmonata, Basommatophora). *Journal of Morphology.* **203**, 107-116.

UNUSUAL H$_2$O$_2$-GENERATING OXIDASES IN LAND GASTROPODS

K R N BAUMFORTH, M POWELL, A T LARGE, N GREWAL, R KAUR,
C J PERRY, M J CONNOCK

School of Applied Sciences, University of Wolverhampton, Wulfruna Street,
Wolverhampton WV1 1SB

ABSTRACT

We report on the detection and subcellular location of peroxide-generating oxidases which, amongst animals, appear to be unique to gastropods. These include: mannitol oxidase which is localised to it's own tubular membrane system; aromatic alcohol oxidase associated with endoplasmic reticulum; a membrane bound FAD-stimulated aliphatic alcohol oxidase most active with C8-12 chain length alcohols. Other oxidases studied include aldehyde oxidase and monoamine oxidase, each of which show particular " molluscan " features.

INTRODUCTION

An ideal molluscicide would be effective against snails and slugs while not impinging significantly on the biology of other organisms. It is likely that such an agent would interact with a target unique to molluscs. We have been studying the activities and subcellular locations of peroxide-generating oxidase enzymes of terrestrial slugs and snails. Amongst these we have identified several putatively unique " molluscan " features which are described below.

MATERIALS AND METHODS

Species studied have included: *Helix pomatia, Helix aspersa maxima*, giant African snails *Achatina fulica* and a Nigerian species provisionally identified as *Achatina achatina*, and the slugs *Arion ater* and *Limax flavus*. Subcellular fractionation in Beckman B XIV zonal rotors employed density gradients of sucrose and Iodixanol. An angled rotor was used for Percol gradients. Substrate dependent H$_2$O$_2$-generation was measured using a peroxidase-linked dye reaction employing amino antipyrene and 2,4,6-tribromohydroxybenzoic acid [TBHBA] synthesised essentially according to Trinder & Webster (1984) and analysed by NMR. Incubations included 1.2 mg/ ml TBHBA, 0.3 mg/ ml amino antipyrene, 0.3 mg/ ml horseradish peroxidase and 15 mM phosphate buffer pH 7.3. Reactions were terminated by addition of an equal volume of 10% SDS and absorbance measured at 510 nm or at 530 nm when FAD was present (at 530 nm the quinoneimine dye extinction was slightly diminished but that of interfering FAD reduced to ~10%). Other enzyme assays were performed by standard procedures and protein estimated using the coomassie blue dye binding method. SDS PAGE was performed on minigels by standard procedures and was followed by silver or Nano Orange fluorescent dye staining.

RESULTS

Mannitol oxidase (mannox): This membrane bound oxidase was first identified in

Helix aspersa by Vorhaben *et al* (1980) and has never, as far as we are aware, been reported in non-molluscs. Vorhaben *et al.* (1986) partially purified the solubilised enzyme and reported a subunit relative molecular mass of ~68 kD. We have found high mannox activity in all species of terrestrial gastropod we have examined and also detected activity in aquatic species *Limnaea stagnalis* and *Planorbis corneus.* The enzyme has highest activity in alimentary tissues but is also found in the kidney. Mannox was initially presumed to be a mitochondrial enzyme responsible for cyanide insensitive mannitol-dependent respiration. Our density gradient subcellular fractionation experiments (Large *et al.* 1993; Large & Connock, 1994) showed it was not localised to peroxisomes or mitochondria (as might be expected) but to an independent membrane we have termed " mannosome ". Electron microscopic examination of isolated pure mannosomes revealed that they are identical to the striking tubular structures previously observed in several electron microscope studies of various gastropod species and tissues. Below is a diagrammatic representation based on their appearance in slug alimentary tissue as reported by Moya & Rallo (1975) and Triebskorn & Kohler (1992).

Electron microscopists have almost universally interpreted these tubules as specialised regions of the endoplasmic reticulum. The elegant micrographs of Triebskorn and Kohler (1992) depict the tubules forming from a smooth parent membrane. However, on ultrastructural grounds without histochemical data one cannot be completely certain of the identity of the parent membrane. Thus this membrane may be SER, peroxisome reticulum, or even an independent membrane.

Subcellular fractionation studies establish that mannosomes : a] Have different physical properties to other cell membranes demonstrated by characteristic velocity sedimentation and equilibrium density in centrifugation experiments and making them separable from all other organelles; b] They completely lack activities of classical ER marker enzymes such as NADPH cytochrome c reductase, non-specific esterase and glucose 6 phosphatase; c] They do not share with ER any polpeptides demonstrable by SDS PAGE; d] They have distinct morphology. By these criteria (enzyme and polypeptide complement, centrifugal behaviour and structure) they represent a distinct organelle. Their origin from the ER, if indeed they are so derived, must involve a very precise mechanism for segregation/ exclusion of proteins.

We have now developed a large scale procedure for the isolation of mannosomes. This method involves equilibration in an "isosmotic" Iodixanol gradient followed by equilibration in a Percoll gradient, rate sedimentation into sucrose and finally density perturbation with digitonin and equilibration in a sucrose gradient. The purest

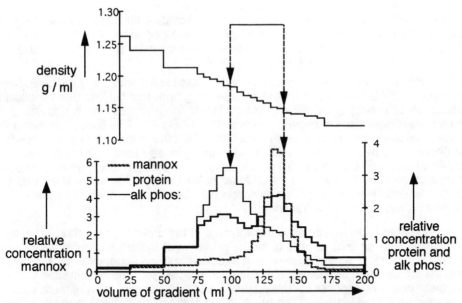

Fig 1. Distribution of protein, mannitol oxidase and alkaline phosphatase after centrifugation (48000 rpm x 80 min in a B XIV zonal rotor) into a linear density gradient of sucrose. The payload was derived from the mannox rich fractions isolated successively from Iodixanol, Percol and rate-sedimentation sucrose gradients. The vertical arrows include those fractions encompassing the two protein peaks and selected for SDS PAGE analysis illustrated in Fig 2.

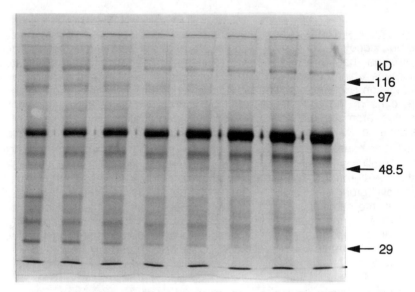

Fig 2. SDS PAGE of fractions enclosed by the arrows in Fig 1 encompassing the two protein peaks. Each track was loaded with ~ 0.5 μg of fraction protein as determined by the coomassie dye binding assay. Position of molecular mass markers is indicated on the right.

preparations are obtained from *Arion ater* and *Achatina achatina*. Figure 1 shows the distributions of mannox, protein and alkaline phosphatase in the final *A. achatina* gradient (alkaline phosphatase was used as a marker for plasma membrane, the major contaminant of mannosomes in this gradient).

The mannox peak at about 1.15 g/ml density is associated with a major protein peak. Analysis by SDS PAGE of the fractions spanning from this peak to that for alkaline phosphatase indicates that most of the protein in the mannosomes is accounted for by a single polypeptide of between 48 and 97 kD (Fig 2). This band is fainter in the peak fraction for alkaline phosphatase, whereas all other bands, except one, become stronger in the fractions toward the alkaline phosphatase peak. Similar results were obtained after gel staining with fluorescent dye. The major polypeptide is also the strongest band in the region of the denser protein peak. It appears likely that this protein is underestimated in the coomassie blue binding assay used to estimate the protein concentration in the fractions.

Since the major polypeptide accounts for most of the protein in the mannosomes we speculate that, as well as acting as an oxidase (if indeed it does represent the mannox subunit) , it must also have a structural role and contribute to the striking tubular morphology. Studies by electron microscopists indicate an increase in these tubules occurs in response to 'stress' stimuli such as administration of molluscicide , starvation and shell damage. The tubular form of these structures makes it tempting to speculate that they may act as conduits for segregation or transport of selected metabolites. The products of the mannitol oxidase reaction are mannose and hydrogen peroxide. If mannose was compartmentalised from the cytosol it might be prevented from entering glycolysis, it could then perhaps be delivered elsewhere and used for 'specialised' purposes (eg. complex carbohydrate / glycoprotein synthesis). Mannose has been detected in gastropod blood.

Aromatic alcohol oxidase (arox): We first reported the presence of arox in digestive gland and kidney of *Arion ater* (Mann *et al.*, 1989). We have found similar activity in all terrestrial gastropods we have examined. The enzyme is associated with membranes which cosediment with ER markers in velocity and density dependent centrifugation experiments. The only other reports of arox activity of which we are aware concern the enzyme secreted by various wood rot fungi. Lignin polymer generates aromatic alcohols on oxidative degradation and these can act as substrates for H_2O_2 production via arox action. This serves a useful purpose because ligninase requires H_2O_2 to function. The fungal arox enzymes contrast with those of molluscs in being soluble and in having slightly different but equally broad substrate preferences. The molluscan enzyme can also oxidise short chain alcohols with an allyl group (eg. crotyl alcohol). Soluble aldehyde oxidase (aldox) is also present in the same tissues (Large & Connock, 1993). The concerted action of these two enzymes could convert aromatic alcohols to their corresponding acids. Terrestrial gastropods are likely to consume plant material containing small amounts of lignin and aromatic lignin precursors together with partially degraded lignin from 'decaying' food.

FAD-dependent aliphatic alcohol oxidase (alox). We have recently found evidence of a further alcohol oxidase in digestive gland and kidney of terrestrial gastropods. In experiments to which various potential oxidase cofactors were added it was observed that FAD stimulated H_2O_2-generation in the absence of added substrate.

Most of this FAD stimulated activity was eventually found to be due to the action of an enzyme upon ethanol present in the TBHBA . During it's preparation TBHBA had been recrystallised from ethanol. NMR spectroscopy showed the presence of 1 mol. of ethanol per 2 mol. TBHBA. NMR indicated complete removal of ethanol when TBHBA was heated at 120 °C overnight. Use of heated TBHBA in the oxidase assay almost completely eliminated FAD-stimulated activity thereby indicating the presence of an aliphatic alcohol oxidase.

Alox is associated with the same membranes as arox, however it is demonstrated to be a different enzyme by it's far greater temperature sensitivity and by it's stimulation by exogenous FAD. The apparent Km for ethanol was found to be ~1 mM. The apparent affinity for longer chain alcohols was much greater with apparent Km values of ~130 μM and ~60 μM for propanol and butanol respectively. In order to investigate longer chain alcohols it was necessary to dissolve them in dimethyl sulphoxide (DMSO), which at 15% in the incubation had no detectable inhibitory effect on the activity with shorter chain substrates. The action of the enzyme from *Helix aspersa maxima* was tested upon different chain length saturated alcohols presented at 440 μM in DMSO. Greatest activity was observed for chain lengths C8-C12. Similar results have been obtained with other species of gastropod. The affinity of the enzyme is greatest with these medium chain substrates but Km values remain to be measured with precision. The enzyme can be solubilised in active form and we are now employing affinity chromatography on an agarose-FAD column in attempt to purify it.

Alcohol oxidation is a wide spread capability in organisms. Aliphatic alcohol oxidases are well known in methyltrophic yeasts. These FAD dependent enzymes prefer methanol as substrate and have greatly reduced activity as chain length increases. The higher plant *Tanacetum vulgare* contains an oxidase for primary medium chain alcohols and shows preference for unsaturated substrate. Alkane utilising yeasts have a long chain aliphatic alcohol oxidase. As far as we are aware an enzyme of this type has never previously been reported in animals. The molluscan enzyme requires further work to determine it's properties and function. The waxes on the surface of plant cells in leaves and fruits could supply substrate for the gastropod enzyme. Gastropod digestive gland is an extremely rich source of ether-linked phospholipids and it is feasible that the oxidase plays a role in the metabolism of alcohol used for, or derived from, these phospholipids. Medium chain aliphatic alcohols are membrane seeking and in mammals have a potent anaesthetic action.

Other oxidases: Gastropod tissues contain a monoamine oxidase active with benzylamine and phenylethylamine as substrates. Most of this activity is membrane bound but not associated with mitochondria. It is far more active and more temperature sensitive than the monoamine oxidase exhibited by gastropod digestive gland mitochondria. It does not correspond exactly in inhibitor response or substrate preferences to the semicarbazide sensitive plasma membrane monoamine oxidase of mammals. It's function and distribution in gastropod tissues requires further investigation.

Lastly we have found an FAD-dependent oxidase activity associated with digestive gland membranes which apparently acts on an endogenous substrate. This activity has similar FAD reliance and temperature sensitivity to the alox described above. Initial results indicate that this activity increases with prolonged storage of the membranes (at -20 °C). These results imply that the substrate for H_2O_2 generation is accumulating during storage of the membrane due to slow degradative or other changes that are taking place.

DISCUSSION

The tissues of terrestrial gastropods contain at least three oxidases which, amongst animals, appear to be unique to molluscs. One of these, mannitol oxidase, is associated with an unusual subcellular membrane which electron microscopists have previously described as increasing in abundance in response to stress stimuli. The physiological function and biological importance of these enzymes and structures is far from clear and remains to be established by future work, however their apparent restriction to terrestrial gastropods means that they represent potential targets of selective molluscicide action. We are presently attempting to investigate the roles of these oxidases by more clearly defining their properties, cellular and subcellular distributions, and by designing inhibitors which may be administered to animals so as to perturb oxidase function.

REFERENCES

Large, A T; Connock M J (1993) Aromatic alcohol oxidase and aldehyde oxidase activities in the digestive gland of three species of terrestrial gastropod. *Comparative Biochemistry and Physiology.* **104B**, 489-491.

Large, A T; Jones C J P; Connock M J (1993) The association of mannitol oxidase with a distinct organelle in the digestive gland of the terrestrial slug Arion ater. *Protoplasma.* **175**, 93-101.

Large, A T; Connock M J (1994) Centrifugal evidence for association of mannitol oxidase with distinct organelles (mannosomes) in the digestive gland of several species of terrestrial gastropod. *Comparative Biochemistry and Physiology.* **107A**, 621-629.

Mann V; Large A T; Khan S; Malik Z; Connock M J (1989). Aromatic alcohol oxidase: a new membrane bound H2O2-generating enzyme in alimentary tissues of the slug *Arion ater. Journal of Experimental Zoology.* **251**, 265-274.

Moya J; Rallo A M (1975). Intracisternal polycylinders: A cytoplasmic structure in cells of the terrestrial slug *Arion empiricorum. Cell and Tissue Research.* **159**, 423-433.

Triebskorn R; Kohler H-R (1992) Plasticity of the endoplasmic reticulum in three cell types of slugs poisoned with molluscicides. *Protoplasma.* **169**, 120-129.

Trinder P; Webster D (1984) Determination of HDL cholesterol using 2,4,6 tribromo 3 hydroxybenzoic acid with a commercial CHOD-PAP reagent. *Annals of Clinical Biochemistry.* **21**, 430-433.

Vorhaben J; Scott J F; Campbell J W (1980) D-mannitol oxidation in the land snail *Helix aspersa. Journal of Biological Chemistry.* **255**, 1950-1955.

Vorhaben J; Smith D D; Campbell J W (1986) Mannitol oxidase: partial purification and characterisation of the membrane bound enzyme from the snail *Helix aspersa. International Journal of Biochemistry.* **18**, 337-344.

HEAVY METAL TOXICITY IN THE SNAIL *HELIX ASPERSA MAXIMA* REARED ON COMMERCIAL FARMS: CELLULAR PATHOLOGY

M D BRADLEY, N W RUNHAM

School of Biological Sciences, University of Wales, Bangor, Gwynedd, LL57 2UW, UK

ABSTRACT

The effects of the three metals copper, zinc and manganese presented in solution, have been investigated on the commonly farmed snail *Helix aspersa maxima*. Copper acts on the external epithelium of the foot, mantle and lung, resulting in the rupture of epidermal cells so allowing blood seepage and facilitating bacterial infection. Zinc pathology occurs internally, primarily by necrosis of the digestive and basophil cells of the digestive gland and disruption of spermatogenesis in the ovotestis. Despite the low toxicity of manganese, transformations were observed in the basophil cells of the digestive gland. The characteristic granules of these cells were transformed by the deposition of micro-crystalline deposits on their outer surface.

INTRODUCTION

The present restrictive economic environment has caused commercial snail farmers to become increasingly reliant on an automated approach to cultivation. The most successful of these automated systems is the Hawkyard Self-Cleaning Cage which reduces labour costs to a minimum and yields marketable snails in 9 - 10 weeks. However the Hawkyard system has been found to have only limited success when introduced into some farms, in extreme cases resulting in high mortalities of livestock. This has been attributed to heavy metal contamination of the water supply, the animals displaying similar characteristic symptoms as described by Runham (1991) for copper induced pathology in *Achatina fulica*.

The animals are cleaned once a day by spraying with water. During a cleaning cycle the snail is extend fully from the shell exposing the entire dorsal and ventral surfaces of the foot and head. These external surfaces are well known to allow the passage of solutes into the body and it is likely that metallic ions are able to exploit this same route, allowing direct entry into the haemolymph (Burbidge *et al.*, 1994).

MATERIALS AND METHODS

Exposure Regime

Three toxicity bioassays were conducted, one for each of the metals under investigation; zinc, copper and manganese. A range of five concentrations were selected for copper and manganese, while six were chosen for zinc. The concentrations used were:
For zinc: $1500\mu g/ml$, $1000\mu g/ml$, $500\mu g/ml$, $300\mu g/ml$, $100\mu g/ml$ and $10\mu g/ml$.
For manganese: $2500\mu g/ml$, $2000\mu g/ml$, $1500\mu g/ml$, $1000\mu g/ml$ and $500\mu g/ml$.
For copper: $10\mu g/ml$, $5\mu g/ml$, $4\mu g/ml$, $3\mu g/ml$ and $1\mu g/ml$.

The control group was exposed to ultra-pure deionised water (conductivity <0.4μs), while a further thirty animals were deprived of food to monitor the effect of starvation on mortality.

Each day, for thirty days the experimental groups were removed from their holding chambers and placed in plastic containers to which a litre of treatment solution was added. This provided sufficient volume to completely submerge the animals. After a fifteen minute treatment period, which was designed to mimic the washing period of the Hawkyard system, the animals were returned to their holding pens.

Each day the number of dead animals was recorded.

Material Preparation

The tissues of animals surviving the experimental period were dissected and prepared either for light microscopy by embedding in Historesin or Transmission Electron Microscopy (TEM) by embedding in Transmit resin.

RESULTS

Mortality

There were no control deaths during the thirty day experimental period. Probit analysis was used to determine the LC50 value for the zinc and copper treatment groups, it was not possible to perform this analysis on the manganese treated animals due to the insignificant mortality rate. The LC50 for copper was 4.22μg/ml (upper limit=4.5μg/ml, lower limit=3.97μg/ml) and for zinc was 563.25μg/ml (upper limit=684.65μg/ml, lower limit=460.02μg/ml).

With zinc (Figure 1) significant mortality occurs between the four highest treatment groups and the control ($p<0.05$ for 300μg/ml; $p<0.01$ for 500 and 1000μg/ml; $p<0.001$ for 1500μg/ml). The concentrations of 4 - 10μg/ml of copper (Figure 2) all result in a mortality rate that is significantly different from the control at the end of the thirty day period ($p<0.01$ for 4μg/ml; $p<0.001$ for 5 and 10μg/ml). Surprisingly, large changes in manganese concentration do not result in a significantly different mortality rate from the control, in any of the treatment groups. Animals starved over the thirty day period do show a degree of mortality which is significantly different from the control ($p=<0.01$). This mortality rate is lower than for the highest two concentrations of copper and zinc.

Pathology

Starvation:
The characteristic vacuole of the glycogen cell, which is found in the connective tissue of most organs, is dramatically reduced. The resulting cell is smaller with a dense cytoplasm that stains intensely with bromophenol blue and contains numerous small granules (1 - 2μm). In the digestive gland the digestive cells are predominantly of the B-type, but appear highly disorganised with large secondary lysosomes (green granules) dominating the cytoplasm. In the kidney there is a tendency towards larger excretory granules which are highly irregular in shape.

Zinc Treatment:

At concentrations above 100µg/ml cellular pathologies become evident, and a sequence of histological and morphometrical changes can be related to the concentration used.

Figure 1   Figure 2

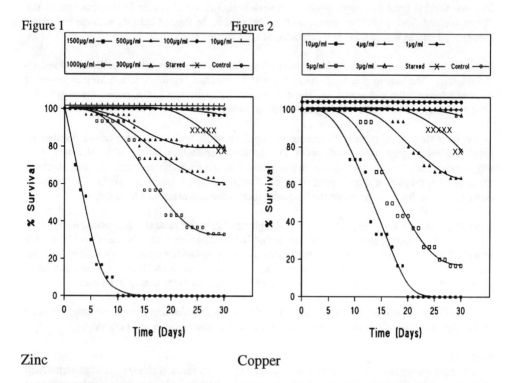

Zinc   Copper

Animals were exposed to copper and zinc solutions by immersion for 15 minutes a day for 30 days. The number of animals surviving 24h after treatment were recorded. Starved animals were deprived of food over the same period and are included as a reference for mortality caused by starvation.

In the digestive gland there is a gradual reduction in the proportion of digestive and basophil cells which is contrasted with a rise in the excretory cell contribution. The digestive cells experience extensive vacuolisation of the cytoplasm to form green granules. At 1000µg/ml the cell loses its structural integrity acquiring a bloated appearance with a swollen apical region, which lacks microvilli. Increasing concentrations lead to an abundance of basophil cells that have few granules. At the highest concentration there is an increase in the incidence of basophil cells which contain a large heterogenous vacuole (autolysosome), in extreme cases the cell may contain several of these vacuoles. The most profound cellular response is displayed by the excretory cells, the small spherical granules coalesce resulting in large irregular concretions which have an internal structure of concentric rings.

Exposure to zinc has a profound effect on spermatogenesis. There is a reduction in the number of developing spermatocytes, the tubules becoming dominated by spermatozoa at the highest levels of zinc. The spermatozoa at all stages of development are randomly distributed throughout the tubule, free of the degenerating sertoli cells. There is a proliferation of haemocytes in the connective tissue which are frequently observed to penetrate the acinus wall. The disorganisation of the male gametes contrasts with that of the female gametes, as oogenesis appears to continue unhindered.

The highest two zinc concentrations result in an observable reduction in the size of the vacuole of the glycogen cell in the connective tissue, but not to the extent recorded in the starved animals.

Copper Treatment:
Only the highest level of copper (4μg/ml) reveals a significant alteration in the structure of the digestive gland. The epithelium resembles that observed in the starved animals with the exception of the blood spaces where there is a profusion of haemocytes.

Exposure to 3 and 4μg/ml results in an abundance of haemocyte aggregations in the blood spaces of the kidney. The haemocytes infiltrate and engulf the excretory epithelium, resulting in localised areas of intense cellular degradation. The nephrocytes lack structure and acquire a dense granular composition which stains an intense purple with toluidine blue.

In the lung there is a general thickening of capillary walls and an increase in the number of circulating haemocytes in the blood spaces. The respiratory epithelium is extensively damaged with a high incidence of cellular sloughing; in places this leaves a naked basement membrane. Bacteria are seen to aggregate in a thick cellular mass of necrotic cells on the outer surface of the lung, eventually these bacteria penetrate the basement membrane and enter the blood spaces.

Similarly, treatment with copper results in a sloughing of both the ventral and dorsal epithelium of the foot, this process being more extreme at the higher concentration. The cells which remain attached to the basement membrane show signs of necrosis, lacking cytoplasmic density. On the outer surface the mucus layers become dominated by cellular debris and bacteria, which penetrate the epithelium and result in the occurrence of large granulomas in the connective tissue.

Treatment with 4μg/ml of copper results in a slight, but definite reduction in the volume occupied by the glycogen vacuole of the glycogen cell in the connective tissue in most organs.

Manganese:
The only pathology resulting from manganese exposure is a structural change in the granules held in the basophil cells of the digestive gland. The granules in these cells lose their regular, spherical shape and adopt a highly irregular, stellate crystalline structure.

DISCUSSION

Mortality

Copper and zinc in solution clearly have a deleterious effect on terrestrial snail survival, while manganese does not. There is relatively little data available on the effect of metals in solution on terrestrial molluscs. Previous authors have assumed that in the terrestrial environment the highest intake of toxic metals will occur via the food. This may be true. However in its bound particulate form the metal is clearly less hazardous to life (Burbidge *et al.*, 1994). In the mortality experiment documented here concentrations as low as 4μg/ml of copper and 300μg/ml of zinc caused significant mortality in a relatively short time period. These concentrations are considerably lower than those used by other authors in feeding trials (Dallinger & Wieser, 1984, Marigomez *et al.*, 1986).

Pathology

The formation of large secondary lysosomes in the digestive cells of the digestive gland are not exclusively a characteristic of metal induced pathology. It is rather a general response to stress occurring readily in animals starved for thirty days. These large secondary lysosomes release powerful hydrolysing enzymes that readily digest the cell.

In zinc treated animals the autolysis of basophil cells is surprising since this is the site of zinc accumulation and detoxification. This suggests that the cell is unable to successfully regulate zinc accumulation at the highest treatment levels. Manganese is also thought to be accumulated by basophil cells, resulting in the dramatic transformation of the enclosed granules. Taylor *et al.*, (1988) proposed that the incorporation of manganese ions into these granules caused the liberation of free calcium ions into the cytoplasm. Such a phenomenon would cause extensive cellular damage (Viarengo *et al.*, 1994), which was not evident in the tissues of *H. aspersa*.

The excretory cell granules have been likened to the type-B granules of terrestrial arthropods. Exposure of the cockroach *Blatella germanica* to mercury, results in an identical transformation of the type-B granules to that observed in the excretory granules of *H. aspersa* in response to zinc exposure (Hopkin, 1989). This may suggest that the excretory cells have a specialised role in the detoxification of certain heavy metals.

The decline in the size of the glycogen cell vacuole is undoubtedly related to stress induced starvation, glycogen utilization being one of the most general indicators of environmental stress (Marigomez & Ireland, 1990).

Enlarged granules in the kidney of zinc treated animals are possibly related to the reduced excretion rate, a similar situation occurring in snails which hibernate over winter (Jeżewska *et al.*, 1963). The swelling of the kidney in response to copper exposure is thought to be the result of an influx of water from the body fluids, possibly as part of an inflammatory reaction. This inflammation may be a direct response to metal toxicity or a secondary response to bacterial invasion of the tissue, facilitated by metal ion suppression of haemocyte phagocytosis (Brown *et al.*, 1992).

Hyperplasia of the ovotestis has been observed in many molluscs as a consequence of pollution. Phagocytosis of spermatocytes is believed to act as a supplement to compliment the utilization of storage reserves (Marigomez *et al.*, 1990). This seems unlikely to be the cause of spermatocyte atresia in zinc treated animals, since no such response occurred in starved animals and autolysis is an immediate response before storage reserves are depleted. The degeneration of sertoli cells appears to be an important contributor to the observed responses of developing sperm to the presence of zinc.

The epithelium of the foot, mantle and lung represent the external surface of the animal which is known to be highly sensitive to copper exposure. Copper is thought to induce necrosis in the epithelial cells because it strongly stimulates lipid peroxidation of the plasma membranes (Viarengo *et al.*, 1990). Sloughing of the epithelial cells creates open lesions and generates an osmotic gradient with the bathing solution. This leads to rapid autolysis, the products of which ooze into the blood spaces and initiate an inflammatory response which is typified by an increase in the number of circulating haemocytes. Bacterial invasion of the open wound intensifies this response resulting in the production of granulomas and extensive tissue necrosis. The invading bacteria enter the general circulation and are translocated to other organs, the bacteria in the kidney probably arrive by this route and intensify the initial effect of copper ions on cellular pathology.

The pathologies as described for zinc and copper toxicity would have serious consequences for a snail farmer, resulting in lower growth rates, reduced fertility, deterioration in the general health of animals and the possibility of infection by bacteria, potentially harmful to both the snail and the consumer (Obi & Nzeako, 1980).

REFERENCES

Brown, R P; Cristini A; Cooper K R (1992) Histopathological Alterations in *Mya arenaria* Following a #2 Fuel Oil Spill in the Arthur Kill, Elizabeth, New Jersey. *Marine Environmental Research*. **34**, 65 - 68.

Burbidge, F J; Macey D J; Webb J; Talbot R (1994) A Comparison Between Particulate (Elemental) Zinc and Soluble Zinc ($ZnCl_2$) Uptake and Effects in the Mussel, *Mytilus edulis*. *Archives of Environmental Contamination and Toxicology*. **26**, 466 - 472.

Dallinger, R; Wieser W (1984) Patterns of Accumulation, Distribution and Liberation of Zn, Cu, Cd and Pb in Different Organs of the Land Snail *Helix pomatia* L. *Comparative Biochemistry and Physiology*. **79C**, 117 - 124.

Hopkin, S P (1989) *Ecophysiology of Metals in Terrestrial Invertebrates*, New York: Elsevier Applied Science Publishers Ltd.

Jeżewska, M M; Gorzkowski B; Heller J (1963) Seasonal Changes in the Excretion of Nitrogen Wastes in *Helix pomatia*. *Acta Biochimica Polonica*. **11**, 309 - 314.

Marigomez, J A; Ireland M P (1990) A Laboratory Study of Cadmium Exposure in *Littorina littorea* in Relation to Environmental Cadmium and Exposure Time. *The Science of the Total Environment*. **90**, 75 - 87.

Marigomez, J A; Angulo E; Saez V (1986) Feeding and Growth Responses to Copper, Zinc, Mercury and Lead in the terrestrial Gastropod *Arion ater* (Linne). *Journal of Molluscan Studies*. **52**, 68 -78.

Marigomez, J A; Cajaraville M P; Angulo E (1990) Histopathology of the Digestive Gland-Gonad Complex of the Marine Prosobranch *Littorina littorea* Exposed to Cadmium. *Diseases of Aquatic Organisms*. **9**, 229 - 238.

Obi, S K C; Nzeako B C (1980) *Salmonella, Arizona, Shigella* and *Aeromonas* Isolated from the Snail *Achatina achatina* in Nigeria. *Antonie van Leeuwenhoek*. **46**, 475 - 481.

Runham, N W (1991) Reproduction and Physiology of Commonly Farmed Snails. In: *Snails: The First UK Snail Festival*, R E Groves (ed), Colwyn Bay: The Snail Centre.

Taylor, M G; Simkiss K; Greaves G N; Harries J (1988) Corrosion of Intracellular Granules and Cell Death. *Proceedings of the Royal Society of London Series B*. **234B**, 463 - 476.

Viarengo, A; Canesi L; Pertica M; Poli G; Moore M N; Orunesu M (1990) Heavy Metal Effects on Lipid Peroxidation in the Tissues of *Mytilus galloprovincialis* Lam. *Comparative Biochemistry and Physiology*. **97C**, 37 - 42.

Viarengo, A; Canesi L; Moore M N; Orunesu M (1994) Effects of Hg^{2+} and Cu^{2+} on the cytosolic Ca^{2+} Level in Molluscan Blood Cells Evaluated by Confocal Microscopy and Spectrofluorimetry. *Marine Biology*. **119**, 557 - 564.

LONG-TERM EFFECTS OF *RICCARDOELLA LIMACUM* LIVING IN THE LUNG OF *HELIX ASPERSA*

F J GRAHAM, N W RUNHAM, J B FORD

School of Biological Sciences, University of Wales Bangor, Gwynedd LL57 2UW, UK

ABSTRACT

Heavy infestations with the mite *Riccardoella limacum* reduce the growth rate and considerably delay reproductive development of the snail *Helix aspersa*. Feeding by the mite in the lung of its host results in pathological changes in the respiratory epithelium, increases dormancy, reduces feeding activity and affects the strength and shape of the shell. The implications of these results for the commercial farming of snails are discussed.

INTRODUCTION

Although the presence of a small mite running over the body of snails and slugs is widely known, its identification has been a source of confusion. It is now clearly established that the mite associated with *Helix aspersa* Müller is *Riccardoella limacum* (Schrank)(Graham *et al.*, 1993). In the UK *H. aspersa* is intensively farmed using a variety of indoor systems (Runham, 1993) and some farmers have stated that *R. limacum* can become a serious pest under these conditions. When *H. aspersa* hibernates in the field in the autumn and winter the mites are present in the lung cavity where they lay a series of eggs (Graham, 1994). During most of the year these eggs hatch in 8-12 days but during hibernation they do not hatch. Only rarely do adult mites survive winter in the lung but in spring the larvae emerge from the eggs. In intensively farmed snails there is no period of hibernation. The mites are therefore free to breed throughout the year, the total life cycle lasting 19-23 days (Graham, 1994), so in conjunction with the raised temperature and humidity large populations can result.

While farmers having a problem with this mite are aware that it has adverse effects on the snail there is uncertainty about the diet of *R. limacum* and whether or not it can affect the physiology of the snail. Using immunological techniques it has now been established that this mite feeds on the blood of its host (Graham, 1994), confirming the histological evidence reported by Baker (1970).

When snails aggregate, as when they seek shelter to become dormant during the day, the mite is readily able to spread from snail to snail. The presence of a large number of mites running in and out of the lung would surely cause a physical problem to the host. André and Lamy (1930) state that the relationship between the snail and the mite is a purely commensal one and the snail does not suffer at all. However Baker (1970) working with the related *R. oudemansi* states "there is no doubt that in large numbers they cause considerable loss of blood in slugs".

The mites must attach to the host tissue in order to obtain nourishment, but living in the sheltered environment of the lung it is likely that the ambulacral claws or small chelicerae are adequate for this purpose (Fain, 1969). If, as is suspected by Baker (1970), a feeding tube or stylostome is formed then considerable tissue disturbance must occur within the host tissues and this structure is thought to persist even after the mite has finished feeding (Evans, 1992). Oldham (1931) discovered that unidentified mites living on *Arianta arbustorum* in nesting boxes caused total mortality in some of the boxes while in others growth was irregular, fecundity lessened and badly deformed shells were a regular occurrence.

Attempts to determine what effect *R. limacum* has on growth and reproduction in *H. aspersa* when relatively large populations of mites are present in the lung, are reported here.

MATERIALS AND METHODS

Juvenile (about 8 weeks old, 1cm diameter) *H. aspersa* were obtained from L'Escargot Anglais, Hereford. A culture of *R. limacum* was maintained at 22°C, on wild collected snails. When required, mites were removed from the snails with a fine paint brush and added to the experimental animals. Snails were fed on commercial snail food (Winstay Cereal Compound Mixture), cleaned weekly and sprayed daily with water to keep a high humidity.

Growth of snails was assessed from the fresh weight of snails, including the shell, and by measurements of the maximum height and diameter of the shell. As all snails were maintained in a constant high humidity the wide variation noted in field animals was probably limited to ±10% (Klein-Rollais & Daguzan, 1990). Shells were measured using vernier calipers.

At the end of the experiment reproductive maturation was assessed by visual examination of the reproductive system, histology of the gonad and the weight and histology of the albumen gland. As feeding by a large number of mites could be expected to affect the lung tissue this was also examined histologically.

For histology, tissues were fixed in Heidenhain's Susa, washed and dehydrated in 95% ethanol, cleared in cellusolve and embedded in Historesin. Sections, 4µm thick, were stained in a polychrome stain.

For biochemical analysis the soft body parts of the snails stored at -70°C were first freeze-dried and then ground to a fine powder using a pestle and mortar and a food blender. Samples from each snail, 2mg and 4mg for protein and carbohydrate analysis respectively, were weighed out to enable assays to be carried out, three times each with three replicates in each run.

The Sigma Diagnostics Protein Assay Kit (Procedure No. P5656) was used to determine total soluble protein and the Boehringer Mannheim Glucose GOD Peridochrome Test Combination P525 was used for glycogen and free glucose.

Results are presented as mean±standard deviation, and significance levels relate to Student t-tests between the means.

Experimental Procedures

Isolated Snails.

To study the effect of high and low incidence of mites snails were kept separately. Mites were added to immature (approximately eight weeks old, 1cm diameter) snails. Snails were housed in plastic plant propagators, 10cm x 7cm x 7cm, with ventilation holes pierced into the lids, and supplied with a small square of high absorbency gamgee to provide a moist environment and a source of fluid. Food was placed in the propagator and cleaning took place approximately every five days. Too frequent cleaning tends to damage shells as a result of handling. The propagators were kept in an incubator at 20-25 °C with a 12h light:12h dark cycle.

Sixty snails were divided into three groups of twenty animals each and mites transferred to them as follows: 1. High incidence; twenty adult mites added initially to each snail, followed by a "top up" of ten mites a month later. 2. Low incidence; five adult mites added to each snail initially, followed by a "top up" of five more a month later. 3. Control group; snails with no mites.

Snails were weighed at the start and then at monthly intervals thereafter. The general condition of the shell was assessed at weighing. The experiment continued for nine months to ensure that the snails had time to achieve sexual maturity(normally this would take about seven months). At the end of the experiment snails were weighed, shell heights and diameters measured and then the animals were killed and dissected. All incidences of mites were noted and snails from all three groups were further divided into two sub-groups to be treated as follows: 1. Whole animals (without shells) were wrapped in aluminium foil and frozen at -70 °C for later biochemical analyses. 2. Gonad, albumen gland and lung tissue were removed and prepared for light microscopy and the development of the reproductive system visually assessed.

Communally reared snails.

Snails were infested with mites as above but using only two groups: 1. High incidence of mites. 2. Control snails with no mites. Snails were kept communally in 30cm x 15cm x 10cm plastic boxes under the same conditions as for the previous experiment. This experiment was run for five months the snails being weighed monthly as before and treated in the same way at the end of the experiment. Additionally the individual weights of the albumen glands were noted.

RESULTS

General Observations.

It was noticed that snails with high mite infestations had a tendency to enter periods of dormancy (they adhered to the side of the box with dry mucus). These periods often lasted for the whole of the five days between cleaning. At any one time approximately 50% of the animals would be dormant, though it was not always the same individuals. Snails with no mites always remained active and ate well while those in the infected groups often left food uneaten.

The shells of the control group were to all appearances "normal" in that they were smooth and shiny with a good colour, they exhibited no damage from breakages and were quite strong. The differences between the low and control groups were negligible when compared to the differences between the control and the high incidence group. All shells in the control group and low incidence group had the thickened rim around the growing edge indicative of sexual maturity. Shells in the high incidence group were dull with weak colouration, they tended to be very brittle and weak, they regularly incurred damage from falling from the sides of their containers and the repairs resulted in irregular ridges. None of these individuals possessed the thickened rim to the shell.

Growth

The average final weight of snails in the control group was significantly greater than the weight of snails in the high incidence group of the same age, whether snails were kept individually(16.35 ± 5.54g and 9.11 ± 5.54g, $p < 0.05$) or in groups(9.13 ± 3.3g and 4.66 ± 1.67g, $p < 0.05$). Low mite infestation snails did not differ significantly from controls. Similarly shell size was greater in the control group. This was very clearly demonstrated in the communally reared animals (shell height 27.11 ± 4.36cm and 23.29 ± 4.04cm, $p < 0.05$; shell diameter 32.1 ± 5.34cm and 25.69 ± 3.61cm, $p < 0.05$). Soluble proteins in snail tissues were significantly reduced in high mite infestations compared to controls(381 ± 111 and 528 ± 129mg.g dry wt^{-1}, $p < 0.05$), but there was no significant change in the carbohydrate content.

Reproductive development

In snails cultured individually control animals had a normal mature reproductive tract and in the last few months of the experiment several of the snails showed a pale coloured and swollen genital pore, characteristic of a sexually mature snail in breeding condition. Many mature sperm and oocytes were present in the gonad and vesicular connective tissue cells scattered between the acini stained well with PAS indicating the presence of large amounts of glycogen. The albumen glands sectioned with considerable difficulty due to the presence of large amounts of galactogen.

In the high incidence group it was very difficult to locate the almost completely undeveloped reproductive tract. The albumen glands were only just visible and were relatively easy to section as they had accumulated so little galactogen. The gonad was extremely difficult to identify indicating that the volume of the gonad was considerably smaller than in the controls. In sections, the gonad was found to consist of very small acini containing a few of the early stages of spermatogenesis and no fully mature sperm, but there were more oocytes present than would normally be expected in a snail at this stage of development. There was also a larger proportion of vesicular connective tissue cells than in the control group.

The reproductive organs in snails in the low incidence group were only slightly less well developed than in the control group.

Snails cultured in groups exhibited the same trends in reproductive development as when cultured individually. Thus in animals with high mite infections the albumen gland only reached half the weight of that in controls(0.46±0.16g and 0.93±0.3g, p<0.05).

Histology of the lung

The respiratory epithelium in control animals was 6-10μm thick, with numerous supporting cells. No holes or tissue canals were seen in the respiratory epithelium of lungs that had contained mites, but there were extended areas which were up to 16μm thick due to considerable vacuolation of the epithelial cells. The surface microvilli had in some cases fused together and fewer supporting elements were present beneath the epithelium.

DISCUSSION

R. limacum when present in large numbers clearly had a profound effect on their host H. aspersa. Low initial infection rates of 5 mites/snail while slowing reproductive development had little effect on growth. High infection rates of 20 mites/snail had a profound effect. The level of infestation over the 5 or 9 months of the experiment could not be determined, but as many as sixty live mites and several hundred eggs, as well as the cast skins from the development of all the nymphal stages, have been found in the lung of a single snail. The threshold for mite damage is unclear since after the introduction of mites there could be no control over the numbers that built up.

Snails with high infection rates were observed to have extended periods of dormancy, this inactivity reduced their feeding activity leading to the consumption and presumably assimilation of less food. This could contribute to the lower growth and slowed reproductive development. The cause of this increased dormancy was not investigated. Physical stress due to the continuous movement of large numbers of mites over the surface of the lung and passing in and out of the pneumostome could lead to a change in behaviour. The amount of blood consumed by a mite and the frequency of feeding are unknown but the frequent puncturing of blood vessels at the lung surface, wound healing and removal of blood are likely to affect the physiology of the animal. Many mites produce saliva when feeding which can produce allergic responses in the host. Increased numbers of amoebocytes are visible in the lung tissue of infected snails but the importance of any allergic reaction remains to be studied. The area of lung surface available for respiration could also have been reduced because of the thickening of the epithelium.

Observations on the shells in the infected snails confirm those of Oldham (1931) in *A. arbustorum* in that when mites are present they affect shell growth and shape. As stated above *H. aspersa* with heavy mite infestations had very weak shells which were constantly being damaged and repaired producing irregular roughened surfaces. The cause of the weak shells was not studied but could have been due to a lowered intake of calcium leading to withdrawal of calcium from the shell or to a change in the protein component.

The most striking results in heavy mite infections were the large reductions in growth rate, up to half the normal rate, whether this was measured as body weight or shell size. Although the growth of all body systems appears to have been affected a major contributor was the considerable retardation of reproductive development.

Visual assessment of the development of the reproductive tract indicated that the tract was very much smaller in heavily infected animals. At the start of the experiment all animals were juveniles with immature reproductive systems. During the course of the experiments the tract developed to a fully mature state in control animals while in heavily infested animals it had developed to a small degree but was still immature. This was confirmed by the finding that the weight of albumen glands in control animals was twice that in heavily infested snails. Histologically, galactogen which accumulates in these glands as the animals mature, was almost completely absent from the smaller glands whereas large amounts had accumulated in control glands. Although the other parts of the reproductive tract were not isolated the differences in volume clearly parallel the differences in the albumen glands. Gamete differentiation was well advanced in the gonads of control animals, with mature sperm and ova being present, while in the heavily infested animals little differentiation of sperm had occurred, certainly there were no mature sperm. Differentiation of the ova did not appear to be as severely affected but this requires further study.

It was not possible to find mites embedded in the surface of the lung in any of the histological sections observed in this study, nor have previous authors reported them (e.g. Baker, 1970), but there was clear evidence for pathological changes to the lung surface. The apparent thickening of the respiratory epithelium at the surface of the lung must reduce the efficiency of the respiratory surface and the capacity for oxygen uptake. This would have an affect on the general health of the animals. Other species of mite e.g. *Trombicula autumnalis* (Jones 1950) secrete a feeding tube which is thought to be permanent (Evans, 1992) and Baker (1970) claimed that *R. oudemansi* produced such a structure. No such structures were observed in *H. aspersa*, but the epithelial layer is so thin(6-10µm) that a simple scraping action by *R. limacum* would be sufficient to effect penetration through the respiratory surface of the lung.

Biochemical analysis indicated that heavy infestation with *R. limacum* resulted in a significant reduction in the levels of soluble protein present but had no effect on carbohydrate levels. Histological observations appeared to confirm this as no reduction in glycogen cells or their contents were detected. Reproductive development was retarded in the heavily infested animals so that the heavy requirement for carbohydrates for the synthesis of galactogen in the albumen gland and complex mucopolysaccharides in the other reproductive glands had not occurred. The experiment needs to be extended to discover if carbohydrate levels in such animals would be affected if reproductive development progressed. Blood contains high levels of the respiratory protein haemocyanin and little other protein. Removal of blood by the parasite will lead to the continual loss of this respiratory pigment and failure to replace it would affect the oxygen carrying capacity of the blood. It is possible that the observed reduction in soluble protein levels in tissues may be the result of replacing this lost protein. Further studies are required to discover the amount of blood removed by the parasite and hence the amount of protein lost. Parasite feeding could affect growth and reproductive development by removing circulating hormones which control these processes, or, since there are large salivary glands in this mite (Graham 1994), injection of saliva could affect the snails physiology.

The reported effects of high mite infestations on the growth, reproductive development and shell strength in *H. aspersa* would have serious results for snail farmers. A farmer optimises the culture conditions to 'fatten' the snails and ready them for market in the shortest possible time. A reduction in growth rate, as reported here, would seriously delay this process, and the effect on reproduction would reduce the production of eggs by the breeders so lowering overall productivity. Weak and distorted shells are of significance to processors of snails. Weak shells make extraction of cooked snails from their shells more difficult. Broken and distorted shells would have to be replaced by purchased shells, as the snail meat is conventionally replaced in the shell with some kind of prepared butter. The development of effective methods for controlling the numbers of *R. limacum* on snail farms could have important economic benefits for the industry.

REFERENCES

André, M; Lamy E (1930) Les acariens parasites des mollusques. *Journal of Conchology.* **74,** 199-221.

Baker, R A (1970) The food of *Riccardoella limacum* (Schrank) (Acari: Trombidiformes) and it's relationship with pulmonate molluscs. *Journal of Natural History.* **4,** 521-530.

Evans, G O (1992) *Principals of Acarology.* Oxford.

Fain, A (1969) Adaptive radiation in parasitic Acari and adaptation to parasitism in mites. *Acarologia.* 7,429.

Jones, B M (1950) The Penetration of the Host Tissue by the Harvest Mite, *Trombicula autumnalis* Shaw. *Parasitology.* **40,** 247.

Graham, F J (1994) The biology and control of *Riccardoella limacum* (Schrank), a mite pest of farmed snails. Ph.D. Thesis, University of Wales.

Graham, F J; Ford J B; Runham N W (1993) Comparison of two species of mites of the same genus, *Riccardoella* associated with molluscs. *Acarologia.* **34,** 143-148

Klein-Rollais, D; Daguzan J (1990) Variation of water content in *Helix aspersa* Müller in a natural environment. *Journal of Molluscan Studies.* **56,** 9.

Oldham, C (1931) Some scalariform examples of *Arianta arbustorum* infested by parasitic mites. *Proceedings of the Malacological Society.* **19,** 240.

Runham, N W (1989) Snail farming in the United Kingdom. In *Slugs and Snails in World Agriculture.* I Henderson (ed.). BCPC Monograph No. 41, pp 49-55.

POPULATION STRUCTURE OF *Deroceras reticulatum* IN GRASSLAND

S L HAYNES, S P RUSHTON, G R PORT

Department of Agricultural and Environmental Science, The University of Newcastle, Newcastle upon Tyne, NE1 7RU, UK

ABSTRACT

A population of *Deroceras reticulatum* in rough grassland was studied and age structure assessed using a range of techniques. At all times of year, the population contained all size or age classes. Mating activity of matched samples was recorded using video in controlled conditions. A weak relationship was found between monthly mating activity and the size or age composition of the population.

INTRODUCTION

One of the principles of integrated pest management is the need for a thorough understanding of the biology and ecology of the pests. In contrast to most arthropod pests, many slugs and snails show a variable life cycle. For example, weather conditions may have a large impact on development rates. Nevertheless, there have been several attempts to identify a typical "life-cycle" for pests such as *Deroceras reticulatum* (Duval & Banville, 1989; Hunter & Symonds, 1971). The general conclusion has been that *D. reticulatum* has an annual life cycle with at some sites a single cohort and at other sites two overlapping or "leap-frogging" sets of generations or "populations".

Previous work on slug life cycles has tended to use a simple morphological characteristic such as body weight or length, however such variables are not strongly correlated with sexual maturation (Abeloos, 1944; Smith, 1966; Runham & Laryea, 1968; Prior, 1983). Sokolove & McCrone (1978) found that *Limax maximus* showed a strongly seasonal relationship between gonad and body weight. Duval & Banville (1989) further developed this type of classification using *Deroceras reticulatum*. They used the hermaphrodite and albumen gland weights to derive two indices, produced by dividing the gland weights by the slug body weight, and classified slugs according to the developmental stages identified through histological study by Runham (1978).

As part of an investigation of reproductive behaviour in *D. reticulatum* (Haynes, 1996), we have data on the mating activity and structure of a population of *D. reticulatum* in grassland and the analysis of these data is presented in this paper.

METHOD

Each month between April 1991 and May 1992 samples of *Deroceras reticulatum* were collected from beneath shelter traps in an area of mixed grassland at Close House, Northumberland, UK. Mating activity was recorded using low-light, time-lapse video in a controlled temperature room maintained at between $10^\circ C$ and $13\,^\circ C$ with a day:night cycle similar to ambient. Slugs were held in arenas, constructed from plastic boxes measuring 57cm x 36cm x 16cm. These were filled to a depth of 15cm with soil. Any large particles of organic matter and large soil animals such as earthworms were removed, but otherwise the soil was not treated in any way prior to use. Moisture content of the soil was maintained at about 22% to 26% w/w. by an automatic misting device. Recording of mating activity went on throughout every month.

A range of slug sizes were used in every trial, selected to match the distribution prevailing in the field. The activity of twenty slugs was recorded for between one week and ten days before slugs and soil were renewed. There were few slug deaths during the trial period. Where slugs were noted as dead on the soil surface, they were removed and an allowance made when calculating the results. Slugs were starved for the duration of the trial, and slug activity was recorded from the first night. All attempts to mate were noted. The number of slugs in the arena showing sexual activity (circling) each night over the recording period divided by the cumulative total of slug-nights was used to describe mating activity (proportion of slugs mating).

When selecting slugs for the first trial of the month, twenty further specimens were also collected. These were paired by size visually with those of the first sample, and were dissected to ascertain the state of reproductive development. Slugs in the sample for dissection were killed by a brief immersion in boiling water, and were then placed in sterile Hedon-Fleig saline (Gatenby, 1937). Slugs were blotted to remove excess mucus and then weighed on a Mettler P163N top-pan balance. The hermaphrodite and albumen glands were dissected in the saline, gently blotted and then weighed on a Cahn 29 automatic electro-balance. Evaporation from the gland could be followed and the effect of this was controlled by standardising a 'settling-time' before the weight of each gland was recorded. Glandweight was then divided by bodyweight weight to produce the Gonad-Somatic-Index (GSI) (hermaphrodite gland), and the Albumen-Somatic-Index (ASI), (albumen gland).

The fuzzy c-mean algorithm of Bezdek (1981) was used to classify slugs into 2-5 groups (possible developmental stages); firstly, using three parameters - the two gland weights and bodyweight; and secondly, using two parameters - the GSI and ASI. The partition coefficient (Bezdek, 1981) was used to assess the optimum number of groups within the sample populations.

RESULTS

In September 1991, equipment failure resulted in the loss of some of the specimens preserved for dissection and weight analysis, thus there are data for only 13 months. The number of groups with the highest partition coefficient was 5 in each analysis. For two parameters, GSI and ASI the partition coefficient was 0.49. This was less efficient than the classification based on three parameters, body weight, hermaphrodite gland weight and albumen gland weight which had a partition coefficient of 0.65. Table 1 shows the median class values for each of the parameters in this second classification.

Table 1 Median class values (mg) for body weight, hermaphrodite gland weight and albumen gland weight from a fuzzy cluster analysis of a *D. reticulatum* population

Class	Body weight	Hermaphrodite gland	Albumen gland
1	85.6	2.3	0.5
2	199.8	6.3	3.6
3	313.0	7.5	10.5
4	428.0	8.2	19.2
5	572.7	10.4	23.0

Figure 1 shows the composition of the *D. reticulatum* population using the results of the classification from three parameters. There was a close correlation between body weight and the results of this classification ($r = 0.9833$, n=265, P<0.001), so another classification was produced, based on body weight alone. The classes were 0-100mg, 101-200mg etc. Figure 2 shows the composition of the same *D. reticulatum* population classified by body weight.

Mating was observed in the video arenas in most months, but always occurred in less than 10% of slug-nights. Data for weight analysis were not available for September 1991, and in January 1992 it was not possible to complete the video observations for mating activity. For the remaining 12 months, possible correlations between the percentage of slugs engaged in mating and other parameters were examined. None were significant, but there was an indication of a weak positive correlation with the Albumen-Somatic-Index (ASI) ($r = 0.5415$, n=12, P<0.1) and also with the proportion of slugs with body weights greater than 400mg ($r = 0.5084$, n=12, P<0.1).

Figure 1 The composition of the *D. reticulatum* population classified into five groups based on body weight, hermaphrodite gland weight and albumen gland weight. The composition for each month is shown derived from a sample of about 20 slugs (data for September 1991 missing).

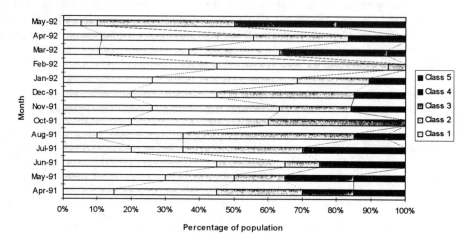

Figure 2 The composition of the *D. reticulatum* population classified into five groups based on body weight. The composition for each month is shown derived from a sample of about 20 slugs (data for September 1991 missing).

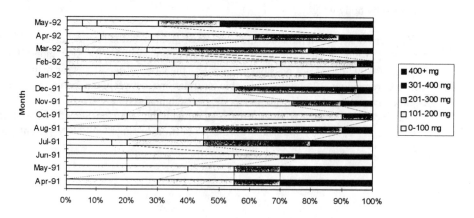

DISCUSSION

Population structure

The fuzzy cluster analysis and the weight class analysis revealed broadly similar compositions of the slug population over the period investigated. Both methods show a peak in the proportion of largest slugs (Class 5) in April and May 1991 and May 1992. There is a smaller peak in December 1991. The smallest slugs (Class 1) show peaks in June 1991 and February 1992. What is more obvious is that all size classes can be found at all times of year. A similar conclusion was reached by Hunter (1968) whose samples came from the same site as ours. Hunter concluded that *D. reticulatum* had two peaks of breeding each year, but subsequently Hunter & Symonds (1971) showed that the situation was more complex with two overlapping generations, a pattern subsequently found by others workers (e.g. Duval & Banville, 1989). Whilst our data support these ideas it is important to note that at any time of year we found slugs of most sizes (Classes) and some slugs in the population can probably breed whenever other conditions are suitable. If mortality factors, such as natural enemies (Ayre & Port, 1996) or environmental constraints act in a size or age specific manner then there will be some slugs susceptible to these factors at any time of year. By the same token there will be some slugs resistant, as a result of their size or age, at any time of year.

The good correlation found between body weight and the fuzzy cluster classification based on gland and body weights suggests that within a population of *D. reticulatum* body weight may act as a good indicator of sexual maturation. However, comparisons between populations may be susceptible to variation due to habitat or genetic differences.

Mating activity

Given suitable environmental conditions (arenas in controlled environments) there was a low level of mating activity at most times of year. The weak correlation between mating activity and the proportion of large slugs or a large ASI was expected and suggests that as the proportion of female phase slugs increases so do the chances of reproductive activity.

ACKNOWLEDGEMENTS

This work was supported by MAFF

REFERENCES

Abeloos, M. (1944) Recherches experimentales sur la croissance des mollusques Arionides. *Bulletin biologique France Belgique*, **78**, 215-256.

Ayre, K; Port, G R (1996) Carabid beetles recorded feeding on slugs in arable fields using ELISA. In: *Slug and Snail Pests in Agriculture*. I F Henderson (ed.). Symposium Proceedings No. 66.

Bezdek, J C (1981) *Pattern recognition with fuzzy objective function algorithms*. Plenum, pp. 256.

Duval, A; Banville, G (1989) Ecology of *Deroceras reticulatum* (Mull.) (Stylommatophora, Limacidae) in Quebec strawberry field. In: *Slugs and Snails in World Agriculture*. I F Henderson (ed.) BCPC Monograph **41**, 147-160.

Gatenby, J B (1937) *Biological Laboratory Techniques* Churchill, London.

Haynes, S L (1996) *Reproductive behaviour and mating patterns of the field slug Deroceras reticulatum (Muller)*. Unpublished Ph.D. thesis, University of Newcastle upon Tyne.

Hunter, P J (1968) Studies on slugs of arable ground II. Life Cycles. *Malacologia*, **6**, 379-389.

Hunter, P J; Symonds, B V (1971) The leap-frogging slug. *Nature*, **229**, 349.

Prior, D J (1983) The relationship between age and body size of individuals in isolated clutches of the terrestrial slug *Limax maximus*. *Journal of experimental Biology*, **225**, 321-324.

Runham, N W (1978) Reproduction and its control in *Deroceras reticulatum*. *Malacologia*, **17**, 341-350.

Runham N W; Laryea, A A (1968) Studies on the maturation of the reproductive system of *Agriolimax reticulatum*. *Malacologia*, **7**, 93-108.

Smith, B J (1966) Maturation of the reproductive tract of *Arion ater*. *Malacologia*, **4**, 325-349.

Sokolove, P G; McCrone, E J (1978) Reproductive maturation in the slug *Limax maximus* and the effect of artificial photoperiod. *Journal of comparative Physiology (A)*, **125**, 317-325.

REPRODUCTIVE DEVELOPMENT IN *ARION ATER*: INFLUENCE OF FOOD AND STORED CARBOHYDRATE.

J M TXURRUKA
University of the Basque Country, Faculty of Sciences, Department of Plant Biology and Ecology, Apdo 644, 48080 Bilbao, Spain

M M ORTEGA.
University of the Basque Country, Faculty of Sciences, Department of Animal Biology and Genetics, Apdo 644, 48080 Bilbao, Spain

ABSTRACT

A multiple regression approach has been used to analyze the relationships between storage tissues and biochemical composition of diet, during gonadial development in *Arion ater* fed on synthetic foods. Starch content is the more important component of food contributing to tissue growth of body wall and genitalia, where carbohydrate deposits show a clear increase. Carbohydrates in diet have a greater influence on the body wall. An inverse relationship between body wall and reproductive tract becomes evident in female stage animals, both in terms of dry weight and carbohydrate content, suggesting that the exponential rise of reproductive demands is partially fmet by glycogen reserves.

INTRODUCTION

Although food is the ultimate source of materials sustaining reproductive demands in animals, cycles of storage and utilization of reserves (particularly glycogen) closely related to annual reproductive processes, are commonly found in molluscs (Gabbott, 1975; Calow, 1981). Some organs experience large fluctuations throughout the year. We have found no information on this in *Arion ater*, but metabolism in Pulmonata has been described as "carbohydrate orientated" (Emerson, 1967) and available data on carbohydrate concentration of different tissues for two species of slugs (Livingstone & De Zwann, 1983) suggests carbohydrate deposition occurs in selected organs. Additionally, dietary increase in starch or glucose increases body carbohydrate in slugs and snails (Wright, 1973; Veldhuijzen & Van Beek, 1976). From our studies on nutrition in *Arion ater* (Txurruka, 1992) a definite pattern for body wall dynamics, in terms of both mass and carbohydrate content, emerges, indicating connections between nutrient availability and reproductive events. The objective of this work has been to study the functional relationships between carbohydrate content in diet and storage tissues, and its coupling with metabolic constraints imposed by genital development.

MATERIALS AND METHODS

Slugs were caught in grassland at Forua (43° 20' N; 2° 41' W, Bizkaia, Basque Country) in early August (male stage) and throughout November (during the egg-laying season of the female stage). Feeding experiments took place over 15 days, prior to which slugs were starved for two days to allow evacuation of formerly ingested wastes. Ten synthetic foods were designed following Wright (1973), their exact chemical composition appearing in

Txurruka (1992); 3 diets are carbohydrate-free, while the rest contain different proportions of cellulose and/or starch (30% or 60% dry weight). On days 0, 5, 10 and 15 slugs were individually weighed, killed by immersion in liquid nitrogen and stored at -18°C. Dissection was performed in a cool chamber (2-4°C) and 298 male stage and 264 female stage slugs were individually divided into four body organs -body wall, digestive gland, genitalia and empty gut. After liophilization, these were stored at -18°C in hermetic flasks with silica-gel under nitrogen atmosphere until analysis. Carbohydrates were extracted with hot trichloroacetic acid according to Newell & Bayne (1980) and colorimetrically quantified by the method of Dubois *et al.* (1956).

Multiple regression procedures have been employed to establish the relationships between genital development and a set of predictors which include both food characteristics and tissue composition of the various body organs. Stepwise multiple regression analyses (forward stepping followed by backward stepping) were initially performed where either the genital weight (mg) or its carbohydrate content (mg) were taken as the dependent variables against the following predictors: feeding period (d), biochemical composition of food (% dry weight), sex (0 = males, 1 = females), body weight (mg), total carbohydrate content (mg), genital index (Dry weight of Genitalia/Dry weight of entire body), and weight (mg), and biochemical content (mg) of the other body organs. Since body wall weight or carbohydrate content explain the greater proportion of genital variation, the analysis has centered on the interrelationships between body wall and genitalia.

RESULTS AND DISCUSSION

The multiple regression approach led to the 4 equations in Tables 1-4, which explain from 82% to 92% of the variation in the total dry weight of body wall and genitalia, and in their carbohydrate contents.

Table 1. Multiple regression equation relating dry weight (mg) of body wall to size, reproductive stage and starch content of food (n=1037; r^2= 0.863; p=0.0001).

Terms	Coefficient	± S. e.	p
Intercept	-234.859		
Total body dry weight (mg)	0.671	0.014	0.0001
Dry weight of genitalia (mg)	-0.540	0.074	0.0001
Genital index	2386.381	230.055	0.0001
Sex x Genital index	-2130.359	196.683	0.0001
t x Starch (day x %)	49.838	3.291	0.0001

As regards the effect of diet, a positive interaction between feeding duration and starch content appears in all equations, indicating food starch contributes to sustained tissue growth and carbohydrate content of both body wall and genitalia. Carbohydrate deposits, in particular, experienced a clear increase. For example, after 15 days feeding on a 60% starch diet, proportions of carbohydrate rose from 29% to 46% dry weight in the body wall of males. Similar findings have been reported for *Arion ater* (Wright, 1973) and *Lymnaea stagnalis* (Veldhuijzen & Van Beek, 1976) where body wall accounts for 70% of total carbohydrate. This accumulating capability could reside in the pore cells of the connective tissue (Skelding & Newell, 1975) and/or muscle cells (Livingstone & De Zwaan, 1983).

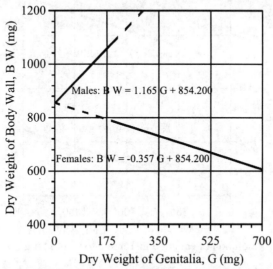

Figure 1. Results from equation in Table 1 for a 1400 mg d.w. slug.

Table 2. Multiple regression relating dry weight (mg) of the genitalia to size, somatic carbohydrate content (CC), reproductive stage and starch content of food (n=902; r^2= 0.828; p=0.0001).

Terms	Coefficient	± S. E.	p
Intercept	-33.202		
t (d)	3.150	0.796	0.0001
Dry weight of body wall(mg)	0.127	0.013	0.0001
CC of body wall(mg)	-2.147	0.057	0.0001
CC of digestive gland(mg)	-1.992	0.246	0.0001
Total carbohydrate(mg)	1.986	0.555	0.0001
Genital code	153.792	10.149	0.0001
t x Starch (d x %)	49.8378	2.278	0.0018

The effect of genital development on body wall weight exhibits opposite trends in male and female stages (Table 1). Figure 1 shows the equation resolved for a mean-size animal of 1400 mg dry weight, whose genital organs range from 50-200 mg for males and 200-600 mg for females (experimental range). In the male stage, before the shift from somatic to germinal growth occurs, growth of body wall and genitalia appears to be isometric (b = 1.16).

Conversely, in the female stage, larger genitalia are associated with smaller body walls, an increase of 100 mg in genitalia implies a decrease of nearly 36 mg in the dry weight of body wall.

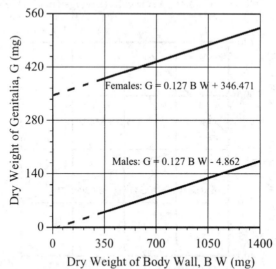

Figure 2. Resolution of equation in Table 2 for a 1400 mg d.w. slug.

The size of genitalia increases with time throughout the experiment (Table 2), and with body wall size in both males and females. For a given body wall size, females' gonads exceed those of males by 350 mg (Fig.2). However, as shown by the negative sign of the coefficients, gonadial growth is associated with a decrease in the carbohydrate content of both body wall and digestive gland.

Table 3. Multiple regression relating carbohydrate content (mg) of the body wall to size, somatic carbohydrate content, reproductive stage and starch content of food (n=902; r^2= 0.818; p=0.0001).

Terms	Coefficient	± S. E.	p
Intercept	-59.283		
DW of body wall (mg)	0.146	0.011	0.0001
CC of digestive gland (mg)	3.102	0.239	0.0001
Genital index	-106.089	45.333	0.0195
Sex x DW of genitalia	-0.101	0.030	0.0009
t x Starch (d x %)	41.697	2.063	0.0001

The relationship between availability of somatic carbohydrate and gonadial development can be seen from the equations in Tables 3 and 4. In Table 3, the negative sign of the genital index means gonadial growth involves a decrement in the carbohydrate content of body wall for both sexes (Fig. 3). Nevertheless, the negative sign of the interaction of sex and dry weight of genitalia, that only takes a value in female stage, results in a higher loss of body wall carbohydrates at this stage: 25 mg vs 11 in males for a 0.1 increase in genital index. This result agrees with views commonly expressed for bivalve molluscs where oogenesis is thought to take place at the expense of glycogen reserves (Gabbott, 1975). Predictions from the model for the variation in carbohydrate content of the gonad (Table 4) are in accord with the preceeding comments: carbohydrates increase in genital tissues throughout gonadial development (positive signs for genital dry weight and Genital Index) coupled to a decrease of sugars within the body wall.

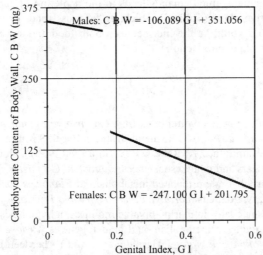

Figure 3. Carbohydrate content of body wall (mg) vs Genital index (from Table 3).

Table 4. Multiple regression relating carbohydrate content (mg) of genitalia to size, somatic carbohydrate content, reproductive stage and starch content of food (n=914; r^2= 0.933; p=0.0001).

Terms	Coefficient	± S. e.	p
Intercept	-21.010		
DW of genitalia (mg)	0.255	0.006	0.0001
CC of body wall(mg)	-0.039	0.004	0.0001
Genital index	34.415	7.697	0.0001
t x Starch (d x %)	6.057	0.424	0.0001

As a conclusion, some suggestions can be made about the fate of foodstuffs, particularly sugars. Growth of both body wall and genitalia are enhanced by dietary carbohydrate, sugar deposits in both tissues accounting for the largest proportion of such increase. However, diet coefficient takes a value close to 42 for the body wall (Table 3) and 6 for genitalia (Table 4). The simplest explanation could be biochemical parsimony: in the somatic tissues stored carbohydrate would be essentially glycogen (Calow, 1981; Livingstone & De Zwaan, 1983), whereas in genitalia, galactogen, a branched polymer of L- and D-galactose in the albumen gland, would account for the largest reserves of sugars. Further, since the albumen gland is involved in glycogenesis and galactogenesis some control of the flow between both pathways should be envisaged.

As regards the metabolic trade-off between body wall and genitalia, sequential commitment of *Arion ater* to first somatic and then reproductive growth (Txurruka et al., 1996) probably ensures that during the immature stage and while reproductive development proceeds at low intensity, surplus energy is channelled to reserve tissues such as body wall. Later, the exponential increase in gametogenic demands leads to the depletion of such stores, either directly or *via* the digestive gland (Table 3) which, at the cross-roads of major metabolic pathways (Gabbott, 1975), would be the link between both food and stored nutrients and biosynthetic requirements of egg production.

REFERENCES

Calow, P (1981) Growth in lower invertebrates In: *Comparative animal nutrition. Vol 4. Physiology of growth and nutrition*. M Recheigl, Jr. (ed.). Basel, S Karger. pp 53-76.

Dubois, M; Gilles, K A; Hamilton, J K; Rebers, P A; Smith, F (1956) Colorimetric method for determination of sugars and related substances. *Analytical Chemistry*. **28**, 350-356.

Emerson, D N (1967) Carbohydrate orientated metabolism of *Planorbis corneus* (Mollusca, Planorbidae) during starvation. *Comparative Biochemistry and Physiology*. **22**, 571-579.

Gabott, P A (1975). Storage cycles in marine bivalve molluscs: a hypothesis concerning the relationship between glycogen metabolism and gametogenesis. *Proceedings of the 9th European Symposium of marine biology*, Harold Barnes (ed.) Aberdeen Univesity Press, pp 191-211.

Livingstone, D R; De Zwaan, A (1983) Carbohydrate metabolism of gastropods. In: *The Mollusca. Vol 1. Metabolic biochemistry and molecular biomechanics*, P W Hochachka (ed.), New York, Academic Press, pp 177-242.

Newell, R I E; Bayne, B L (1980) Seasonal changes in the physiology, reproductive condition and carbohydrate content of the cockle *Cardium* (=*Cerastoderma*) *edule* (Bivalvia: Cardiidae). *Marine Biology*. **56**, 11-19.

Skelding, P F; Newell, P F (1975) On the functions of the pore cell in the connective tissue of terrestrial pulmonate slugs. *Cell and Tissue Research*. **156**, 381-390.

Txurruka, J M (1992) Studies on the nutrition of *Arion ater* (L.): Influence of synthetic foods on growth and reproduction. Ph D Thesis. University of the Basque Country.

Txurruka, J M; Altzua, O; Ortega M M. (1996).Organic matter partitioning in *Arion ater*: Allometric growth of somatic and reproductive tissues throughout its life span. *Proceedings of the Synposium Slug and Snails Pests in Agriculture*.

Veldhuijzen, J P; Van Beek, G (1976) The influence of starvation and of increased carbohydrate intake on the polysaccharide content of various body parts of the pond snail *Lymnaea stagnalis. Netherlands Journal of Zoology*. **26**, 106-118.

Wright, A A (1973) Evaluation of a synthetic diet for the rearing of the slug *Arion ater* L. *Comparative Biochemistry and Physiology*. **46A**, 593-603.

THE CONTROL OF SLUG DAMAGE TO SEEDLING BEET, IN IRELAND, WITH SPECIAL REFERENCE TO BAND PLACEMENT OF SLUG PELLETS

T F KENNEDY

Teagasc, Oak Park Research Centre, Carlow Ireland

ABSTRACT

Slugs and slug/leatherjacket combinations were found to cause significant reductions in beet seedling numbers with most damage taking place up to the two leaf stage of plant growth. Molluscicides applied at the time of sowing gave the best control of damage by slugs while damage by mixed infestations of slugs and leatherjackets were best controlled by applying methiocarb at the time of sowing. Higher plant numbers were recorded where molluscicides were applied at 0.3 normal rate as a band treatment at sowing compared with 0.5 rate incorporated into the seedbed. The Armer-Salmon pesticide applicator used gave excellent delivery and no crushing of slug pellets.

INTRODUCTION

Slugs are normally associated with wetter soils that have high clay or silt contents (Gould, 1961). In Ireland, however, slug damage to seedling beet can often be found on light and medium textured soils. The average annual rainfall in the beet growing areas is 927 mm.

Slug damage to beet is often accompanied by 'leatherjacket' (*Tipulidae*) damage with which it can be confused. Climatic and cultural conditions favouring slug activity also favours leatherjackets.

This paper reports on some investigations where the emphasis was on reducing the amount and cost of molluscicides in controlling damage by slugs to seedling beet.

MATERIALS AND METHODS

The following trials were undertaken :
Trial 1
Plot size :   3.4 m x 5 drills, each 56 cm wide
Plant spacing :   14 cm
Replication :   4 fold randomised block
Soil type :   Free draining, light textured, moderately deep brown earth
Sowing date :   30 March 1987
Applied :   By hand 14 days post sowing

Trial 2

Plot size :	10 m x 5 drills, each 56 cm wide
Plant spacing :	14 cm
Replication :	4 fold randomised block
Soil type :	Moderately deep, medium to heavy textured, well drained grey-brown podzolic soil
Sowing date :	8 April 1988

Trial 3

Trial 3 was adjacent to Trial 2 and had similar plot dimensions and sowing date.

Consult Tables 1, 3 and 4 for treatment details.

In Trials 2 and 3 in-furrow and surface band treatments at sowing were applied using the Armer-Salmon pesticide applicator (Fig. 1). This applicator uses a stainless steel auger to dispense slug pellets (or granule insecticides) and can be used to incorporate slug pellets into the seedbed or to place them in a narrow band on the soil surface. All other treatments were applied by hand. Test baiting using 'shelter traps' and methiocarb slug pellets prior to sowing showed that Trial sites 2 and 3 were infested with both slugs and leatherjackets. Shelter traps were constructed from weather-proof wallboard, 30 x 30 x 0.7 cm, and had wooden legs to penetrate the soil and hold the trap approximately 2 cm from the soil surface. Chlorpyrifos was applied to site 2 at 672g a.i./ha and lightly harrowed into the soil before sowing to eliminate leatherjackets

Assessment methods : The percent plant establishment was measured in Trials 1, 2 and 3. In addition, root and sugar yields were obtained in Trials 2 and 3. The mortality of slugs, leatherjackets and earthworms was recorded in five $0.25m^2$ quadrats in treatments 4 and 6 of trial 3 twenty six days after sowing. Rainfall records for the area during April and May were recorded.

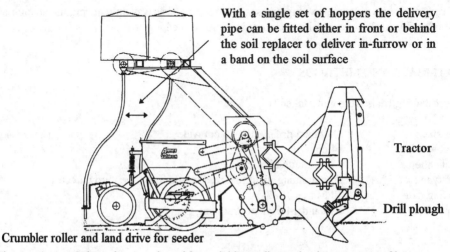

With a single set of hoppers the delivery pipe can be fitted either in front or behind the soil replacer to deliver in-furrow or in a band on the soil surface

Tractor

Drill plough

Crumbler roller and land drive for seeder

Fig. 1: Armer-Salmon beet seeder with pesticide applicator having two sets of hoppers. Hoppers are fitted with stainless steel augers.

RESULTS

Following application of molluscicides only slugs were found dead or dying in Trial 1. No leatherjackets were found. The trial results are presented in Table 1. All four molluscicide treatments increased plant establishment relative to untreated plots. These increases were significant for three treatments. The highest plant establishment was 69.8% recorded for treatment 2, i.e. methiocarb applied in a band at 0.3 normal rate, compared with 56.9% for the untreated plots.

Table 1. The percent plant establishment in sugar beet treated with full and reduced rates of the molluscicides methiocarb and metaldehyde recorded 16 days after application.

Treatments				Field establishment
Compound	Rate	Placement	Applied	(%)
Methiocarb[1]	full	broadcast	14 days post sowing	66.0*
Methiocarb	0.3	band	14 days post sowing	69.8*
Metaldehyde[2]	full	broadcast	14 days post sowing	65.3*
Metaldehyde	0.3	band	14 days post sowing	60.1
Untreated				56.9
L.S.D. (5%)				7.3

$* = P \leq 0.05$
[1]Full rate = 5.5 kg/ha of 4% methiocarb ('Draza')
[2]Full rate = 15 kg/ha of 6% metaldehyde ('PP' Mini slug pellets)

The results of test baiting carried out pre and post sowing at the site of Trials 2 and 3 are given in Table 2 and show that slugs and leatherjackets occurred at this site. *Deroceras reticulatum* was the more plentiful slug, the remainder being *Arion hortensis*. Leatherjackets were *Tipula paludosa*.

Table 2. Numbers of slugs and leatherjackets per 3 refuge traps at the site of trials 2 and 3.

	18-23 March	23-30 March	30Mar- 5 April	20-27 April	27Apr- 4 May	4 - 11 May	11 - 18 May	18-26 May	26 May - 3 June
D.reticulatum	20	4	-	-	1	-	-	-	-
A. hortensis	11	2	-	-	6	-	-	-	3
T. paludosa	17	9	2	-	3	1	-	-	-

The results of Trial 2 are given in Table 3. Ten of the twelve treatments had significantly greater plant establishments, one had significantly greater root yield and two had significantly greater sugar yield when compared with the untreated plots.

Of the commercially available molluscicides, i.e. methiocarb and metaldehyde, the in-furrow application of methiocarb increased final plant establishment by 6.8% relative to untreated

Table 3. The effect of molluscicide and method of application on the percent plant establishment in sugar beet in a slug infested site.

Compound	Rate	Placement	Applied	Two leaf stage	Six leaf stage	Root	Sugar
		Treatments		Field establishment (%)		Yields t/ha	
Methiocarb[1]	0.5	in-furrow	sowing	73.0	72.5	71.0	-
Methiocarb	0.5 + 0.3	in-furrow + band[3]	sowing + sowing	82.4**	82.4**	71.7	-
Methiocarb	0.5 + 0.3	in-furrow + band	sowing + emergence	84.3**	84.4**	72.0	-
Methiocarb	full	broadcast	sowing	84.0**	84.2**	75.0*	12.83*
Methiocarb	full	broadcast	emergence	81.7**	82.6**	72.2	12.18
Methiocarb	full + 0.3	in-furrow + band	sowing + emergence	83.1**	83.5**	-	-
Metaldehyde[1]	0.5 + 0.3	in-furrow + band	sowing + sowing	73.0	72.5	73.3	-
Metaldehyde	0.5 + 0.3	in-furrow + band	sowing + emergence	78.9**	80.2**	-	-
Bensultap[2]	0.5 + 0.3	in-furrow + band	sowing + sowing	85.0**	85.8**	73.0	12.5*
Bensultap	0.5 + 0.3	in-furrow + band	sowing + emergence	82.2**	82.6**	-	-
Bensultap	0.25 + 0.17	in-furrow + band	sowing + sowing	82.5**	83.1**	-	-
Bensultap	0.25 + 0.17	in-furrow + band	sowing + emergence	81.2**	82.5**	-	-
Untreated				68.9	67.9	71.8	11.97
L.S.D. (5%)				6.1	6.6	2.8	0.52

* = P ≤ 0.05
** = P ≤ 0.01

[1]See footnotes 1 and 2, Table 1
[2]Full rate = 10 kg/ha of 5% a.i.
[3]Applied to the soil surface

plots while the in-furrow plus band treatment at sowing gave an additional 14.6% plant establishment over the in-furrrow treatment alone.   Methiocarb at 0.5 rate in-furrow and 0.3 rate at emergence gave the highest plant establishment.   Methiocarb broadcast at normal rate at sowing gave the highest root and sugar yield

At the two leaf stage in Trial 3 all treatments had significantly greater plant numbers when compared with untreated plots (Table 4).   Highest plant numbers were recorded for methiocarb at 0.5 rate in-furrow plus 0.3 rate surface band applied at sowing.   The in-furrow treatment increased final plant establishment by 31.9% relative to untreated plots while the addition of the band treatment gave a further 35% increase in plant establishment.

Plant counts recorded at the six leaf stage differed only slightly from those at the two leaf stage showing that most damage had taken place when plants were at the two leaf stage.

The methiocarb treated plots had higher seedling counts than the γHCH treated plots, which in turn had significantly higher counts than those of the untreated plots.   The reduction in plant numbers in the untreated plots were mainly due to damage by leatherjackets.   In plots receiving a broadcast application of methiocarb a mean of 15.2 dead or dying slugs, 0.8 dead earthworms and 26.4 dead or moribund leatherjackets per m^2 were recorded 9 days following treatment but no dead soil animals were observed on untreated plots.

Root and sugar yields were improved in treated plots but these did not differ significantly from untreated plots.

Table 4.   The effect of methiocarb slug pellets and γHCH used to control slugs and leatherjackets on field establishments and yields of sugar beet.

Treatments				Field establishment %		Yields t/ha	
Compound	Rate	Placement	Applied	Two leaf stage	Six leaf stage	Root	Sugar
Methiocarb	0.5	in-furrow	sowing	55.8**	55.8**	72.5	-
Methiocarb	0.5 + 0.3	in-furrow + band	sowing + sowing	70.5**	70.6**	74.7	-
Methiocarb	0.5 + 0.3	in-furrow + band	sowing + emergence	68.8**	68.5**	74.3	-
Methiocarb	full	broadcast	emergence	65.1**	65.1**	77.0	12.87
γHCH	full	soil incorporated	prior to sowing	65.9**	64.7**	75.4	12.81
Untreated				43.0	42.3	71.5	11.85
L.S.D.(5%)				9.0	8.3	N.S.	N.S.

** = $P \leq 0.01$
N.S. = not significant.

DISCUSSION

The results from Trial 1 indicate that the amount and cost of molluscicide needed to control slug damage to seedling beet could be substantially reduced. They also show that significant damage by these pests can occur even on light textured soils. Arising from this trial the suitability of the Armer-Salmon pesticide applicator to apply molluscicide pellets in a band treatment at sowing was investigated. No modifications were required to deliver in-furrow treatments and only the addition of a steel bar to hold the pesticide delivery tube in place behind the seeder was required to enable the placement of pellets in a narrow band on the soil surface. It was possible to vary the rate of molluscicide applied by varying the diameter of the applicators drive-wheel pulley. The applicators stainless steel auger was found to deliver pellets without crushing or blockages.

Monitoring of plots in the days following sowing in Trial 2 showed considerable leatherjacket mortality occurring in the chlorpyrifos treated area. No leatherjackets were found subsequently in the vicinity of damaged plants and damage was therefore attributed to slugs. Slug activity and damage to beet seedlings in Trial 2 as well as slug/leatherjacket damage in Trial 3 coincided with rainfall. Precipitation for the five days from 30 April to 4 May inclusive was 9.5 mm, 9.5 mm, 4.5 mm, 5 mm and 2.5 mm, respectively. A dry period followed from 4 May. Field observations recorded damage to seedlings and reduction in plant numbers as being confined to these five days, (Tables 3 and 4).

In Trial 2 the highest root and sugar yields were obtained for methiocarb broadcast at normal rate at sowing. Broadcast applications of slug pellets following sowing have also been found to give best protection against slug damage to cereals (Anon, 1984; Kennedy & Connery, 1990).

The results of Trial 3 show that good control of damage to seedling beet by mixed populations of slugs and leatherjackets can be obtained using methiocarb pellets. The band treatment of methiocarb had the highest plant numbers.

In general the results indicate that where slugs are known to occur in beet fields, prior to sowing, that reduction in plant numbers by these pests can be significantly lessened by using reduced rates of molluscicide in a band treatment at sowing. Where slug/leatherjacket combinations occur methiocarb should be used. The practise of applying molluscicide in a band treatment at sowing in beet fields known to harbour these pests is now carried out by some Irish beet growers particularly in the south of the country.

REFERENCES

Anon.(1984) Slugs and snails. *Leaflet 115* Ministry of Agriculture, Fisheries and Food, London, 12pp.

Gould,H G(1961) Observations on slug damage to winter wheat in East Anglia, 1957 - 1959. *Plant Pathology.* **10,** 142 - 146.

Kennedy,T; Connery J(1990) Slug control in cereals. *Teagasc Research Report* 182-183

IMPLICATIONS OF SUBLETHAL EFFECTS ON THE EFFICACY OF METHIOCARB BAITS TO TWO PEST SPECIES OF SLUG

C R KELLY
Huntingdon Life Sciences Ltd, PO Box 2, Huntingdon, Cambs, PE18 6ES

S E R BAILEY
School of Biological Sciences, Manchester University, Oxford Rd, Manchester, M13 9PL

ABSTRACT

Using the acoustic pellet technique developed at Manchester University, the feeding behaviour of two species of slug, *Arion distinctus* and *Deroceras reticulatum*, were compared in a laboratory trial. Key meal parameters from 80 individual meals on pellets containing either 2 or 4 % (w/w) of methiocarb or an untreated control, were compared. Important differences were found in the basic feeding strategy between the two species on control pellets although proportionally meal sizes were similar. In comparison meal sizes were highly reduced with methiocarb present and the response to the onset of poisoning for both slug species was quite different. *D. reticulatum* took more, smaller, quicker bites, whilst *A. distinctus* merely stopped feeding sooner. There appeared to be no beneficial effects to meal size by halving the amount of methiocarb in pellets, as this appeared to merely halve the amount of methiocarb consumed. The displayed sub-lethal effects on feeding behaviour lead us to examine the probable two-stage mode of action of methiocarb. The possibility of improving slug baits by reformulation is examined.

INTRODUCTION

Methiocarb slug baits are the most widely applied molluscicide, in agriculture in the UK. Slugs are a serious pest of winter sown arable crops and potatoes and are notoriously difficult to control. The slug *Deroceras reticulatum* (Müller), the commonest pest species, is said to be less susceptible to methiocarb baits than slugs of *Arion hortensis* agg. (Kelly and Martin, 1989).

Although a highly successful slug bait with over 25 years on the market, slugs do not always ingest a lethal dose of methiocarb slug pellet. It is known that the presence of methiocarb significantly reduces meal size compared to a non-toxic meal (Wright and Williams,1980). Early meal termination has been suggested as the reason for slugs not consuming a lethal dose (Bailey *et al.*, 1989). Various explanations; the distastefulness of the active ingredient, paralysis of the gut wall and sublethal levels of blood anticholinesterases affecting ingestion, have been proposed (Wright and Williams, 1980, Bailey *et al.*, 1989). Standard commercial bait contains 4% methiocarb (w/w) although a 2% pellet is also available. The study reported here investigated differences in slug species susceptibility by comparing the feeding behaviour of *D. reticulatum* and *Arion distinctus* using specially prepared non-toxic control pellets and similar pellets containing either 2 or 4% methiocarb (w/w).

METHODS AND MATERIALS

Slugs of both species were collected from Jodrell Bank Experimental Grounds (Cheshire, UK). Each species was held separately and kept in plastic holding boxes (17 x 11.5 x 6 cm) on moist peat in groups of ten. Holding boxes were placed in conditions of 10 h of subdued light at 12 ± 1°C and 14 h darkness at 7 ± 1°C in incubators. Fresh lettuce and carrots were provided *ad libitum* and any dead or sick animals were removed. Slugs were starved for 48 h prior to being presented with a pellet, to ensure that the digestive tract was empty and to heighten food arousal. All slugs for each species in this test were of similar weights to reduce response variability. The mean weights for *D. reticulatum* and *A. distinctus* (± SD) were 0.4972g (± 0.1701g) and 0.3377g (± 0.0814g) respectively. Pellets were manufactured from wheat flour. The required amount of active ingredient was initially mixed with the dry flour to improve homogeneity. A fixed volume of 1% gelatine in water was then added, to give the pellets similar hardness, before being mixed to a paste. Pellets were extruded through a spaghetti maker and air dried.

The meal recording equipment consisted of a weighed pellet (c. 1.5 cm long) stuck on a short length of wire with quick-set epoxy resin. This was attached to a piezo-electric transducer (sub-miniature microphone insert) with a small crocodile clip. Signals caused by the vibrations produced by each bite of a feeding slug were amplified and displayed on an oscilloscope and converted by a signal conditioner to a five volt square wave pulse. This was fed into the user port of a BBC microcomputer programmed to record the number of bites, meal duration and mean and median bite frequency. Each rasp produced a series of pulses. The time that the computer ignored incoming signals was pre-set to correspond with the rasp rate of the test species so that the series of signals caused by each rasp were recorded as one bite. Each bite was plotted on a graph of bite number (x axis) against bite rate per minute (y axis) on the computer screen. The end of a biting period (a 'meal') was defined as a pause in biting of longer than two minutes. The number of periods greater than eight seconds between rasps was also recorded.

After 48 h of food deprivation, slugs were randomly selected and weighed (± 0.1 mg) with minimal disturbance to ensure meal initiation. Each slug was released singly into a washed small plastic Petri dish with a lid, containing a disc of moist filter paper. The test pellet was moistened with a few drops of distilled water before being presented to a slug, to increase initial palatability. If the animal rejected the test pellet three times the slug was removed from the test. If the animal sampled the pellet and continued feeding (> five bites) the meal was recorded. Immediately after a meal, the animal was reweighed (± 0.1 mg), and the pellet was retrieved and air dried to constant weight. The state of intoxication was assessed immediately following the meal by turning the animal on its right side and timing its ability to resume crawling (Pessah and Sokolove, 1983). Each slug was returned to its Petri dish and on fresh moist filter paper. A slice of carrot was provided as food in each dish. Each chamber was then placed in an incubator in similar conditions to those during acclimation and slugs were assessed every other day for mortality. Death was defined by evidence of decay.

RESULTS

Results for 20 meals on dummy pellets and 10 meals on each of the 2 and 4 % pellets were collated for both species, giving a total of 80 single meals. Key meal parameters, number of bites, duration of meal, mean rasp rate, mean rasp period and the number of pauses per meal greater than 8 seconds were compared for treatment group and between each species. Other factors, mean rasp size and the amount of pellet consumed compared to the size of the slug were calculated from the data.

Toxicity of methiocarb

The amount of non-toxic pellet consumed expressed in mg/g of slug was remarkably similar for both species, with the *Arion* slug eating slightly more than *Deroceras*. The presence of methiocarb reduced the amount of pellet eaten considerably for both species of slug. For *D. reticulatum* the amount of both active pellet types eaten was reduced to approximately 40% of a meal on non-toxic bait. This was further reduced by 4 and 7% for *A. distinctus* with 2% and 4% pellets respectively.

Both species of slug ingested proportionally the same amount of methiocarb during a meal on a 2% pellet (Table 1). However during a meal on a 4% pellet *D. reticulatum* ingested proportionally more methiocarb. The actual amounts ingested for both species were very much lower than the calculated theoretical values, assuming early meal termination did not occur, based on control meals.

Table 1 The theoretical amount of methiocarb that would be ingested in a single meal on non-toxic pellets, compared to the actual amount ingested by both species of slug with 2 or 4 % (w/w) pellets.

| | Mean amount of methiocarb ingested (µg/mg slug) ± SD | |
Pellet type	*Arion distinctus*	*Deroceras reticulatum*
Theoretical if a 2% pellet	0.94 ± 0.28	0.88 ± 0.26
Theoretical if a 4% pellet	1.88 ± 0.56	1.77 ± 0.52
Actual with a 2% pellet	0.34 ± 0.10	0.37 ± 0.11
Actual with a 4% pellet	0.62 ± 0.13	0.82 ± 0.45

The righting test immediately post-meal, showed that *A. distinctus* took considerably longer than *D. reticulatum* to initiate crawling, irrespective of the meal type. On dummy pellets, mean righting times were 55 ± 25s for the former and 17 ± 14s for the latter. On 2% pellets, four *Arion* slugs made no attempt to right themselves (within three minutes) though the rest were comparable to the controls. After a 4% meal one *Arion* did not right within three minutes, though the mean of the rest had increased to 84 ± 26s. Two *Deroceras* failed to right in both the 2 and 4% groups and means were about three times the control value for the rest, for both pellet types.

Percentage mortality of slugs, however, did not reflect differences in the total ingestion of methiocarb (Table 2). Nevertheless mortality rates in ideal recovery conditions after a methiocarb meal were all 80% or more, irrespective of pellet concentration, and all *A. distinctus* died after feeding on 4% pellets. Mortality of some control animals occurred but at low levels after almost double the time and was distinct from treatment effects.

Table 2 Percentage mortality of two species of slug after feeding on either a 2 or 4% methiocarb pellet or a non-toxic control pellet (mean time to death in days ± SD).

Species	Control pellet	Pellet type 2% Methiocarb	4% Methiocarb
A distinctus	30	80	100
	(14 ± 9)	(7 ± 4)	(8 ± 3)
D. reticulatum	10	90	80
	(18 ± 9)	(7 ± 3)	(10 ± 6)

Effects of methiocarb on feeding behaviour

The mean values ± SD for meal parameters with *D. reticulatum* for non-toxic, 2% and 4% (w/w) methiocarb were: 3155.8 ± 632.8s, 2666.7 ± 904.9s and 2019.4 ± 884.7s for duration; 720 ± 200, 732 ± 275 and 515 ± 232 for rasp number; 13.6 ± 2.9, 16.1 ± 3.1 and 15.8 ± 2.1 for rasps per minute; 463 ± 109 centiseconds (cs), 391 ± 81cs, 402 ± 62cs for rasp period, and 0.067 ± 0.024, 0.028 ± 0.008 and 0.042 ± 0.011 for rasp size in µg/mg slug.

D. reticulatum exhibited a negative dose response to increasing methiocarb in a number of key meal parameters. The meal duration, the total number of bites, and the mean number of bites between pauses (> 8 seconds) reduced progressively as methiocarb concentration increased. Interestingly mean rasp size was higher in a meal on a 4% pellet compared to a 2%, though reduced compared to the mean for control meals . The mean rasp rate was higher in meals on both methiocarb pellets than in a control meal. This is linked to the onset of toxic effects.

When feeding on non-toxic pellets *A. distinctus* took longer meals with slower, longer and larger bites than *D. reticulatum*. The duration of dummy meals for *A. distinctus* ranged from 1591 to 4974s with a mean rasp rate of 11.1 ± 2.3 rasps per minute. These subtle differences in feeding behaviour were amplified on baits containing methiocarb. The largest differences were seen in meals on 2% pellets. Although the number of rasps, rasps rate and meal duration were lower than in a comparable *Deroceras* meal, rasp size and rasp duration were very much larger. These two similar parameters were markedly reduced for *Arion* at the higher rate of methiocarb. The differences are highlighted in Figure 1.

DISCUSSION

In the controlled conditions of this laboratory experiment on single meals, both 2 and 4% methiocarb pellets were effective in killing the majority of slugs. Although the populations were small, all *Arion distinctus* slugs were dead within 5 to 11 days after a 4% methiocarb meal. In comparison, 100% mortality with *D. reticulatum* was not achieved with either toxic pellet type though mortality was 80 to 90%. In this single meal experiment it appears that *A. distinctus* was slightly more susceptible to methiocarb baits than *D. reticulatum*. However, in arena studies *D. reticulatum* were observed to take multiple meals on methiocarb pellets whereas *A. distinctus* were immobilised after just one meal (Bailey and Wedgwood, 1991).

Fig. 1. Relative key meal parameters for *A. distinctus* compared to *D. reticulatum* for three pellet types (untreated control, 2 and 4 % methiocarb (w/w)).

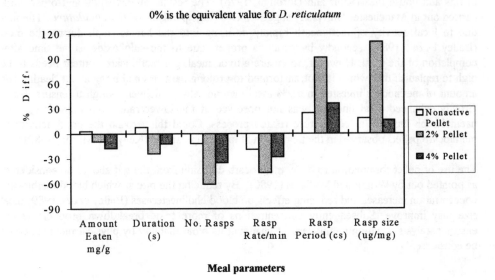

0% is the equivalent value for *D. reticulatum*

Meal parameters

It is remarkable that when feeding on non-toxic pellets both slug species consumed proportionally the same amount although subtle differences in their feeding behaviour were observed. *A. distinctus* rasped more slowly and took larger rasps than *D. reticulatum*. On toxic pellets these differences were amplified. The even larger rasp size and duration, however, are phenomena created by very short meal duration. It follows that at initiation of any meal, when food arousal is at its strongest, the rasps are larger than towards the end of a long meal when the animal becomes sated. For a long meal, on a dummy pellet, the mean rasp period reflects the long 'tail' of declining food arousal. However during toxic meals, which were much shorter, initial arousal is reflected. *A. distinctus* showed few signs of the onset of methiocarb toxicity before feeding came to an abrupt halt. However rasping up to the end of a meal was very efficient and methiocarb intake sufficient to ensure relatively rapid ingestion of a lethal dose. This behaviour did not seem to equate to repellency or distastefullness.

D. reticulatum feeding behaviour displays more clear signs of the onset of paralysis with a higher rasping rate compared to control meals. It is not likely that this is a product of increased food arousal at the beginning of feeding, as meals on 2% pellets were only 15% shorter than those on a non-toxic pellet while rasp rate was nearly 20% higher. However this increased rasp rate was less efficient and bite size was also reduced. Perhaps feeding stops as a result of fatigue of the over-stimulated muscles controlling radula action. Again this behaviour is not commensurate with repellency of the active ingredient. Briggs and Henderson (1987) attributed the success of methiocarb as a molluscicide to its low repellency.

Although early meal termination occurred slugs were only partially paralysed at the end of the meal, as indicated by the righting tests. Pessah and Sokolove (1983) found that methiocarb's sulphoxide derivative had speedier effects at a lower molar concentration, than methiocarb alone, on the activity of foot and haemolymph cholinesterase activity extracted from *Limax flavus*. The sulphoxide is the major product of microsomal sulphoxidation of similar carbamates in rats and house flies (Kuhr and Dorough, 1976). The results of this study and other work carried out at Manchester University suggest a two stage paralysis of *D. reticulatum*. The first due to local activity of methiocarb rapidly diffusing into the haemolymph during the meal (Bailey *et al.*, 1989), secondly by paralysis proper, due to the sulphoxide, some time after completion of the meal. However, to ensure a toxic meal, gut methiocarb content needs to be high to replenish that which is biotransformed to prolong paralysis and bring about death. The amount of methiocarb ingested in a 2% meal may not always deliver enough to ensure death. Greatly enhanced meal duration was not observed at the lower rate. Sulphoxidation in the haemolymph must be a temperature related process. Could this explain the good activity of methiocarb pellets observed in the field, even at low temperatures (Kelly and Martin, 1989)?

The use of pellet formulation to delay methiocarb diffusion from the gut should be considered, as pointed out by Wright and Williams (1980). By reducing the rate at which blood methiocarb concentration increases and reducing effects on blood cholinesterases (Bailey *et al.*, 1989) meal size may improve. Subsequently concentrations of more toxic breakdown products would ensure paralysis. Coating methiocarb or changing its bioavailability by physical methods could be considered.

ACKNOWLEDGMENTS

This work was carried out as PhD research project at Manchester University, funded by SERC (Total Technology Case Award) and sponsored by Bayer plc, Crop Protection Business Group.

REFERENCES

Bailey, S E R; Wedgwood, M A (1991) Complementary video and acoustic recordings of foraging by two pest species of slug on non-toxic and molluscicidal baits. *Annals of Applied Biology*. **119**, 163-176.

Bailey, S E R; Cordon, S; Hutchinson, S (1989) Why don't slugs eat more bait? A behavioural study of early meal termination produced by methiocarb and metaldehyde baits in *Deroceras caruanae*. In *Slugs and snails in world agriculture*. I F Henderson (ed.). B C P C Monograph No. 41, pp. 385-390.

Briggs, G G; Henderson, I F (1987) Some factors affecting the toxicity of poisons to the slug *Deroceras reticulatum*. *Crop protection*. **6**, 341-346.

Kelly, J R; Martin, T J (1989) 21 years' experience with methiocarb bait. In: *Slugs and snails in world agriculture*. I F Henderson (ed.). B C P C Monograph., No 41, pp. 131-146.

Kuhr, R J; Dorough, H W (1976) Carbamate Insecticides: Chemistry, Biochemistry & Toxicology. C R C Press, Ohio.

Pessah, I N; Sokolove, P G (1983) The interaction of organophosphate and carbamate insecticides with cholinesterases in the terrestrial pulmonate, *Limax maximus*. *Comparative Biochemistry and Physiology*, **74C (2)**, 291-297.

Wright, A A; Williams R (1980) The effect of molluscicides on the consumption of bait by slugs. *Journal of Molluscan Studies* **46**, 265 -281.

OPTIMISING THE CHEMICAL CONTROL OF THE RICE PEST *POMACEA CANALICULATA* (LAMARCK)

J S ARTHUR, E J TAYLOR, I D BOWEN

Mollusc Research Group, Department of Pure and Applied Biology, University of Wales Cardiff, PO Box 915, Cardiff, CF1 3TL United Kingdom

ABSTRACT

The control of pests in the aquatic environment using chemical methods are discussed with reference to the Golden Apple Snail (*Pomacea canaliculata* Lamarck), an increasingly important pest of rice crops in the Far East.

The "optimum" chemical treatment - based on the exposure period and the concentration of the agent - must be effective against *P. canaliculata* whilst minimising the impact on non-target biota. Treatments used against pest species are often chosen on the basis of the predicted mortality, however, in the case of this crop pest a sub-lethal dose which inhibits feeding may be more suitable.

Extracts of tissues from Nigerian plants are being screened in this laboratory according to their relative toxicity to *P. canaliculata* and a reference non-target freshwater species, *Daphnia magna*, in acute lethality tests. Extracts which prove more toxic to the target species are subsequently investigated for their potential use in sub-lethal control strategies. This is quantified using a behavioural bioassay which determines the feeding rate of apple snails after a specified exposure period.

The suitability of chosen plant extracts for the control of *P. canaliculata* are discussed with respect to commercially available products and other existing methods.

INTRODUCTION

The Golden Apple Snail, *Pomacea canaliculata* Lamarck, has become a major problem pest of rice crops in several Asian countries since it's relatively recent introduction (Halwart, 1994). The use of chemical agents is still the most widely used method in the control of this aquatic snail pest, but these may be costly to both the farmer and the environment. There is a need to identify readily available agents which specifically target the pest snails whilst minimising adverse effects on non-target biota.

Plant extracts have been investigated for molluscicidal activity in a number of laboratories (Adewunmi & Saforwora, 1980; Adewunmi, 1991; Belot *et al.*, 1993; Marston *et al.*, 1993, Perrett & Whitfield, 1996) since they may provide cheaper, safer and more effective alternatives to commercially available products. In an ongoing study in this laboratory various Nigerian plants, which have been shown to possess molluscicidal properties (Adewunmi & Safowora, 1980; Kela, 1992; Kela & Bowen, 1995; Kela, pers. comm.), are being assessed for their potential suitability as control agents. The relative toxicity of plant extracts to both the target species and non-target reference species is being investigated using a screening procedure - those proving more toxic to the target species than to the non-target species will be further investigated in order that the active compound(s) may be isolated and identified.

Two freshwater species are used as the reference non-target organisms in the screening procedure. The cladoceran, *Daphnia magna* Strauss, and *Gammarus pulex* Linnaeus

(Amphipoda), a freshwater detritivore, are known to be important components of freshwater ecosystems. They were chosen on the basis of the large toxicity database which exists for these species and because they are known to be sensitive to a range of pollutants (McCahon & Pascoe, 1988a).

Exposure of aquatic organisms to sub-lethal concentrations of toxicants can be important since they may reduce feeding rates, growth, reproductive success (Pascoe *et al.*, 1991), and induce post exposure mortality (Abel, 1980). The use of plant extracts at a sub-lethal level may therefore prove to be sufficient to control *P. canaliculata* by reducing or preventing feeding on the rice crops during the period for which the crops are vulnerable. In order to assess likely control potential and also environmental impacts the screening process focuses on sub-lethal effects of plant materials to both target and non-target species using a modified feeding bioassay.

MATERIALS AND METHODS

Candidate plant species

The plants listed below were chosen on the basis of previously documented molluscicidal activity reported by the cited authors.

1. *Ximenia americana* - leaves and stem bark (Kela, 1992)
2. *Detarium microcarpum* - stem bark (Kela & Bowen, 1995)
3. *Tatum spp.* - (Kela, pers. comm.)
4. *Ankubrang spp.* - (Kela, pers. comm.)
5. *Polygonum limbatum* - leaves (Adewunmi & Safowora, 1980)

Culturing and maintenance of test species in the laboratory

Pomacea canaliculata: A static-with-weekly-renewal culture system (28°C and photoperiod 16 h light) is used with 8 - 15 adult snails maintained in tanks of dimensions 600 mm x 300 mm x 300 mm half filled with 6 litres of dechlorinated mains water (pH 7.3 ± 0.2, hardness 170 mg/l as $CaCO_3$, and conductivity 312 ± 10 S/cm) to allow snails to lay eggs (positioned above the surface of the water). Eggs are carefully removed form the tanks and transferred to "egg hatching" tanks (dimensions 300 mm x 150 mm x 150 mm) where they are placed on wire mesh which is suspended above the water, this allows hatched snails to fall through to the water below. The snails are fed every 2 days on a mixture of lettuce and Tetramin[®].

Daphnia magna and *Gammarus pulex:* A static-with-weekly-renewal culture system (photoperiod 16 h light and 20°C for *Daphnia*, 15°C for *Gammarus*) is used for both species maintained as described by Taylor *et al.* (1995) and McCahon & Pascoe (1988b).

Preparation of plant extracts and stock solutions

Extracts of the plant species were prepared using either water or acetone, chosen as non-polar and polar solvents, respectively, with the potential to dissolve different fractions from the original plant material. Tissues from each plant species (leaves or stem bark) were ground to a fine powder, 5 g of which was then suspended in 75 ml of solvent (water or acetone) and left to stir for 24 h at room temperature. The resulting supernatant was filtered off and the solvent was evaporated under reduced pressure in a rotary evaporator. The extract obtained was used to make "stock" solutions for use in the toxicity tests.

Stock solutions were prepared by the ultrasonication of a known mass of plant extract (in solid form) in dechlorinated water (dilution control water) made up to a known volume. The required nominal concentrations were then made by diluting this stock. The concentration of the plant extract stock and subsequent dilutions were verified colorimetrically using spectrophotometric techniques at maximum absorption wavelength for the extract in question.

Acute lethality tests

A range of five plant extract concentrations, selected on the basis of results from preliminary toxicity tests, and a control of dechlorinated mains water were used in order that the pattern of mortality and, if appropriate, median lethal concentrations (LC50's) could be determined for particular exposure periods. The exposure periods currently under investigation include 3, 6, 24 and 96 h. Juvenile life stages of *P. canaliculata* (3-5 mm shell height) and *Daphnia magna* (24 h old) were sampled from the laboratory stock for use in the bioassay. 10 animals were exposed as a group in 100 ml of control or test solution at 28°C and 20°C for snails and daphnids respectively, and a photoperiod of 16 h light.

After the specified exposure period the animals were transferred to 100 ml of control water for a 24 h period. This was performed in order that snail mortality could be accurately determined since the snails often do not respond to mechanical stimulation during toxicant exposure. Mortality was recorded after this additional 24 h period - both daphnids and snails were deemed to be dead if there was no response to mechanical stimulation. This method determines the post exposure mortality (after a 24 h period) which has been shown to occur with other aquatic invertebrates after various toxicant exposures (Abel 1980).

Sub-lethal tests

Feeding behaviour was chosen as the criterion of sub-lethal toxicity since this has been widely recognised as a sensitive indicator of toxicant induced stress. The method described by Taylor *et al.* (1993), used to quantify the feeding activity of the freshwater amphipod *Gammarus pulex*, is a rapid and non-destructive bioassay which allows the suitable quantification and statistical comparison of feeding rates using time-response analysis of the feeding on eggs of *Artemia salina*. This method was modified for use with the target species, *P. canaliculata*, and the original method was used with the reference non-target species, *Gammarus pulex*, using the following protocol:

1. Juvenile life stages of *P. canaliculata* (3 - 5 mm shell height) and *G. pulex* (5 - 6 mm body length) were sampled from the laboratory stock for use in the bioassay.

2. 10 animals of each species were exposed as separate groups in 500 ml of either control or test solution. Sub-lethal plant extract concentrations were chosen on the basis of the results from the acute lethality tests. Food was provided during the exposure - lettuce for the snails and conditioned leaves of horse chestnut, *Aesculus hippocastanum* (L.), for the gammarids.

3. After the exposure period the test animals were carefully transferred to individual pots (base area 18 cm^2) each containing 10 shell-less *Artemia salina* eggs and 18 ml of the relevant treatment solution (i.e. exposure continued).

4. The number of eggs eaten by each individual was recorded frequently to allow the cumulative percentage of eggs consumed by each treatment group to be calculated during the test.

5. Time response analysis (based on the methods of Litchfield, 1949) was used to determine the median feeding times, (FT50's, time at which 50 % of the eggs have been consumed) for each treatment group. The FT50's were statistically compared ($p = 0.05$) to identify the relative toxicities of the plant extracts.

Three plant extracts, *X. americana* leaf-water, *D. microcarpum* stem bark-water, and *Ankubrang spp.* stem bark-water, were chosen for use in the bioassays based on their differential toxicity in the acute mortality tests, being more toxic to the target snails than to non-target *D. magna*. The water temperature throughout the 24 h exposure period was maintained at 28°C for the snails and 15°C for the gammarids.

Water quality and toxicant analyses

Temperature, pH, dissolved oxygen and conductivity were determined at regular intervals by portable meters, while total hardness was measured by Atomic Absorption Spectrophotometry.

RESULTS

The means (95% confidence limits) for pH, conductivity, and hardness of the solutions during the tests were 8.12 ± 0.28, 319.5 ± 31.5 μS/cm and 176.9 ± 3.7 mg/l as $CaCO_3$, respectively. The dissolved oxygen exceeded 70 % air saturation in all cases. Mean temperatures were 27.2 ± 1.31°C for the snails, 18.1 ± 0.32°C for the daphnids, and 15.7 ± 0.78°C for the gammarids.

24 h acute lethality tests

Table 1 shows the results of 24 h exposure acute lethality tests using 5 plant species in the screening procedure. There were no mortalities in any of the control treatments.

Water extracts: concentrations of plant extract of 300 mg/l or more effected high mortality in both species, except this extract concentration of *X. americana* leaf and *P. limbatum* stem bark. The 300 mg/l concentration of *X. americana* leaf was markedly more toxic to *Pomacea* than to *Daphnia* whilst both species were apparently tolerant of this concentration of *P. limbatum* stem bark. The relative toxicity of the water extracts other than *X. americana* leaf and *P. limbatum* stem bark becomes apparent at lower concentrations and generally 100 mg/l causes greater than 80 % mortality of snails and less than 40 % mortality of daphnids. The water extract of *Ankubrang spp.* proved to have the largest differential in toxicity with a concentration of 30 mg/l still producing high mortality (70 %) of snails whilst only 10 % mortality of daphnids.

Acetone extracts: of the acetone extracts tested, *Tatum spp.* and *Ankubrang spp.* produced a similar pattern of mortality to that generally found with the water extracts (described above). *Tatum spp.* produced the largest differential in toxicity, similar to that shown by *Ankubrang spp.* water extract. However, the acetone extract of *X. americana* stem bark was more toxic to *Daphnia* than to *Pomacea* e.g. at concentrations of 3 - 30 mg/l there was at least 50 % mortality of *Daphnia* whereas there was at most 10 % mortality of *Pomacea*.

Sub-lethal feeding tests

Median feeding times (FT50's) for *P. canaliculata* and *G. pulex* after 24 h exposure to control and plant extract solutions, together with 95 % confidence intervals where these could be calculated, are given in table 2. Statistical differences (p = 0.05) between FT50's of animals in the reference water and FT50's in plant extract solutions are also given in table 2. It should be noted that an increase in FT50 represents a decrease in feeding rate.

Exposure to certain plant extract concentrations markedly reduced the feeding rates such that 50 % of the eggs were not eaten. The FT50 could not be calculated in these cases and statistical comparison was obviously precluded, however it is clear that these extracts had a dramatic effect on feeding behaviour. In these cases the number of eggs consumed during the bioassay was noted together with the time taken.

Table 1. Mortality of *P. canaliculata* and *D. magna* after 24 h exposure to plant extracts and a further 24 h period in control water.

Plant Species and Tissues Used	Solvent	Test Species	Nominal Plant Extract Concentration (mg/l) and Percentage Mortality after 24 h Exposure						
			1000	300	100	30	10	3	1
Ximenia americana leaves	water	*Pomacea*	-	100	90	0	0	0	-
		Daphnia	-	40	30	0	0	0	-
	acetone	*Pomacea*	not sufficiently soluble to cause mortality						
		Daphnia							
Ximenia americana stem bark	water	*Pomacea*	100	100	100	60	0	-	-
		Daphnia	100	80	30	30	20	-	-
	acetone	*Pomacea*	-	100	100	10	0	0	0
		Daphnia	-	100	80	50	60	50	0
Detarium microcarpum stem bark	water	*Pomacea*	-	100	80	0	0	0	-
		Daphnia	-	80	40	0	0	0	-
Tatum spp. stem bark	acetone	*Pomacea*	-	100	100	70	0	0	-
		Daphnia	-	90	30	10	0	0	-
Ankubrang spp. stem bark	water	*Pomacea*	-	100	90	70	0	0	-
		Daphnia	-	100	20	10	0	0	-
	acetone	*Pomacea*	-	100	100	40	0	0	-
		Daphnia	-	100	30	0	0	0	-
Polygonum limbatum leaves	water	*Pomacea*	-	*10	0	0	0	0	-
		Daphnia	-	0	0	0	0	0	-

* indicates not sufficiently soluble above this concentration

The median feeding times indicate that in general the feeding rates of both *P. canaliculata* and *G. pulex* are decreased (FT50's increased or 50 % of eggs not consumed) by 24 h exposure to the three tested plant extracts in a concentration-related manner. However, an exception to this is seen for snails exposed to *Ankubrang spp.* water extract where there was a significant increase ($p = 0.05$) in feeding rate of animals exposed to 5 mg/l of extract - the feeding rate of snails was reduced at concentrations higher than 5 mg/l. It is evident that the feeding rate of *G. pulex* is significantly reduced ($p = 0.05$) at lower concentrations of *X. americana* leaf water extract and *Ankubrang spp.* water extract than affects *P. canaliculata*.

Table 2. Median feeding time (FT50) of *P. canaliculata* and *G. pulex* after 24 h exposure to plant extracts (95 % confidence limits are given in parentheses). Where FT50's could not be calculated the number of eggs consumed during the experiment is displayed as a percentage of the initial number (duration of test given in parentheses).

Plant Species	Solvent	Concentration (mg/l)	Median feeding time (FT50) after 24 h exposure (min.)	
			Pomacea canaliculata	*Gammarus pulex*
Ximenia	water	Control	35.3 (22.2-56.2)	60.7 (46.48-79.26)
americana		5	42.6 (25.1-72.5)	108.2 (69.2-168.6)*
leaves		10	-	133.1 (170.5-103.9)*
		30	32% in 173 min.[a]	-
		50	0% in 173 min.[a]	27% in 267 min.[a]
Detarium	water	Control	29.5 (22.5-38.5)	46.1 (37.3-57.1)
microcarpum		10	52.0 (34.6-78.3)*	133.3 (85.5-207.9)*
stem bark		30	60.2 (43.8-82.8)*	189.2 (129.8-275.7)*
		50	15% in 65 min.[a]	169.7 (129.5-222.4)*
Ankubrang spp.	water	Control	30.4 (22.8-40.5)	71.4 (53.6-95.0)
stem bark		5	16.7 (13.7-20.4)*	174.9 (135.4-225.8)*
		10	24.3 (15.8-37.4)	229.0 (131.2-399.8)*
		30	5 % in 45 min.[a]	10 % in 120 min.[a]

* indicates significant difference from control (p = 0.05)

[a] indicates animals not consuming 50 % of the eggs provided

DISCUSSION

Results from the screening procedure using the criteria of mortality of *D. magna* and *P. canaliculata* indicate that several of the plant extracts tested are worthy of further investigation. After 24 h exposure these extracts were found to cause higher mortality in the target species, *P. canaliculata*, than the non-target reference species, *D. magna*. Such extracts may be useful in lethal doses (at specified concentration and exposure period) as control agents. The shortlist of suitable extracts is given below:

Ximenia americana	leaves	water extract
	stem bark	water extract
Detarium microcarpum	stem bark	water extract
Tatum spp.	stem bark	acetone extract
Ankubrang spp.	stem bark	water extract
		acetone extract

A 24 h exposure to sub-lethal concentrations of plant water extracts was seen to reduce the feeding rates of both the target snail and the reference species, *G. pulex*. This suggests that a relatively short exposure to low levels of plant extract may be sufficient to reduce feeding for the period when rice crops are vulnerable, however this may also affect the feeding of non-target species. This may be an acceptable alternative to the use of higher concentrations to effect the mortality of this crop pest.

It was apparent that at a low concentration of *Ankubrang spp*. water extract the snails were stimulated into feeding at a faster rate than in the reference water. This may be because *P. canaliculata* is stimulated to feed in the presence of dissolved plant material. However, at higher concentrations the feeding rate is reduced, probably because the effect of the toxic ingredient in the extract reducing feeding outweighs any stimulant effect.

In order to increase the database for the choice of the most suitable chemical the screening tests must be extended to include indigenous non-target species including invertebrates and fish, as well as investigating other aspects such as the impacts on the soil metabolism. It is envisaged that the active ingredient(s) will be isolated and identified from the short listed plant extracts and using the methodology detailed above the efficacy of the plant isolates in the control of snail pests will be compared with other commercially available chemical applications such as Niclosamide (Bayluscide®) and Metaldehyde (Optimol®). If the plant materials tested prove to be more or equally as effective as those agents already in use, whilst causing minimal environmental impact even at sub-lethal levels, then they will provide important alternatives in the management of pests such as *P. canaliculata*.

REFERENCES

Abel, P D (1980) Toxicity of γ-hexachlorocyclohexane (Lindane) to *Gammarus pulex*: mortality in relation to concentration and duration of exposure. *Freshwater Biology* **10**, 251-259.

Adewunmi, C O (1991) Plant Molluscicides: Potential of Aridan, *Tetrapleura tetraptera*, for Schistosomiasis control in Nigeria. *The Science of the Total Environment* **102**, 21-33.

Adewunmi, C O; Safowora, E A (1980) Preliminary Screening of some Plant Extracts for Molluscicidal Activity. *Planta Medica* **39**, 57-65.

Belot, J S; Geerts, S; Sarr, S; Polderman, A M (1993) Field Trials to Control Schistome Intermediate Hosts by the Plant Molluscicide *Ambrosia maritima* L. in the Senegal River Basin. *Acta Tropica* **52**, 275-282.

Halwart, M (1994) The Golden Apple Snail, *Pomacea canaliculata*, in Asian Rice Farming Systems: Present Impact and Future Threat. *International Journal of Pest Management* **40 (2)**, 199-206.

Kela, S L (1992) Molluscicidal Property of Polygonum limbatum (L.). In: *Proceedings of the 16th Annual Conference of the Nigerian Society for Parasitology*, Ahmadu Bello University, Zaria, Nigeria, September 1992.

Kela, S L; Bowen, I D (1995) Control of Snails and Snail Borne Diseases. *Pesticide Outlook*, February 1995.

Litchfield, J T (1949). A method for the rapid graphic solution of time percent effect curves. *Journal of Pharmacology and Experimental Theory* **97**, 399-408.

McCahon, C P; Pascoe, D (1988a) Use of *Gammarus pulex* (L.) in Safety Evaluation Tests - Culture and Selection of a Sensitive Life Stage. *Ecotoxicology and Environmental Safety* **19**, 245-252.

McCahon, C P; Pascoe, D (1988b) Culture Techniques for Three Freshwater Macro-invertebrates Species and their use in Toxicity Tests. *Chemosphere* **17**, 2471-2480.

Pascoe, D; Gower, D E; McCahon, C P; Poulton, M J; Whiles, A J; Wulfhorst, J (1991) Behavioural Responses to Pollutants - Applications in Freshwater Bioassays. In: Jeffrey, D W and Madden, B (eds) *Bioindicators and Environmental Management*. pp. 245-254. Academic Press.

Perrett, S; Whitfield, P J (1996) Currently Available Molluscicides. *Parasitology Today* **12 (4)**, 156-159.

Taylor, E J; Jones, D P W; Maund, S J; Pascoe, D (1993) A new method for measuring the feeding activity of *Gammarus pulex* (L.). *Chemosphere* **26**, 1375-1381.

Taylor, E J; Morrison, J E; Blockwell, S J; Tarr A; Pascoe, D (1995) Effects of Lindane on the Predator-Prey Interaction between *Hydra oligactis* Pallas and *Daphnia magna* Strauss. *Archives of Environmental Contamination and Toxicology* **29**, 291-296.

REFORMULATION STUDIES WITH METHIOCARB

I D BOWEN, S ANTOINE
School of Pure and Applied Biology, University of Wales Cardiff, PO Box 915, Cardiff, CF1 3TL, UK

T J MARTIN*
Cydia, Gislingham Road, Finningham, Stowmarket, IP14 4HZ, UK

ABSTRACT

The efficacy of methiocarb at 1.0%, 2.0% and 4.0% has been tested in the laboratory against the grey field slug *Deroceras reticulatum* and compared with methiocarb reformulated with a range of potential synergists including piperidine and veratrylamine. Using a battery of tests including simple arena repellency/attractancy trials, voluntary feeding trials and terraria trials, a number of useful additives which appear to increase slug mortality and enhance crop protection have been identified.

INTRODUCTION

It is generally accepted that slugs are serious pests on arable farms in the UK (Glen, 1989). Slugs cause serious damage to both wheat and potato crops (Port & Port, 1986). Grasslands, horticultural crops such as Brussels sprouts, lettuce, carrots, sugar beet and ornamental flowers may also be damaged (Martin & Kelly, 1986). It appears that fashionable farm practices may contribute to the slug problem (Glen *et al.*, 1984) and the use of molluscicides containing metaldehyde or carbamate (especially methiocarb) has been increasing, particularly where slug damage is incipient due to the rotation of vulnerable crops with oilseed rape. Chemical control is not always easy to obtain (Henderson & Parker, 1986) and, it is claimed, may be harmful to non-target organisms including predatory carabid beetles (Kennedy, 1990; Purvis & Bannon, 1992). However, because arable crops are essentially transient, most resident organisms are mobile and thus populations generally return to "normal" in a relatively short space of time following the impact of pesticides, according to Martin (1993).

Bowen & Antoine (1995) argued that whilst many molluscicides as currently formulated may pose a hazard to the environment and non-targeted animals, biological agents were not yet a practicable alternative for use on a large scale; thus further improvements or reformulation of molluscicides would be required to produce acceptable preparations of minimal impact on non-targeted species. Bowen & Jones (1985a) suggested that a number of surfactant additives could be successfully formulated with molluscicides, resulting in a reduction of the concentration of active ingredient required coupled with an increased efficacy.

The relative merits of metaldehyde and methiocarb as terrestrial molluscicides were investigated by Bourne *et al.* (1988, 1990) and formulations containing combinations of metaldehyde and methiocarb were also tested. In this paper we look for synergism between methiocarb and a selected number of additives, some derived from the beehive product "propolis". In general much potential still remains to be exploited in terms of chemical reformulation. New slug attractants, repellents and antifeedants are emerging along with mammal and avian repellents which promise to improve pest targeting (Bowen *et al.*, 1993; Henderson *et al.*, 1992). The use of mammalian and avian repellents could prove particularly important in improving the targeting of conventional molluscicides such as metaldehyde and methiocarb (Mason *et al.*, 1991).

---

* formerly Bayer plc

Preliminary work undertaken at Cardiff has demonstrated that a number of chemicals derived from the complex mixture that is found in the bee-product "propolis" are potential slug control compounds. In particular derivatives such as glutaramide and piperidine were found to be molluscicidal. In this study derivatives including piperidine, veratrylamine, methyl morpholine and two commercially sensitive candidates A and B, were tested for their synergism with methiocarb. The methodology employed for comparing the efficacy of the additives was based on that of Wright & Williams (1980) where a chromium marker is introduced into homogeneously produced baits in order to accurately measure bait and active ingredient consumption by means of atomic absorption. These data are coupled with accumulated laboratory data on mortality and crop protection. Separate arena trials on attractancy and repellency were also carried out.

MATERIALS AND METHODS

Voluntary feeding tests

Fresh grey field slugs, *D. reticulatum* (Müller) were collected from the wild and stored in glass covered trays lined with damp filter paper, held at 13°C. Animals were starved for 48 hours before being presented with laboratory prepared baits for the next 24 hours in individual chambers (30 ml bottles) where the relative humidity was kept high with moist cotton wool. Slugs of a narrow weight range (400-600 mg) were chosen to minimise any differences that might arise due to larger individuals consuming more bait than smaller ones. The bait was removed after 24 hours and replaced by a single wheat grain as an alternative food source. The condition of the slugs was then monitored for the next 7 days, and each slug was scored as being either normal, paralysed, or dead using an electrode linked to a 12 volt D.C. pulse generator to stimulate slugs to move (Henderson, 1968). Slugs which showed only a slight reflex contraction when stimulated were considered to be paralysed and were unable to crawl away from the point of stimulation. Slugs showing no response were dead, and those that demonstrated none of the above symptoms were considered normal. Elimination of faeces (containing chromic oxide marker), slug weight and seed hollowing were also monitored over the experimental period. Bait (and hence a.i.) consumption was determined by atomic absorption spectroscopy of chromium in the slug tissue and any faeces produced.

Preparation of baits

Baits for laboratory feeding trials were prepared as follows:

Plain flour (balance)

Chromic oxide (to give a final concentration of 2% chromium wt/wt)

4-nitrophenol: mould repellent (0.06% wt/wt)

Active ingredient at the desired concentration, i.e. 1%, 2% and 4% methiocarb

15 ml of double distilled water per 50 g mix

Adjuvant additives at 1%, namely piperidine, veratrylamine, methyl morpholine, compound A and compound B

50 g batches were mixed in a domestic food mixer and the pH of the aqueous phase adjusted to 6.5 with 10 drops of 0.1 M sodium hydroxide solution. Baits were pelleted through a 1 ml syringe and dried at 37°C for 48 hours to produce pellets weighing approximately 300 mg. Measurements of bait and active ingredient consumed were calculated and associated slug mortality and grain consumption data accumulated.

Arena attractancy/repellency

Simple arena attractancy, repellency and phagostimulation tests were carried out in moist petri dishes as described by Bowen & Antoine (1995).

Terraria trials

Terraria, measuring 0.2 m², were used in this experiment, and lined with a double thickness of filter paper as a base on which to sow wheat grains. Winter wheat grains were sown at a rate of 60 per terrarium - 6 rows of 10 grains each. Experimental pellets, methiocarb 4% pellets*, and blank pellets were introduced at a rate of 5.5 kg/ha and randomly distributed.

Five slugs pre-starved for 48 h were introduced into each terrarium representing a rate of 400,000 slugs per hectare, that is a heavy slug infestation.

Six replicates of each treatment were undertaken. Terraria were watered regularly to maintain the required humidity, and were kept in a CT room at 14° ± 1°C, and moderate humidity.

Slug mortality, paralysis, grain loss, and seedling damage were monitored daily for 14 days.

RESULTS

Data on slug mortality and grain loss from the voluntary feeding experiments are presented in Figure 1 and indicate that 1% methiocarb combined with 1% additive A is superior to 4% methiocarb on its own. The same advantage is seen at 2% (not presented) and 4% methiocarb. Advantages in mortality are reflected also in the crop protection figures. Thus 4% methiocarb with additive A results in only 6.7% grain loss. Additive B had no advantageous effect whilst both piperidine and methyl morpholine at 1% did appear to bestow an advantage over methiocarb alone.

The range of bait and methiocarb ingested are presented in Figures 2 and 3 respectively. It is clear that methiocarb at 4% is less palatable to slugs than at 1% and again addition of additive A enhances ingestion of bait which at 1% methiocarb almost approached that achieved by blank pellets. Interestingly piperidine and methyl morpholine seem to suppress ingestion of bait and the enhanced levels of slug mortality which they appear to induce may be due to an inherent molluscicidal action. Veratrylamine marginally enhanced the uptake of bait at 1% methiocarb level but bait uptake was generally suppressed at higher methiocarb concentrations. The pattern of methiocarb ingestion is similar to that for bait ingestion.

Data produced from the direct comparison of methiocarb 4%, methiocarb 4% + additive A (Exit) and a blank show clearly that the additive bestows an advantage both in terms of slug mortality induced, crop protection in terms of grain loss and the amount of active ingredient ingested (Table 1). Terraria trials also show clearly that a formulation containing additive A leads more quickly to a higher level of mortality under simulated field conditions (Fig. 4).

The simple arena repellency/attractancy trials showed that methiocarb at 4% on its own was more repellent than at 1%. Additive A appeared mildly attractive and phagostimulant at 1 and 2% whilst additive B seemed repellent to slugs at concentrations above 0.05%. Veratrylamine, methyl morpholine and piperidine were repellent at 1% whilst the latter two compounds were also molluscicidal at this concentration.

---

* Draza

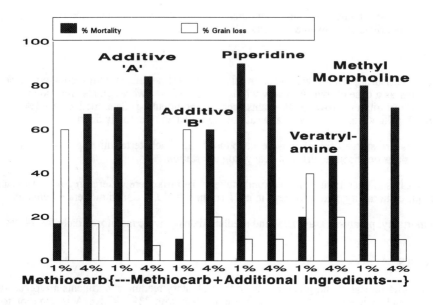

Fig. 1. Mortality and grain loss at the end of the experiment.

Table 1. Laboratory tests and voluntary feeding studies with *D. reticulatum.*

	Mortality after 7 days	% Grain loss	Ingested a.i μg/slug
methiocarb 4%	70	20	564
methiocarb 4% + additive A	90	10	840
blank pellet	30	70	-

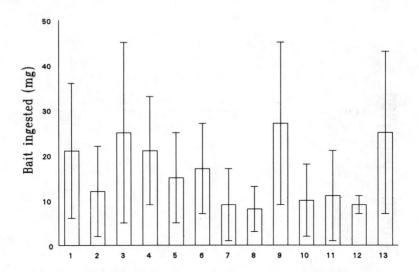

Fig. 2. Ranges and means of bait ingested.

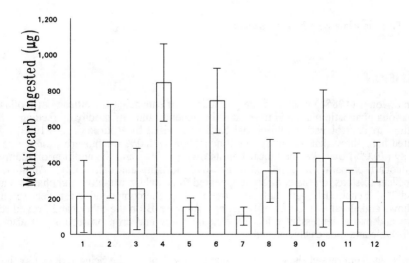

Fig. 3. Ranges and means of methiocarb ingested

Key to Figs. 2 and 3
1 = 1% methiocarb
2 = 4% methiocarb
3 = 1% methiocarb + 1% additive A
4 = 4% methiocarb + 1% additive A
5 = 1% methiocarb + 1% additive B
6 = 4% methiocarb + 1% additive B
7 = 1% methiocarb + 1% piperidine

8 = 4% methiocarb + 1% piperidine
9 = 1% methiocarb + 1% veratrylamine
10 = 4% methiocarb + 1% veratrylamine
11 = 1% methiocarb + 1% methyl morpholine
12 = 4% methiocarb + 1% methyl morpholine
13 = blank with chromium

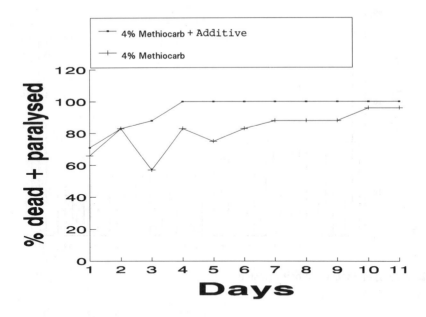

Fig. 4. Terraria trials with *D. reticulatum*

DISCUSSION

Bowen & Jones (1985b) discussed the possibilities of enhancing the efficacy of molluscicides with various phagostimulant additives, and the general utility of modifying feeding behaviour in relation to control with molluscicidal baits was raised by Bourne *et al.* (1988). Results presented here show that additive A in combination with 1% methiocarb produced a higher mortality (70.0%) than 4% methiocarb on its own (66.7%) and the addition of additive A to all methiocarb preparations appeared to improve mortality and palatability. Additive A was not repellent to slugs, on the contrary it appeared to be mildly attractive and phagostimulant at 1% and 2%. Felicitously, although attractive to slugs the compound is repellent to birds and thus shows added value in terms of targeting. Additive B on the other hand proved repellent to slugs and had a depressant effect on slug mortality and crop protection. It also proved highly soluble and thus easily leached.

Veratrylamine suppressed the efficacy of methiocarb, bait uptake being reduced at the higher methiocarb concentrations. Addition of veratrylamine to methiocarb resulted in a reduced slug mortality and lower grain protection.

Combination of piperidine or methyl morpholine with methiocarb resulted in the enhancement of slug mortality and grain protection although accompanied by a reduction in bait eaten and active ingredient ingested. The enhanced mortalities seen here were probably due to the inherent molluscicidal properties of the additives. Both compounds at 1% proved to be strongly repellent to slugs and slugs that experienced contact with filter paper soaked with the additive at 1% usually died within a short time of contact. Both thus appear to be contact molluscicides in their own right and may act in a synergistic manner with methiocarb.

CONCLUSIONS

1. Additive A appeared to enhance ingestion and uptake of methiocarb resulting in improved performances in terms of slug mortality and crop protection. Results in the field are presented by Meredith (1996).

2. The addition of either 1% methyl morpholine or 1% piperidine also seemed to improve the efficacy of methiocarb as a molluscicide despite inducing a reduction in bait ingestion.

REFERENCES

Bourne, N B; Jones, G W; Bowen, I D (1988) Slug feeding behaviour in relation to control with molluscicide baits. *Journal of Molluscan Studies.* **54**, 327-338.

Bourne, N B; Jones, G W; Bowen, I D (1990) Slug feeding behaviour in relation to control with molluscicidal baits containing various combinations of metaldehyde and methiocarb. *Annals of Applied Biology.* **117**, 455-468.

Bowen, I D; Jones, G W (1985a) Molluscicide compositions. UK Patent GB2098869B.

Bowen, I D; Jones, G W (1985b) Getting pesticides into cells. *Industrial Biotechnology.* **5**, 29-32.

Bowen, I D; Mendis, V W; Symondson, W O C; Liddell, J E; Leclair, S (1993) The integrated management of terrestrial slug pests. *Proceedings of the Second Malacological Convention,* Diliman, The Philippines, pp. 26-36.

Bowen, I D; Antoine, S (1995) Molluscicide formulation studies. *International Journal of Pest Management.* **41**, 74-78.

Glen, D M (1989) Understanding and predicting slug problems in cereals. In: *Slugs and Snails in World Agriculture,* I F Henderson (ed.). BCPC Monograph No. 41, pp. 253-262.

Glen, D M; Wiltshire, C W; Milsom, N F (1984) Slugs and straw disposal in winter wheat. British Crop Protection Conference - Pests and Diseases, **2**, pp. 139-144.

Henderson, I F (1968) Laboratory methods for assessing the toxicity of contact poisons to slugs. *Annals of Applied Biology.* **62**, 363-369.

Henderson, I F; Parker, K A (1986) Problems in developing chemical control of slugs. In: *Crop Protection of Sugar Beet and Crop Protection and Quality of Potatoes. Aspects of Applied Biology* **13**, Part II, pp. 341-347.

Henderson, I F; Martin, A P; Perry, J N (1992) Improving slug baits: the effects of some phagostimulants and molluscicides on ingestion by the slug *Deroceras reticulatum* (Müller). *Annals of Applied Biology.* **121**, 423-430.

Kennedy, P J (1990) The effects of molluscicides on the abundance and distribution of ground beetles (Coleoptera: Carabidae) and other invertebrates. Ph.D. Thesis, University of Bristol.

Martin, T J (1993) The ecological effects of arable cropping including the non-target effects of pesticides with special reference to methiocarb pellets used for slug control. *Pflanzenschutz-Nachrichten Bayer.* **46**, 49-102.

Martin, T J; Kelly, J R (1986) The effect of changing agriculture on slugs as pests of cereals. British Crop Protection Conference - Pests and Diseases, **2**, 411-424.

Mason, J R; Avery, M L; Glahn, J F; Otis, D L; Matteson, R E; Nelms, C O (1991) Evaluation of methyl anthranilate and starch-plated dimethyl anthranilate as bird repellent feed additives. *Journal of Wildlife Management.* **55**, 182-187.

Meredith, R H (1996) Testing bait treatments for slug control. *These proceedings.*

Purvis, G; Bannon, J W (1992) Non-target effects of repeated methiocarb slug pellet application on carabid beetles (Coleoptera: Carabidae) activity in winter-sown cereals. *Annals of Applied Biology.* **121**, 201-422.

Port, C M; Port, G R (1986) The biology and behaviour of slugs in relation to crop damage and control. *Agricultural Zoology Reviews.* **1**, 255-299.

Wright, A A; Williams, R (1980) The effect of molluscicides on the consumption of bait by slugs. *Journal of Molluscan Studies.* **46**, 265-281.

A STUDY ON OUTBREAKS OF *BRADYBAENA RAVIDA* IN CHINA

CHEN D, ZHANG G Q
Institute of Zoology, Academia sinica, Beijing, China

XU W, LIE Y
Shaanxi Plant Protection Station, Xian City, Shaanxi Province, China

WANG M, YAN Y
Shaanxi Plant Protection Institute, Yanglin Town, Shaanxi Province, China

ABSTRACT

Bradybaena ravida has become a serious crop pest in some locations. The snail has a high reproductive capacity, but appears to have few natural enemies. Environmental conditions, especially drought and temperature extremes are probably important in reducing populations, but where there are no breaks in the cropping cycle the snail population is protected from these extremes.

INTRODUCTION

The Huiba snail, *Bradybaena ravida* has become a serious crop pest in some locations. This paper reports a series of investigations into reasons for its increasing importance.

MATERIALS AND METHODS

Captive snails were put into small feeding pots after mating, and the timing and frequency of egg laying, the numbers of eggs laid, the duration of incubation, growth and development were all recorded. A similar approach was used to investigate the effects of differing water availability. From field observations, food preferences for different crops were noted, the effects of intercropping and interplanting, moisture, fertilizer and crop rotation were also observed. Snails in experimental plots were observed every day to determine the effects of environmental temperature, moisture and wind strength. The effects of temperature on snail mortality were observed in the field and using captive populations in the greenhouse.

Soil samples from outbreak areas were analysed to determine calcium content and pH. Systematic observations were made every five days in major crop plots to monitor the number of snails and egg mass and to analyse the pattern of outbreaks.

RESULTS

Life cycle

The Huiba snail, *B. ravida* snails reproduce once every year, overwintering as adult snails, larva or eggs. Some of the snails overwinter in the crop rhizosphere, among roots and stubble, within piles of corn culm and under fallen leaves. Overwintering adult snails start to move about and feed in the last ten days of March when the average temperature rises above 8°C and there is adequate moisture. Between the second ten days in April and June there is widespread feeding damage. The peak period of mating and egg laying is between the first ten days and the last ten days in May. Snails oversummer mostly within loose soil layers of the crop rhizosphere between the last ten days in June and the first ten days in August (when the average temperature rises above 25°C and the relative humidity decreases to 65%). After the second ten days in August there is a second period feeding and egg laying. Snails overwinter from the second ten days of November when the average temperature drops to -10°C and the relative humidity to 76%. The snail can secrete mucus and form a soft film at the shell mouth to protect itself from adverse conditions during overwintering and oversummering. The snails may lay eggs twice a year, in spring and in autumn. All the offspring should be mature in the following spring and autumn respectively, after one year. Every adult snail may pass through 2-3 laying seasons although some adults die after every laying period. The life span is 1.5 to 2 years.

Biological habits

Snails are usually nocturnal, they begin activity at dusk and stop at sunrise. However, snails may remain active through the day if it is overcast and raining. When conditions are unfavourable, such as strong wind, high temperature, and drought, snails draw back into their shells and secrete mucus at the shell mouths.

The snails are polyphagous feeding on a range of green plants including wheat, cotton, rape, soybean, kidney bean, hot pepper and cabbage. The exception was cowpea which was not fed on. Results from feeding cotton leaves to snails show that average consumption is 0.0709 g leaf / g snail.

Mating occurs mostly between 2300 and the following sunrise, but can also be seen in the daytime during wet weather. Mating usually lasts 5-6 h with the maximum recorded at 16 h. Table 1 shows data regarding egg laying. Snails usually lay their eggs in small holes 2-4 cm deep in the soil. The newly laid eggs are milky and gradually become light yellow. Several eggs are laid together to form an egg mass. An egg mass takes 14-20 days from laying to the end of incubation. Eggs are rapidly destroyed in direct sunlight therefore cultivation in the laying season is an effective measure for eliminating eggs.

Table 1. Observations of reproductive activity of *Bradybaena ravida*, Jingyang 1991

Mating date	Mating to laying period (d)		Laying period (d)		Egg masses laid		Max. eggs / mass		Total eggs		Grand total
	A1	A2	A1	A2	A1	A2	A1	A2	A1	A2	
28 March	20	40	19	12	3	5	27	53	61	140	201
28 March	20	30		28	1	8	24	68	24	371	395
28 March	33	36	21	22	5	6	62	58	238	260	448
5 April	30	39	23	20	5	6	101	69	187	336	523
27 April	4	4	29	37	7	7	122	143			392
10 May	5	8	52	49	7	7	140	85	638	346	984
Average					5.6				260.1		

A1 and A2 represent each of two snails after mating

Environmental relations

The most suitable temperature for snails is 16-20°C. Snails still can move about normally between 10-35°C. During the winter, observations showed 33% of snails can survive a temperature of -29°C for a day. Snails were exposed during the afternoon to high temperatures in a greenhouse. The following morning all snails exposed to 46°C and above had died, but 73.3% of the snails exposed to 44 °C had survived

Moisture has a great effect on the snails. Outdoor observation showed that activity decreased when the weather was dry from April to October in 1991. A great number of snails became active when it rained and the moisture increased. Several hundred thousand larvae in early incubation could be seen between April and May 1991. They all died during a period of higher temperature and low moisture (the average relative humidity was 75.4%) between the first ten days in July and the second ten days in August. In Xichencun village and Zhoujiadao village, snail damage is more serious because the area is irrigated. Soil moisture content influences egg laying and incubation. When moisture content is below 10% adult snails are unable to lay eggs, and when moisture is about 20% adult snails can lay eggs normally. Incubation of eggs is normal when soil moisture content is 17-20% and stops when moisture content is below 15%.

Soil was sampled in eight different types of plot in the area of outbreak in Jingyang county in July 1991. Calcium content was 4.69-5.54%. Abundant calcium may meet the demand for snail growth. The pH was between 7.74 and 8.72.

Population patterns

Regular and systematic investigation was made in crop fields (wheat, corn), and cotton fields in Xichencun village and Zhoujiadao village where damage of snails was serious from 1989 to 1991. Figure 1 shows that snails start to be active in the last ten days of March and stop moving in the first ten days of November. Two stages are apparent. The first stage is from the second ten days in April to the last ten days in May in wheat fields. The second stage is from the second ten days in August to the last ten days in September in corn and cotton fields. It can be seen from Figure 2 that only a handful of snails appear in cotton field from seedling to squaring stage mainly because of the low moisture. When wheat is harvested, snails move into cotton field in great number in the first ten days of June. This leads to a rapid increase in the numbers of snails.

Figure 1
Numbers of snails recorded in wheat and corn fields at Xichencun village 1989-1991.

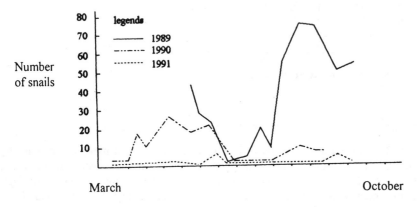

Figure 2
Numbers of snails recorded in cotton fields at Xichencun village 1989-1991.

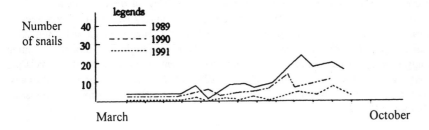

The timing and amount of egg laying differs with different crop types. Figures 3 and 4 show that the last ten days in April is the beginning of laying season in crop field and the first laying season is mainly between the first ten days and the last ten days in May in wheat fields. The second widespread laying season appears between the second ten days in August and the second ten days in September in corn fields and cotton fields. The laying season may last till the first ten days in October.

Figure 3
Numbers of eggs and snails in different crops at Zhoujiadao village, 1991.

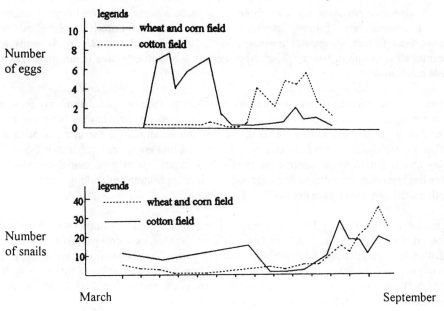

Figure 4
Numbers of eggs in different crops at Xichencun village, 1989-1991.

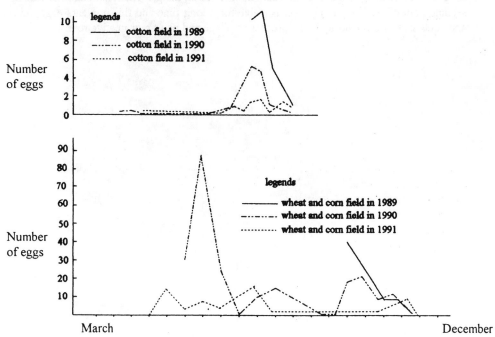

DISCUSSION

Bradybaena ravida snails are serious crop pests in some areas of Jingyang county. This is a consequence of many interacting factors, including historical, biological and meteorological factors. A general investigation in 1990 showed that *B. ravida* snails are distributed all over the province. That crop damage was not extensive is due to a lack of suitable conditions.

The snail has the capacity to produce numerous offspring, thus the population may increase and spread rapidly if conditions are favourable. Natural enemy populations have not been observed in large numbers and this leads to a favourable condition for the increase of snails. *Carabus brandti* (Faldam) has been observed, but this natural enemy can only catch 2-5 snails per day and is not likely to control the outbreak of snails. It may be that the ecological balance has been disturbed due to pesticide and fertilizer application over a long period of time and other natural enemies have become locally extinct.

The annual precipitation in Jingyang county is above 500 mm. Rainfall was fairly well distributed from April to June in the last 5-6 years and serious drought does not occur. Rainfall in July, August and September is usually greater. The temperatures are suitable for snails to grow and develop and reproduce. However, the area of the outbreak may be irrigated through both canals and wells. Thus moisture content of the soil remains high which provides favourable conditions for the increase of snail numbers.

Due to a decrease in cotton growing the main crop rotation involves transplanting corn before the ripening of wheat and to plant wheat by hoeing the soil after the harvest of corn. This long-term cycle of growing prevents cultivations during the growth season and deep-tillage in seeding period. Soil in plough layer is stable for a long time and this prevents eggs, juvenile and adult snails from injuring in the process of tillage and plays a protective role.

CARABID BEETLES RECORDED FEEDING ON SLUGS IN ARABLE FIELDS USING ELISA

K AYRE, G R PORT

Department of Agricultural and Environmental Science, The University of Newcastle, Newcastle Upon Tyne, NE2 7RU, UK

ABSTRACT

Carabid beetles were collected by pitfall trapping in fields of oilseed rape and winter wheat in Northumberland and subjected to ELISA analysis of their gut contents to detect slug proteins. A proportion of some species contained slug remains, other species were consistently negative. However carabids in this latter group were present at low densities. The more abundant carabids all fed on slugs to some extent. The proportion containing slug remains varied over time. There was some evidence of increased consumption of slug prey during periods when prey density increased. The carabid species involved in feeding on slugs were compared with those identified from other studies.

INTRODUCTION

Many carabid species occur on arable land (e.g. Sunderland, 1975; Jones, 1976 and 1979; Attah, 1986). Some species are known to feed on molluscs (e.g. Davies, 1953; Tod, 1973; Luff, 1974; Symondson, 1992) and therefore have the potential to exert an impact on populations of slugs. There is increasing interest in assessing the contribution such polyphagous predators may make to slug population reduction in integrated management systems (e.g. Symondson *et al.*, 1996) and the objective of this study was to investigate slug predation by carabids in the north of England.

MATERIALS AND METHODS

Carabid beetles were collected from three commercial fields in the Tyne valley, Northumberland. In 1992 all three fields were sown with oilseed rape. The rape was harvested at the end of July, winter wheat was sown in September and harvested in August 1993.

Plastic beakers (72mm diameter and 105mm deep) were used as pitfall traps to catch ground active beetles. Twenty pitfall traps were placed at five meter intervals in each field in a five by four block. Traps were filled to 75 percent capacity with a weak detergent (e.g. Sunderland, 1975).

In 1992, pitfall traps were set on 15 June and visited most working days until 27 July. Trapped beetles spent a maximum period of 72 hours in the trap before they were collected. After the oilseed rape was harvested and winter wheat sown (September)

traps were set again on 26 October. The traps were visited weekly until 21 December. No further beetles were captured until 12 April 1993, when weekly visits resumed until August. In this second phase of trapping, beetles may have spent a maximum of seven days in the traps before they were collected.

On each visit, beetles were removed from the traps and transferred to a freezer box for transportation to the laboratory. The beetles were washed, sorted, identified following the nomenclature of Kloet & Hincks (1945) and frozen at -20°C. An ELISA (Ayre, 1995) was used to identify slug tissues (antigens) in the guts of field-caught carabids.

Twenty ceramic tiles measuring 15x15 cm were placed at five meter intervals within the sampling area and used to assess the activity-density of *Deroceras reticulatum* over the summer periods in 1992 and 1993. Other molluscs were noted. On 30 June and 8 July 1992, slugs were collected from the 3 sites, weighed and returned to their respective site.

RESULTS

A total of 2184 carabid specimens, representing forty species, were collected from the three study sites. Not all specimens could be assessed by ELISA and three criteria were used to choose species for the analysis. The first was predator size, as larger species are more likely to eat slugs than smaller species (e.g. Tod, 1973). Secondly, predator abundance in the field was considered. Species which were rare or occurred infrequently were considered to be unlikely to exert an effect on slug populations. Exceptions to this were predator species which were very large and/or known slug-specialist species. Finally, food specialisation was considered. *Loricera* are specialised collembolan feeders (Hengeveld, 1980) and the predatory behaviour of *Loricera pilicornis* is adapted to overcome their collembolan prey (Bauer, 1982). Tod (1973) collected 56 *L.pilicornis* specimens from the field and found none contained mollusc remains. Although *L.pilicornis* was abundant at two of the sites, only one specimen was assessed as this species was not considered to be a slug predator. Nineteen of the forty species, a total of 1311 specimens, were assessed by ELISA.

Carabid species containing slug tissues

The beetle species assessed by ELISA are shown in Table 1. Thirteen of the nineteen beetle species assessed contained molluscan tissues. As snails were not found in the sites the beetles were considered to have fed on slugs. The beetle species were compared with each other by calculating the proportion of each species containing slug tissues (Table 1).

Nebria brevicollis apparently fed frequently on slugs, thirty-seven percent of beetles containing slug tissues. Two *Pterostichus* species fed on slugs in similar proportions. Ten percent of all *Pterostichus madidus* and nine percent of all *Pterostichus melanarius* had fed on slugs. Five larger species were assessed. All four specimens of *Carabus violaceus* and one of two *Cychrus caraboides* had fed on slugs. A single *Abax parallelepipedus* was recovered which had eaten a slug meal and a single *Carabus problematicus* was recovered which had not fed on slugs. Only one of thirteen *Pterostichus niger* had fed on slugs. In general, a proportion of most beetle species assessed contained slug tissues. However some species tested consistently negative. None of the twenty-three *Harpalus aeneus* had fed on slugs and none of the *Pterostichus cristatus* and *Amara aenea* had fed on slugs.

Table 1 The number of each beetle species tested for slug remains by ELISA in 1992 and 1993 at each of the three sites is given in the 'No.' columns. The percentage of beetles of each species found to contain slug tissues is given in the '%+' column.

1992	Site 1		Site 2		Site 3		Overall	
	No.	%+	No.	%+	No.	%+	No.	%+
Pterostichus melanarius	25	0	47	10	4	0	76	6
Pterostichus madidus	48	6	4	0	5	0	57	5
Pterostichus niger	5	0	5	20			10	10
Pterostichus cristatus	3	0	3	0			6	0
Abax parallelepipedus	1	100					1	100
Harpalus rufipes	238	29	3	33	2	0	243	29
Harpalus latus	2	50					2	50
Harpalus aeneus	15	0					15	0
Amara similata	465	30	80	18	26	0	571	27
Amara aenea	4	0			1	0	5	0
Amara aulica	5	20					5	20
Amara lunicollis	4	75	6	16			10	40
Amara plebeja	1	0					1	0
Carabus violaceus	4	100					4	100
Cychrus caraboides	1	0			1	100	2	50
Loricera pilicornis					1	0	1	0
Nebria brevicollis	9	44	95	40	11	9	115	37
Overall 1992	830	27	243	25	51	4	1124	26
1993								
Pterostichus melanarius	14	14	29	13			43	13
Pterostichus madidus	34	14	7	28	1	0	42	16
Pterostichus niger	2	0	1	0			3	0
Pterostichus cristatus			1	0			1	0
Pterostichus nigrita	2	50					2	50
Harpalus rufipes	85	14	1	0			86	14
Harpalus latus	1	0					1	1
Harpalus aeneus	8	0					8	0
Carabus problematicus			1	0			1	0
Overall 1992	146	13	40	15	1	0	187	14
Overall 92 and 93	976	25	283	23	52	4	1311	32

Site differences in beetles feeding on slugs

Contingency tables of the numbers of beetles found containing/without slug remains were used to determine if there were any differences in slug feeding between the three sites. A significantly lower percentage of beetles fed on slugs at site 3 than at site 1 ($X^2 = 539.3$, d.f. = 1, P < 0.001) and site 2 ($X^2 = 20.6$, d.f. = 1, P < 0.001). No other significant differences were found.

Each common species was also considered separately. Similar proportions of *N.brevicollis* beetles fed on slugs at sites 1 and 2. The largest proportions of *P.madidus* and *P.melanarius* beetles containing slug tissues were from site 2, but there was no significant difference between this site and the other sites. The largest proportion of *Amara similata* beetles containing slug tissues was from site 1. A significantly greater proportion of *A.similata* fed on slugs at site 1 than site 2 ($X^2 = 4.8$, d.f. = 1, P < 0.05) and site 3 ($X^2 = 11.2$, d.f. = 1, P < 0.001). A significantly greater proportion of *A.similata* fed on slugs at site 2 than at site 3 ($X^2 = 5.6$, d.f. = 1, P < 0.05). *Pterostichus madidus*, *P.melanarius*, *Harpalus rufipes* and *A.similata* beetles did not feed on slugs at site 3, where only a single *N.brevicollis* and *C.caraboides* beetle had fed on slugs. Slugs were similar in weight at sites 1 and 2 but heavier at site 3 (Table 2). Slugs were significantly heavier at site 3 than at site 2 (t = 2.036, d.f. = 70, P < 0.05).

Table 2 The mean relative activity-density of *D.reticulatum* at each site in 1992 and 1993. The mean is calculated from the weekly densities of slugs found under tile traps over the two summer seasons at each site (n = number of weekly samples). The slug weight are calculated from slugs collected from each site on 30 June and 8 July, 1992 (n = number of slugs weighed).

| | Activity-density | | | | | | Weight | | |
| | 1992 | | | 1993 | | | 1992 | | |
	Mean	n	S.E	Mean	n	S.E.	Mean	n	S.E.
Site 1	0.31	6	0.08	0.84	5	0.54	0.465	10	0.076
Site 2	1.38	6	0.38	0.22	5	0.09	0.468	54	0.039
Site 3	0.30	6	0.12	1.00	5	0.31	0.646	18	0.093

Effect of slug activity-density on beetles feeding on slugs

The mean slug activity density at sites 1 and 3 were similar in 1992 (Table 2) but the proportion of beetles feeding on slugs at site 3 was lower than at site 1. *Amara similata* fed on slugs at site 1 but not at site 3 and *N.brevicollis* beetles ate slugs in lower proportions at site 3 compared to site 1. Insufficient *P.madidus* were recovered to detect any differences in slug feeding between the sites. However, *A.similata* was numerous at all sites and the differences in slug feeding are probably real.

In 1992, slug activity-densities at site 2 were significantly higher than those at site 1 (t=2.997, d.f.=5, P<0.05). Despite this, the proportion of beetles feeding on slugs was similar at sites 1 and 2.

Response of beetles to slug activity-density

The effect of weekly slug activity-density on the proportion of beetles feeding on slugs was investigated using regression analysis.   Each site and year were considered separately. The analysis was not possible at site 3 in 1993. None of the correlations were significant. Similarly, each beetle species was analyzed separately. The only possible positive correlation was found in *N.brevicollis* at site 2 in 1992 (r^2=0.75, d.f.=4, P<0.1). In the summer period of 1992, two peaks in slug activity densities occurred at site 2, on 6 July and 27 July. *Pterostichus melanarius* was caught throughout the summer period at this site, but fed on slugs only on the two occasions when slug activity-density peaked.

DISCUSSION

The results of this study, in which thirteen carabid species fed on slugs, are compared with four other studies in Table 3. The proportion of beetles recorded as feeding on molluscs differs between the four studies. The relatively low proportion of beetles containing mollusc remains in the investigation by Davies (1953) may reflect the technique (microscopic gut analysis) used to identify mollusc remains, as slugs leave few visually identifiable remains.

Fourteen of twenty-six species fed on slugs in Tod's (1973) study and nine of these species were tested in this study. Eight of these nine species fed more frequently on slugs in Tod's (1973) study compared with this study. The proportion of beetles feeding on slugs in the four studies will partly reflect the ecosystem from which the beetles were trapped and the relative availability of slugs and alternative prey. In the other studies, beetles were collected from a walled garden (Luff, 1974), hillside, wood and grass pasture (Tod, 1973), woods (Symondson, 1989) and unspecified inland sites (Davies, 1953).

Carabid species containing slug tissues

It is likely that all of the carabid predators investigated scavenged dead slugs to some degree. However, the extent of scavenging may vary from species to species.

Most of the large *Carabus* and *Cychrus* beetles fed on slugs in this study and in Tod's (1973) study. In this study, all *C.violaceus* beetles and one of two *C.caraboides* assessed had fed on slugs. Slugs are considered to be part of the diet of *Carabus* and *Cychrus* species (e.g. Gruntal & Sergeyeva, 1989). The large beetle species *A.parallelepipedus*, *P.melanarius* and *P.madidus* ate molluscs in this study and in other studies (Table 4). A high proportion of *H.rufipes* beetles contain liquid food (Sunderland, 1975) which may represent the remains of snails (molluscs) (Hengeveld, 1980). Therefore, *H.rufipes* may feed on slugs extensively. A large proportion of *H.rufipes* beetles fed on slugs in this study.

Nebria brevicollis has not been recorded as having fed on slugs to a great extent in other studies (Tod, 1973). However, a large proportion of beetles fed on slugs in this study. The Amarini are generally considered to be phytophagous (Aubrook, 1949; Davies, 1953). However, a large proportion of another member of this tribe, *A.similata*, fed on slugs in this study. These two smaller species can kill small slugs under laboratory conditions (Ayre, 1995) and small *D.reticulatum* slugs are often abundant in arable land. Therefore these smaller species have an opportunity to prey on slugs in the field. *Deroceras reticulatum* breeds throughout the year, (Haynes *et al.*, 1996). Autumn breeding may occur too late in the season for many carabid species to utilize the abundance of small, newly hatched slugs, as many beetle species are becoming inactive at this time. Slug generation intervals are affected by prevailing weather conditions and vegetation (Bett, 1960) and the availability of small slugs may change from year to year. This will affect the extent to which some carabid species feed on slugs. The extent of carabid predation of other pests such as aphids is dependant on the degree of synchronisation between the life cycle of the predator and the phenology of the aphid species (Sunderland & Vickerman, 1980).

Table 3   Comparison of the percentage of each beetle species containing mollusc remains in this study with four other studies. * indicates that ten or fewer beetles were assessed.

	This study	Tod (1973)	Davies (1953)	Symondson (1992)	Luff (1974)
Pterostichus melanarius	9	35	0*		
Pterostichus madidus	10	20	5	44	23
Pterostichus niger	7	43	0*		
Pterostichus nigrita	50*	0*	0*		
Pterostichus cristatus	0*				
Abax parallelepipedus	100*		16	92	
Harpalus rufipes	25		0		
Harpalus latus	33*	0*	0*		
Harpalus aeneus	0		0*		
Amara similata	27		0*		
Amara aenea	0*		0*		
Amara lunicollis	40*				
Amara aulica	20*		0*		
Amara plebeja	0*		0*		
Carabus violaceus	100*	100*	0*		
Carabus problematicus	0*				
Cychrus caraboides	50*	73	0*		
Loricera pilicornis	0*	0	0*		
Nebria brevicollis	37	2	0		

Site differences in beetles feeding on slugs

Differences in the carabid fauna of the three fields partly explain the differences in the proportion of beetles feeding on slugs. Slug feeding species were more abundant at sites 1 and 2 than at site 3. Some species occurred at each site, but ate slugs differentially between sites. The size of individual slugs may affect the number of carabid species able to prey on them and large slugs may deter smaller beetle predators. The mean size of slugs was greatest at site 3 where few beetles fed on slugs.

Effect of slug activity-density on beetles feeding on slugs

There was no evidence to suggest that differences in slug activity-density affected the proportion of beetles feeding on slugs between the sites in 1992. Slug activity-density was significantly higher at site 2 than at site 1 in 1992, but the overall proportion of beetles feeding on slugs at the two sites was similar. Slug activity-densities at sites 1 and 3 were similar but the proportion of *P.melanarius*, *P.madidus*, *N.brevicollis* and *A.similata* feeding on slugs was always lowest at site 3. This indicates that differences in slug activity-density between fields did not affect slug feeding by the predators.

Two species exhibited an increased feeding response to slug activity-density over the summer periods. More *N.brevicollis* beetles fed on slugs at site 2 as slug density increased. *Nebria brevicollis* shows a similar response to increasing aphid densities (e.g. Sunderland & Vickerman, 1980).

Pterostichus melanarius fed on slugs at site 2 in 1992 on only two occasions, when the highest slug densities were recorded. This species has been shown to feed on invertebrate prey which are most abundant at a particular time (Pollet & Desender, 1985), indicating that *P.melanarius* may change its diet according to prey availability and may feed on slugs if they are abundant.

This study has shown that a number of carabid species feed on slugs in the field. Differences in the carabid fauna may influence the extent of slug predation in different locations but many widely distributed species feed on slugs. Slug weight and slug density (e.g. Symondson *et al.*, 1996) may also affect the extent that some carabid species feed on slugs at particular locations. Clearly more studies are needed to reveal the extent to which carabid feeding results in the death of slugs and whether slug populations are significantly affected by these predators.

ACKNOWLEDGEMENTS

This work was funded by a MAFF studentship to Kevin Ayre.

REFERENCES

Attah, P K (1986) *The insect pests and polyphagous arthropod predators associated with crops of oilseed rape in North Yorkshire and Humberside.* Unpublished Ph.D.

thesis, University of York.

Aubrook, E W (1949) *Amara familiaris* Duft. (Col., Carabidae) feeding in flower heads. *Entomologist's Monthly Magazine.* **85**: 44.

Ayre, K (1995) *Evaluation of carabids as predators of slugs in arable land.* Unpublished Ph.D thesis, University of Newcastle upon Tyne.

Bauer, T (1982) Predation by carabid beetle specialised for catching Collembola. *Pedobiologia.* **24**: 169-179.

Bett, J A (1960) The breeding season of slugs in gardens. *Proceedings of the Zoological Society of London.* **135**: 559-568.

Davies, M J (1953) The contents of the crops of some British carabid beetles. *Entomologist's Monthly Magazine.* **89**: 18-23.

Gruntal, S Y; Sergeyeva, T K (1989) Food relations characteristics of the beetles of the genera *Carabus* and *Cychrus. Zoologicheskii Zhurnal.* **58**: 45-51.

Haynes, S L; Rushton, S P; Port, G R (1996) Population structure of *Deroceras reticulatum* in grassland. I F Henderson (ed.). BCPC Monograph No 66

Hengeveld, R (1980) Polyphagy, oligophagy and food specialisation in ground beetles (Coleoptera, Carabidae). *Netherlands Journal of Zoology.* **30**: 564-584.

Jones, M G (1976) The carabid and staphylinid fauna of winter wheat and fallow on a clay with flint soil. *Journal of Applied Ecology.* **13**: 775-791.

Jones, M G (1979) The abundance and reproductive activity of common Carabidae in a winter wheat crop. *Ecological Entomology.* **4**: 31-43.

Kloet, G S; Hincks, W D (1945) *A check list of British insects.* Buncle and Co. Ltd.

Luff, M L (1974) Adult and larval feeding habits of *Pterostichus madidus* (F.) (Coleoptera; Carabidae). *Journal of Natural History.* **8**: 403-409.

Pollet, M; Desender, K (1985) Adult and larval feeding ecology in *Pterostichus melanarius* Ill (Coleoptera, Carabidae). *Mededelingen van de Fakulteit Landbouwwertenschappen Rijksuniversiteit (Gent),* **50/2b**: 581-594.

Sunderland, K D (1975) The diet of some predatory arthropods in cereal crops. *Journal of Applied Ecology.* **12**: 507-515.

Sunderland, K D; Vickerman, G P (1980) Aphid feeding by some polyphagous predators in relation to aphid density in cereal aphids. *Journal of Applied Ecology.* **17**: 389-396.

Symondson, W O C (1989) Biological control of slugs by carabids. 1989 BCPC Mono. No. 41 *Slugs and Snails in World Agriculture,* pp 295-300. Ed; Henderson, I.F. British Crop Protection Council, Thornton heath.

Symondson, W O C (1992) *Biological control of slugs by carabid beetles.* Unpublished Ph.D. thesis, University of Wales college of Cardiff.

Symondson, W O C; Glen, D M; Wiltshire, C W; Langdon, C J; Liddell, J E (1996) Effects of cultivation techniques and methods of straw disposal on predation by *Pterostichus melanarius* (Coleoptera: Carabidae) upon slugs (Gastropoda: Pulmonata) in an arable field. *Journal of Applied Ecology.* **33**: (in press).

Tod, M E (1973) Notes on beetle predators of molluscs. *The Entomologist.* **106**: 196-201.

FIELD TRIALS WITH *PHASMARHABDITIS HERMAPHRODITA* IN SWITZERLAND

B SPEISER

Research Institute of Organic Agriculture (FiBL), CH-4104 Oberwil, Switzerland

M ANDERMATT

Andermatt Biocontrol AG, CH-6146 Grossdietwil, Switzerland

ABSTRACT

Three field experiments on slug control in organic farming were carried out with the nematode biocontrol agent *Phasmarhabditis hermaphrodita*. The aims were to determine the efficiency of *P. hermaphrodita* under Swiss field conditions, and to collect additional data for the registration process for *P. hermaphrodita* in Switzerland.

In the so-called pilot experiment, we compared slug damage to several vegetables and an ornamental species two weeks after application of either *P. hermaphrodita* or metaldehyde pellets and in untreated plots. In the so-called lettuce experiment, we compared slug damage to lettuce in plots treated with either a high or a low dose of nematodes (1 million or 0.1 million nematodes per m^2) or in plots without molluscicidal treatment. In the so-called rape experiment, we counted the numbers of intact or slug-damaged rape plants two weeks after the application of either *P. hermaphrodita* or metaldehyde pellets and in untreated plots.

In the pilot experiment, both the nematode and the metaldehyde treatment provided good protection against slug damage. In the lettuce experiment, however, neither the high nor the low nematode dosage protected the lettuce from being eaten by slugs. In the rape experiment, the number of intact rape plants was highest in the nematode-treated plots, intermediate in the metaldehyde-treated plots and lowest in the untreated plots. The total number of rape plants (intact or damaged) was higher in both molluscicidal treatments than in the untreated plots.

In conclusion, *P. hermaphrodita* is a promising alternative to chemical molluscicides. More research is needed to determine the conditions under which it can be successfully used. In particular, we should know whether it is infective for all species of pest slugs occurring in Switzerland.

INTRODUCTION

In Switzerland, slugs cause damage in a large number of crops: among the arable crops, oilseed rape and sugar beet are regularly damaged, while maize, winter wheat and potato suffer slug damage less frequently. In horticultural production, a large number of crops are attacked by slugs, particularly lettuce, strawberries and ornamentals. However, slug damage is an economically serious problem only in certain areas, where the soil is heavy or the climate is wet. In addition, slugs are notorious pests in home gardens.

In arable crops, *Deroceras reticulatum*, the grey field slug, causes most of the damage. In horticulture and home gardens, two other slugs are also harmful: *Arion distinctus* and *A. lusitanicus*. *A. lusitanicus* has been introduced to central Europe in the past decades. Because it was confounded with *A. rufus* for a long time, its pest status is not well known. Fechter & Falkner (1990) call *A. lusitanicus* «the most frequent and the only truly harmful species of pest slug» and Sulzberger (1996) estimates that *A. lusitanicus* is responsible for 90 % of all slug damage in home gardens.

Organic farmers are not allowed to use unnatural pesticides, including chemical molluscicides. As a result, there is great interest in non-chemical methods of slug control. The methods presently available («slug fences», «beer traps» and hand-collecting) are expensive, labour-intensive and often ineffective. Recently, a method for biocontrol of slugs with the nematode species *Phasmarhabditis hermaphrodita* has been developed (Wilson *et al.* 1993). In England, this method has been successfully used to protect winter wheat, oilseed rape, Chinese cabbage, lettuce and strawberries (Wilson *et al.*, 1994a, b, 1995a, b, Glen *et al.*, 1996).

We carried out three field experiments to investigate whether *P. hermaphrodita* provides adequate protection against slug damage under Swiss field conditions, and to facilitate the registration process for *P. hermaphrodita* in Switzerland.

MATERIALS AND METHODS

For the field trials, we used the commercial formulation of *P. hermaphrodita* known as 'Nemaslug' or 'BioSlug' (supplier: MicroBio, UK). The nematodes were stirred in a watering can full of water and applied to the soil at a rate of 10^6 per m^2, unless otherwise stated. The purpose of the 'pilot experiment' was to verify the effect of *P. hermaphrodita*. This experiment was done in plots with a mixture of several vegetables and an ornamental plant. The 'lettuce experiment' and the 'rape experiment' simulated the agronomic conditions prevailing in horticulture or in arable farming, respectively (crop, soil cultivation, season etc.).

The pilot experiment

The pilot experiment was carried out in autumn 1993, at two sites: Grossdietwil and Oberwil (for details see Speiser & Andermatt, 1994). Both the climate and the soil are more favourable for slugs at Grossdietwil than at Oberwil. At Grossdietwil, we installed four randomized blocks of three plots (a nematode-treated plot, a metaldehyde-treated plot and an untreated plot). Each plot measured 2.5 x 2.5 m and was surrounded by a slug fence (supplier: Thomas Pfau, Switzerland). Inside, all of the surface was covered with a living grass sward except for the central area of 1 x 1 m, where we planted lettuce, chinese cabbage and kale (three plants each). A mixed population of 1 *D. reticulatum*, 5 or 15 *A. distinctus* and 4 or 14 large arionids (*A. rufus* or *A. lusitanicus*) was introduced into the plots at Grossdietwil. At Oberwil, three nematode-treated plots and three untreated plots were installed. Here, lettuce, kohlrabi and tagetes were planted in the centre of each plot. In Oberwil, the slug population resident in the living grass sward inside the slug fence was high; no additional slugs were therefore introduced into the plots. Both molluscicides were applied to the entire plots, including the grass. Throughout the experimental period, the weather was unusually wet.

To quantify slug damage, we estimated the percentage leaf area removed from each plant twice a week at Grossdietwil, and daily (except for weekends) at Oberwil. The relative performance of the different treatments was very similar throughout the experiment. Therefore, we only present in detail the results after two weeks. We compared treatments pairwise with the Wilcoxon signed-rank test, treating data for different crops and from different blocks as separate pairs.

The lettuce experiment

The lettuce experiment was carried out in summer 1994 at Oberwil. It consisted of four rows with eight plots, measuring 1 x 1.5 m each. The central two rows of plots were completely surrounded by a slug fence, while the plots in the outer two rows were open for slugs immigrating from the surrounding meadow. Seven specimens of *D. reticulatum* were introduced into each plot five days prior to the start of the experiment. Twelve lettuce seedlings were planted in each plot. Eleven plots were treated with 10^6 nematodes per m^2, ten plots with 10^5 nematodes per m^2, and eleven plots were left untreated. To quantify slug damage, we estimated the percentage leaf area removed on four dates after treatment. Again, the relative performance of the different treatments was very similar throughout the experiment. Therefore, we report only the results of the last census (at harvest). In this experiment, the weather was very dry initially and humid later.

The rape experiment

The rape experiment was carried out in autumn 1994 at Grossdietwil. In a rape field, twelve plots measuring 2 x 2 m were fenced off at the time of emergence of the rape plants. Thirty *D. reticulatum* were introduced into each plot. Four plots were then treated with nematodes, four with metaldehyde and four left untreated. After two weeks, we counted the numbers of undamaged and damaged rape plants.

RESULTS

The pilot experiment

At both sites, slugs heavily attacked the plants in the untreated plots. At Grossdietwil, damage was low during the first week but increased strongly afterwards. At Oberwil, damage in the untreated plots increased steadily throughout the experiment; although in lettuce, it levelled off after two weeks. In both molluscicidal treatments, nematodes and metaldehyde, damage was low from the beginning of the experiment and increased more slowly than in the untreated plots. In the top 5 cm of the soil, we found dead *D. reticulatum* showing the typical signs of nematode infection (Wilson *et al.*, 1993). We also observed large arionids which seemed to be nematode infected, but we could not follow their fate.

After two weeks, leaf loss averaged for all crops was 36.2 % in the untreated plots, 3.9 % in the nematode-treated plots and 6.9 % in the metaldehyde-treated plots (Fig. 1). The difference between the nematode-treated and the untreated plots was highly significant (Wilcoxon signed-rank test: N=21, p<0.001). Probably because there were only four metaldehyde-treated plots (at site Grossdietwil), both comparisons involving the metaldehyde treatment were not significant (metaldehyde vs. untreated: N=12, p>0.1; metaldehyde vs. nematodes: N=12, 0.05<p<0.1).

The lettuce experiment

Slug damage to the lettuce plants increased steadily throughout this experiment, until most of the plants were eaten completely. The nematode treatment had no effect on slug damage (Fig. 2). On the other hand, slugs damaged the lettuce significantly less within the slug fence than in the open plots (ANOVA with arcsine-transformed data; nematode treatments: p>0.5; fence: p<0.01).

The rape experiment

Two weeks after treatment, we found several dead or moribund *D. reticulatum* in the metaldehyde-treated plots. We also found at least two slugs on the soil surface or closely below with a swollen mantle characteristic for nematode infection. We counted an average (±SE) of 108 (±39) undamaged rape plant in the untreated plots, 337 (±13) plants in the nematode-treated plots and 221 (±25) plants in the metaldehyde-treated plots (Fig. 3, black bars). The total number of (damaged and undamaged) rape plants averaged 279 (±35) in the untreated plots, 376 (±16) in the nematode-treated plots and 380 (±16) in the metaldehyde-treated plots (Fig. 3, total height of black and hatched bars). For the number of intact plants, there were significant differences among treatments (ANOVA; p<0.001), and all treatments differed significantly trom each other (Fisher PLSD test, p<0.05). For the total number of plants, there were also significant differences among treatments (ANOVA; p<0.05). Here, the untreated plots differed significantly from both the nematode- and the metaldehyde-treated plots, but the nematode and the metaldehyde treatment were not significantly different from each other (Fisher PLSD test, p<0.05).

Figure 1. Slug damage (leaf loss in %) found on the different crops in the pilot experiment.

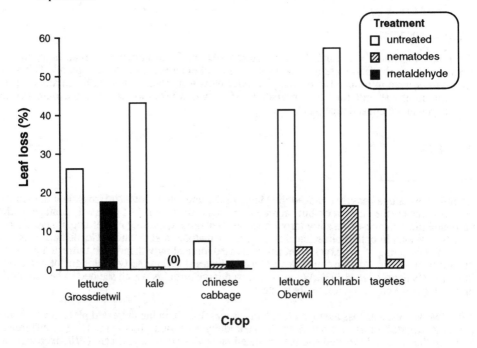

Figure 2. Slug damage (leaf loss in %) in the lettuce experiment. Bars indicate one standard error.

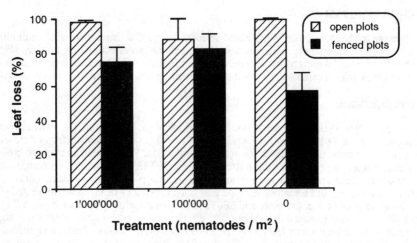

Figure 3. Average number of damaged and undamaged rape plants counted per plot in the rape experiment.

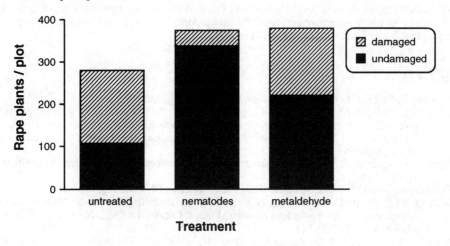

DISCUSSION

The nematode treatment clearly reduced slug damage in the pilot experiment and the rape experiment. The effect of the nematodes was similar to metaldehyde, or better. This is confirmed by the literature: Treatment with *P. hermaphrodita* was superior to methiocarb treatment in several experiments reported by Wilson *et al.* (1994a, b) and by Glen *et al.* (1996).

On the other hand, the nematodes failed in the lettuce experiment. Possible explanations are: (I) The nematodes might have become inviable in the hot weather prevailing at the time of application. Shortly before application, we had checked under the binocular microscope that the nematodes were alive, but we did not perform any other quality control; (II) The nematodes might have been ineffective against the species of slug which was causing the damage. Although *D. reticulatum* was released into the plots, we suspect that *A. lusitanicus* might have caused much of the feeding damage to the lettuce. *A. lusitanicus* was frequent in the surrounding areas and had easy access to the unfenced plots; some individuals may also have entered the fenced plots (the fences prevent slug migration to a high degree, but not completely). To test whether the nematodes are indeed differentially infective for *D. reticulatum* and *A. lusitanicus*, we carried out a preliminary laboratory trial shortly after the lettuce experiment was finished. This trial suggested that, unlike *D. reticulatum*, *A. lusitanicus* is not very sensitive to infection by the nematode *P. hermaphrodita* (unpublished data). Further tests are necessary to corroborate this preliminary finding. In particular, the sensitivity of *A. lusitanicus* towards *P. hermaphrodita* might be age or size-dependent, as reported for the snail *Helix aspersa* by Glen *et al.* (1996). Juvenile *A. lusitanicus* often hide in soil crevices, where nematodes could easily infect them. By contrast, larger animals, such as those used in this laboratory trial, tend to stay on the soil surface.

The nematode *Phasmarhabditis hermaphrodita* is a promising alternative to chemical molluscicides. It is known to infect all major pest slugs occurring in England, but is no threat for non-target snails (Wilson *et al.*, 1993, Glen *et al.*, 1996). The present results demonstrate that it can be successfully used for slug control in Switzerland. Whether it is infective for all pest slugs prevailing in Switzerland, however, remains to be clarified.

ACKNOWLEDGEMENTS

We warmly thank A. Winkler, U. Gujer and D. Niederhauser for help with the field work, and F. Weibel for his constructive advice on the manuscript. This work was financially supported by the canton Basel-Landschaft and by K. Züst.

REFERENCES

Fechter, R; Falkner, G (1990) *Weichtiere*. Mosaik Verlag, München.
Glen, D M; Wilson, M J; Hughes, L; Cargeeg, P; Hajjar, A (1996) Exploring and exploiting the potential of the rhabditid nematode *Phasmarhabditis hermaphrodita* as a biocontrol agent for slugs. In: *Slug and Snail Pests in Agriculture*, I F Henderson (ed.) BCPC Monograph 66
Speiser, B; Andermatt, M (1994) Biokontrolle von Schnecken mit Nematoden. *Agrarforschung*. **1**, 115-118.
Sulzberger, R (1996) *Wenn Schnecken zur Plage werden*. BLV Verlag, München.
Wilson, M J; Glen, D M; George, S K (1993) The rhabditid nematode *Phasmarhabditis hermaphrodita* as a potential biological control agent for slugs. *Biocontrol Science and Technology*. **3**, 503-511.
Wilson, M J; Glen, D M; George, S K; Pearce J D; Wiltshire C W (1994a) Biological control of slugs in winter wheat using the rhabditid nematode *Phasmarhabditis hermaphrodita*. *Annals of Applied Biology*. **125**, 377-390.
Wilson, M J; Glen, D M; Wiltshire, C W; George, S K (1994b) Mini-plot field experiments using the rhabditid nematode *Phasmarhabditis hermaphrodita* for biocontrol of slugs. *Biocontrol Science and Technology*. **4**, 103-113.
Wilson, M J; Glen, D M; George, S K; Hughes, L A (1995a) Biocontrol of slugs in protected lettuce using the rhabditid nematode *Phasmarhabditis hermaphrodita*. *Biocontrol Science and Technology*. **5**, 233-242.
Wilson, M J; Hughes, L A; Glen; D M (1995b) Developing strategies for the nematode, *Phasmarhabditis hermaphrodita*, as a biological control agent for slugs in integrated crop management systems. In: *Integrated Crop Protection: Towards Sustainability?* R G McKinlay & D Atkinson (eds). BCPC Monograph No 63, pp.33-40.

A STUDY OF COMPLEMENTARY TECHNIQUES FOR SNAIL CONTROL

CHEN DE-NIU, ZHANG GUO-QING
Institute of Zoology, Academia Sinica, Beijing, China

XU WENXIAN, LIU YANHONG
Shaanxi Plant Protection Station-in-chief, Xian, Shaanxi Province, China

CHENG XINGMIN, WU JIQING
Shaanxi Plant Protection Institute, Yanglin town, Shaanxi Province, China

ABSTRACT

Laboratory and field studies revealed a number of effective methods of controlling snail pests (*Bradybaena ravida*). These methods were deep ploughing in autumn or spring; application of ammonium carbonate fertilizer in early morning or evening; applying the pesticide metaldehyde (Chuwote) mixed with edible bait or soil; exposure of eggs by cultivation; hand removal of snails. Adults and larvae of the carabid beetle *Carabus brandti* were shown to feed on snails. Deep ploughing, exposure of eggs and hand picking of snails should be the basis of control measures, supplemented by pesticide (metaldehyde) where necessary. Extract of tobacco leaves may have potential value for snail control, but further studies are needed.

INTRODUCTION

Snails (*Bradybaena ravida*) (Gastropoda: Bradybaenidae) are important pests of wheat, maize, cotton and other crops in Shaanxi Province, China. This paper summarises detailed investigations of a wide range of techniques for controlling snail damage.

MATERIALS AND METHODS

The effect of deep ploughing

The effects of deep ploughing on snail survival were tested using artificial simulated soil cover, in autumn and spring.

Deep ploughing in autumn. A tile basin with 34 cm diameter opening and height 30 cm was buried in the field, then snails collected from the field were buried in the basin with a covering of 10 cm or 20 cm of soil. The soil surface in the basin was kept at the same level as the field surface. A quantity of crop leaves was supplied, to serve as food and shelter for snails. The opening of the basin was covered with nylon mesh. Snail survival and death were recorded at intervals.

<u>Deep ploughing in spring.</u> Three cylindrical holes with a diameter of 30 cm were dug, with depths of 10 cm, 20 cm and 30 cm, and snails collected from the field the same day were buried into the holes, then the soil was replaced and crop leaves were placed on top. Finally, the holes were covered with nylon mesh which was buried in the soil around the edges of the hole. Each treatment was replicated three times. Snail survival and death were recorded at regular intervals.

Chemical Control

<u>Effect of chemical fertilizer.</u> Ammonium carbonate, ammonium chloride, ammonium nitrate, urea, ammonium phosphate and calcium superphosphate were applied to soil at the manufacturers' recommended rate of usage. The test was made both in basins buried in the field and in baskets (1 x 1 m) placed in the field.

<u>Effect of tobacco water.</u> On 22 November 1991, hibernating snails collected in the field the previous day were soaked in water at 16°C to make them emerge from behind the operculum. Whole tobacco leaves were soaked in cold water (1 volume of leaves to 20 volumes of water) for 24 hours. Snails were soaked in tobacco water, then placed in a basin and observed at regular intervals.

<u>Effect of insecticides.</u> Tests were made using more than 20 pesticides including metaldehyde (Chuwote). Generally, tests were done in basins buried in the field. Snails were collected from the field on the same day. Each treatment was repeated two or three times. Effects were assessed 24, 48, 72 and 96 hours after treatments. To find an economic bait from a convenient source, we tested poison bait consisting of metaldehyde and carrier in the proportion of 1:20 and we compared this with commercial metaldehyde (techniucal) powder, and metaldehyde bait mixed with soil.

<u>Biological control.</u> In the laboratory, adults and larvae of *Carabus brandti*, which is a natural enemy of this snail, were placed individually in glass jars, with a layer of fine wet soil in the base. At least five adult and juvenile snails were placed in each jar each day. Jars were inspected each morning.

Field experiments.

<u>The effect of deep ploughing before autumn sowing.</u> On 5 October 1989, a field in Xi-chen village, Qiaode Town was cultivated to a depth of 13 cm using a rotocultivator. This field was compared with a hoe-ploughed field which is common at present. Before sowing and 15 days later, the numbers of snails in both fields were assessed by taking 5 samples, each 1 m² examined to 4 cm depth of soil.

<u>Use of ammonium carbonate during autumn sowing.</u> On 7 October 1990, when wheat was sown, 50 kg ammonium carbonate was applied as fertilizer to an experimental plot of 0.07 ha (1 mu) in area (1 ha = 15 mu) in Group 2, Zhangjiadeo, Zhongzhang township. This was compared to a neighbouring field in which 25 kg urea/0.07 ha was used. Both fields were sown using the hoe planting method. Next spring, when snails began to resume activity, the number of snails in these two fields was assessed.

Integrated control of snails: large scale field experiment. Field sites were chosen in Group 2, Zhangjiadao, Zhongzhang township, where serious damage had been caused by snails in two previous years. An area of 6.5 ha (98 mu) was planted, with 3.3 ha of wheat and 1.2 ha of cotton. In order to reduce the number of snails, the techniques of killing snails by ammonium carbonate application or by deep ploughing followed by ammonium carbonate application was used. In spring 1991, when snail numbers increased dramatically, Metaldehyde was used to kill snails in the cotton fields which were not deep ploughed in the previous year. Snails were also collected by hand and deep or medium ploughing was adopted to dessicate eggs in cotton fields. The cotton fields left unsown were not deep ploughed until snails became active the following year. Before the autumn harvest, the number of snails in both demonstration fields and conventional fields were assessed.

RESULTS

Effect of deep ploughing

When snails were covered with 20 cm layer of soil in the autumn, 73% died and no living snails emerged from the soil (Table 1). When soil cover was 10 cm thick, the death rate of snails was just above 60%, and most of the survivors emerged from the soil.

Table 1. Effect of covering snails with soil in the autumn (17 September 1990) on survival at 6 July 1991, Jingyang.

Depth of soil cover (cm)	Initial No. of snails	Total No. snails recovered	No. snails on soil surface	No. snails surviving in soil	No. dead snails in soil	Death rate (%)
10	50	50	18	0	32	64
10	50	50	14	5	31	62
20	100	100	0	27	73	73

Table 2 Shows that the death rate of snails covered with soil in spring reached 80% or more, irrespective of whether the soil cover was 30 cm, 20 cm, or 10 cm thick. The death rate of snails with a soil cover of 10 cm was a little lower than snails with a soil cover of 20 cm and 30 cm, and more snails emerged from soil after burying to 10 cm depth than after burying to 20 cm or 30 cm.

Table 2. Effect of covering snails with soil in spring (April 27 - July 3, 1991, Jingyang). Results are for three replicates of 50 slugs at each soil depth.

	Depth of soil cover (cm)		
	10	20	30
Initial no. of snails	150	150	150
Total no. recovered	143	140	150
No. on soil surface	15	0	2
No. surviving in soil	14	19	20
No. dead	114	121	128
Death rate (%)	79.7	86.4	85.3

Comparing the results in Table 1 with Table 2, more time was needed to kill snails by covering them with soil in autumn than in spring, and the rate of kill was lower following autumn burial. The main reason is that after covering with soil in autumn, the temperature declines and snail metabolism declines, and as the snails have sufficient food reserves to survive the winter, it took longer to kill them in the autumn. On the contrary, after covering with soil in spring, soil temperature and snail metabolism and food requirements increase, so that more snails were killed in a short period.

Chemical Control

Chemical fertilizers. Urea and diammonium phosphate were ineffective. Where 750 kg/ha ammonium carbonate was applied, more than 90% of snails were dead 48h later, but where 450 kg/ha ammonium chloride was used, the death rate was only 78% 48 hours later (Table 3). The contrast with the untreated control is very clear. In the test where ammonium carbonate was applied at different times of day, better results were achieved by applying in early morning or after 4 pm when autumn sowing, with snail mortality as high as 92%. Where ammonium carbonate was used at noon, when snails had contracted into their shells, efficiency did not reach 20%, because ammonium carbonate did not readily make contact with snails and was lost by volatilisation.

Table 3. Effect of different types of chemical fertilizer on snail survival (18-20 April, 1990, Jingyang).

Treatment	Rate of application (kg/ha)	No. snails in test	% snails dying during test
Ammonium carbonate	750	50	98
	1125	50	100
Ammonium chloride	375	50	72
	450	50	78
Untreated	-	50	0

Effect of tobacco water. Initial tests indicate that exposure to tobacco water may kill snails: eight out of 15 snails exposed to tobacco water had died after 6 days, whereas there were no deaths in the untreated controls.

Effect of insecticides. Tests showed that metaldehyde (Chuwote), produced in Tongshan Insecticide Plant, Jiansu Province, China, is effective in killing snails. Though calcium arsenite is also effective, it is no longer produced in this country. In order to investigate the optimum dose of metaldehyde, we tested poison bait consisting of a mixture of metaldehyde commercial powder and cornflour, in different proportions, (Table 4). Treated snails still survived after 24h but showed obvious poison symptoms, did not eat or move, and excreted light yellow mucus from the aperture. After 48 hours, the average death rate of treated snails reached 93-96%. In tests of different baits (Table 5), all baits were effective.

Table 4. Effects of different dosages of metaldehyde on snail mortality (9-12 April, 1991, Jingyang).

Ratio of metaldehyde to maize flour	No. of snails tested	Death rate (%) after		
		24h	48h	72h
1:10	300	0	92.7	97.7
1:15	300	0	95.7	98.0
1:20	300	0	93.7	94.7
1:25	300	0	93.0	94.0
maize flour alone	300	0	0	0

Table 5. Effect of incorporating metaldehyde with different bait carriers (9-12 April, 1991, Jingyang).

Bait	No. of snails tested	Death rate (%) after		
		24h	48h	72h
Maize flour	300	27.7	99.3	100
Wheat bran	300	13.0	98.3	99.0
Cotton seed cake	300	13.3	95.7	98.6
Grass	300	18.2	97.3	97.3

Table 6. Comparison of metaldehyde baits and metaldehyde/soil formulation (1 part: 30 parts soil).

Formulation	No. of snails	Death rate (%) after	
		24h	48h
Metaldehyde dust bait	150	76.5	98.0
Metaldehyde wheat bran bait	50	51.0	98.0
Metaldehyde mixed with soil	150	56.8	97.0
Untreated control	50	0	2.0

Table 6 shows that all three methods of using metaldehyde were effective, with 97-98% mortality after 48 hours.

Effect of natural enemies. Replicated experiments showed than an adult beetle (*C. brandti*) could eat 1 adult snail and 4 to 5 young snails per day, while a larva ate 1 to 2 adult snails and 2 to 3 young snails per day, under the conditions of the test.

Tests of field control

In the experiment in the wheat field and cotton field in Zhoujadeo village, Zhong-zhang township, 105 kg/ha poison bait or 450 kg/ha poison soil (1: part of metaldehyde to 20 parts wheat bran or soil) was applied, along the cotton plant rows or broadcast in the wheat field. After 72 hours, the results were assessed (Table 7).

Table 7. Effect of method of application of metaldehyde in field trials (22-24 May, 1991, Jingyang).

Crop	Method of application of metaldehyde	No. of snails tested	Death rate (%)
Cotton	mixed with bait	100	98
	mixed with soil	100	99
Wheat	mixed with bait	100	43
	mixed with soil	100	62
Cotton	untreated	30	0
Wheat	untreated	30	0

In 1989 and 1991, egg piles exposed on the surface of the soil by cultivation were observed. When eggs were exposed in the sun in clear weather in summer and autumn, they were ruptured in 5 minutes, while in cloudy or rainy weather, it took as long as 24 hours.

On 9, 13 and 20 May 1990, snails were removed by hand from an area of over 25.3 ha wheat fields. The total weight of snails collected was 1551.6 kg and snail numbers were reduced by 45% from $29.4/m^2$ before collection to $16.2/m^2$ after hand collection.

Field experiment and demonstration of integrated control

Before deep ploughing to 22 cm, a mean number of 56.8 snails/m^2 were recorded on 10 May 1989. After ploughing, numbers declined to $1.6/m^2$ on 20 October. In contrast, numbers on the field to be hoe ploughed were estimated to be $25.8/m^2$ on 10 May 1989 before cultivation, and $18.0/m^2$ on 20 October, after hoe-ploughing to a depth of 7 cm. Where ammonium carbonate was applied to the soil in October 1990, snail numbers declined by 94% by 26 May 1991. Where urea was applied to soil on the same date, snails declined by only 23% over the same period.

In the demonstration fields, snail numbers declined by 88% and 76% in cotton and maize fields, respectively, between 7 and 25 September 1991. In the corresponding untreated fields, snail numbers declined by only 9% and 18% in cotton and maize fields, respectively, over the same period.

DISCUSSION

Deep ploughing is of obvious value in killing snails and the area deep-ploughed before autumn sowing, should be expanded. Fields which are not sown in autumn should be deep ploughed before winter. Moreover, soil should be deep-ploughed before sowing maize in summer. The depth of ploughing should exceed 20 cm. This is beneficial both for the growth of the crops and in greatly reducing the number of snails.

A large number of snails could be killed by applying ammonium carbonate at the time of autumn sowing. This technique was previously employed in many areas in Jingyang County but has since declined as urea or diammonium phosphate has been adopted as the base fertilizer. We suggest that in areas where snails are a problem, we must encourage the use of ammonium carbonate at a rate of 750 kg/ha, as a fertilizer. Metaldehyde is an effective pesticide for snails, applied as a poison bait or by admixture with soil. Extract of tobacco leaves shows possible potential as a method of killing snails, but insufficient experimental data are available and further study is needed.

It is important not to rely on a single method for controlling snails, but to take every appropriate measure depending on the numbers of snails and the state of crop growth. In addition to killing snails by deep ploughing, application of ammonium carbonate and pesticide, we should try to take measures such as hand-picking snails. Cultivation, exposing eggs, and hand-picking of snails, should be the basis for control measures, with pesticides used to supplement these methods in order to optimise the economic, social and ecological benefits.

REFERENCES

Chen De-niu; Gao Jia-xiang (1980) Investigation and control of land snails in XinJiang Province, China. *Agricultural Science*, **3,** 35-38.
Chen De-niu (1984) On the land snails from farmland of China, Agricultural Publishing House, 1-221.
Chen De-Niu (1987) Economic Fauna Sinica of China (Terrestrial Mollusca), Science Press Beijing, 1-186.
Daxl T (1972) Das Verhaaltem der molluskiziden Verbindungen Metaldyhyd Isolan und Ioxynil gegen Nacktschnecken unter Frielandedingungen *Zeitschrift für angewande Zoologie*, **58,** 203-241.
Ranaivosoa H (1972) Lutte biologique contre les escargots phytpphages à Madahascar, **299,** 359-368.

DIFFERENTIAL TOLERANCE OF WINTER WHEAT CULTIVARS TO GRAIN HOLLOWING BY THE GREY FIELD SLUG (*DEROCERAS RETICULATUM*): A LABORATORY STUDY

K A EVANS & A M SPAULL
SAC, West Mains Road, Edinburgh EH9 3JG, UK

ABSTRACT

Differences in the susceptibility of twelve winter wheat cultivars to grain hollowing by the field slug (*Deroceras reticulatum*) were tested in the laboratory. Differences were apparent with the most popular UK cultivars in recent years, Riband, Mercia and Avalon, being significantly more damaged than the cultivars Buster, Hunter, Parade, Hussar and Brigadier. The damage to grain could be related to sugar content in the ungerminated seed and release of sugars and other solutes during germination at both extremes of slug damage to winter wheat cultivars. The potential for assessing the risk of damage to winter wheat cultivars based on the mobilisation and release of sugars in germinating seeds is discussed.

INTRODUCTION

Slug damage to cereals in the UK, especially winter wheat has increased in the last few years due in part to the increase in set-aside and other changes in agronomic practice. Despite the wide range of molluscicidal products currently available for slug control, and the knowledge gained on cultural methods for reducing slug problems (Glen & Wiltshire, 1992; Spaull & Davies, 1992), slugs are still a major problem to many growers of winter wheat.

Spaull & Eldon (1990) demonstrated differences in the degree of grain hollowing of winter wheat cultivars by the field slug (*Deroceras reticulatum*) in the laboratory. Differences in damage were thought to be due to the rate of release of water soluble sugars and other solutes by the seed. The cultivars tested in 1990 by Spaull & Eldon are nearly all outclassed now, apart from one (Mercia), which is still on the 1995 Fully Recommended List of winter wheat cultivars produced by SAC and NIAB. This paper reports on results obtained using cultivars on the current Full and Provisional Recommended list in order to determine whether varietal susceptibility to grazing by the grey field slug is present in modern winter wheat cultivars, and whether susceptibility can be related to the sugar content and rate of leaching of water soluble sugars by seed.

MATERIALS AND METHODS

Assessment of slug damage to grain

Adult slugs (*D. reticulatum*) were obtained from field populations by trapping with a bait of grain and bran. The slugs were kept on a diet of lettuce for a week after collection at 10°C before being used in any experiments. Adult slugs were starved for 24h before use in feeding tests on winter wheat grain.

Untreated seed of twelve cultivars of winter wheat were used, seven of which are on the 1995 Fully Recommended List of wheat cultivars (cvs. Brigadier, Hereward, Hunter, Hussar, Mercia, Riband and Spark), two from the Provisionally Recommended List (cvs. Buster and Dynamo), two cultivars which were the most susceptible (cv. Avalon) and resistant (cv. Parade) to slug damage in Spaull & Eldon's previous work (1990), and cv. Zodiac which is a cv. Parade cross. The cultivars Buster and Dynamo are also Parade crosses.

The protocol of Spaull & Eldon (1990) was followed where four seeds were placed on well-moistened filter paper equidistant from each other and the edge of a 9cm diameter petri-dish. A single adult grey field slug was placed in the centre of the dish, and kept in an incubator at 10° C. Ten replicates were used for each cultivar, and grain hollowing was assessed after 1, 2 and 5 days. Grain hollowing was scored as a crude percentage of individual seeds.

Assessment of sugar content in seed

The water-soluble carbohydrate content of non-germinated seed was determined using the methods of Spaull & Eldon (1990). Two hundred seeds (approx. 10 g) were boiled in 25 ml distilled water for three minutes, macerated and centrifuged at 4000 g for three minutes. The supernatant was tested for total water-soluble sugar content using a Boehringer-Mannheim test kit, and expressed as glucose equivalents.

Assessment of electrical conductivity of germinating seed

The electrical conductivity of seed (a crude estimate of leaching of electrolytes) was undertaken from a dish containing 200 seeds (10 g fresh weight) on saturated filter paper after 1, 2 and 3 days at 10°C. The conductivity was measured using a conductivity meter, and results expressed as electrical conductivity (μS/g) of seed.

RESULTS

Assessment of slug damage to grain

Differences in slug damage between cultivars were apparent after 1 day (Fig. 1a); cvs. Parade, Dynamo, Hunter and Buster had been grazed significantly less than Riband, Mercia and Avalon. Over the next four days, grain hollowing increased on all cultivars until the experiments were terminated on Day 5 (Figs. 1b-c). After 5 days, Buster was the least damaged cultivar, with 18% of grain hollowed, compared to Riband which had a mean score

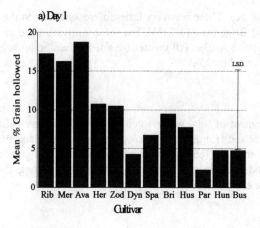

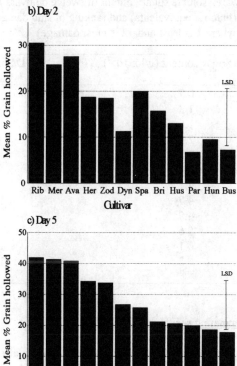

Fig. 1. Percentage grain hollowed by *Deroceras reticulatum*
L.S.D = Least Significant Difference (P = 0.05)

of 42% grain hollowed (Fig. 1c). There was very little difference between the cultivars Buster, Hunter, Parade, Hussar and Brigadier (18-21%), but these cultivars were all significantly less grazed than Riband, Mercia and Avalon (all greater than 40% grain hollowing). The remaining cultivars had had grain hollowed by 25-35%.

Sugar content in seed

The results from the assessment of total water-soluble sugar content are indicated in Table 1. The cultivars Parade, Buster and Hunter had the lowest sugar content, and were also the three least damaged cultivars in the slug feeding experiments (Fig. 1). The greatest sugar content was in the cultivars Riband, Mercia and Hereward, which were among the most damaged cultivars in the slug grazing experiments.

Table 1.   Total water-soluble sugar content of twelve cultivars of winter wheat (glucose equivalents) and ranking of slug damage (1-12, where 1 is least and 12 is most damage)

Cultivar	Sugar content (g/l x 10^{-2})	Damage ranking
Parade	0.7	3
Buster	1.3	1
Hunter	1.6	2
Hussar	2.2	4
Zodiac	3.5	8
Spark	3.8	6
Dynamo	4.1	7
Avalon	4.7	10
Brigadier	4.9	5
Hereward	5.6	9
Mercia	6.1	11
Riband	6.2	12

Electrical conductivity of germinating seed

The measurements of electrical conductivity differed between cultivars after 1 day (Table 2), but the differences became less marked after 2-3 days. The lowest conductivity measurements after 1 day were obtained from cvs. Buster, Hussar, Spark, Mercia and Parade (all < 10 μS), and these were also among the cultivars with the lowest conductivity after 3 days. The highest measurements after 1 day were obtained from Dynamo, Riband and Avalon.

Table 2. Electrical conductivity (µS) of leachate from 12
winter wheat cultivars (expressed per g of seed)

Cultivar	Day 1	Day 2	Day 3
Avalon	12.6	17.4	23.8
Brigadier	11.4	19.6	26.2
Buster	8.4	15.1	19.8
Dynamo	14.2	20.2	24.3
Hereward	12.3	17.2	22.3
Hunter	10.1	13.6	18.6
Hussar	8.4	14.4	19.7
Mercia	9.6	12.2	18.8
Parade	9.4	13.1	16.8
Riband	13.8	20.6	27.2
Spark	8.8	12.6	18.6
Zodiac	11.2	17.9	23.7

DISCUSSION

The results obtained from these experiments confirm some of the earlier findings of Spaull &
Eldon (1990); particularly the relative tolerance of the cultivar Parade to slug damage and the
susceptibility to damage of the cultivar Avalon. The hypothesis proposed by Spaull & Eldon
(1990), that the exudation of sugar solutes during the early stages of seed germination makes
the seed more attractive to slugs, is to some extent upheld based on many of the results
obtained during this study. The four cultivars with the lowest sugar content prior to
germination (Parade, Buster, Hunter and Hussar) were also the cultivars that were least
damaged in the slug grazing experiments. The opposite is also true to some extent; cvs.
Riband, Mercia, Hereward and Avalon had high sugar content in the seed, as well as being
most heavily grazed by slugs.

The rate of mobilisation of sugars by seed, as measured by the change in the electrical
conductivity of the leachates from germinating seed, suggests that cultivars that rapidly begin
production of sugars, are more prone to being damaged by slugs. Rapid sugar producers such
as Riband and Avalon suffered the heaviest grazing by slugs within the first 2 days of the
feeding experiments. Slower sugar producers such as Parade and Hunter suffered lower
damage from slugs. This hypothesis of 'high seed sugar content and rapid rate of sugar
production' does not explain some of the variability in results obtained from cultivars such as
Zodiac (low sugar, high grazing), Brigadier (high sugar in seed and rate of production, low-
moderate grazing) and Spark (moderate sugar content, low rate of sugar production, moderate
grazing). However the hypothesis fits well at the extreme ends of the spectrum, namely low
slug grazing when sugar content in seed is low and sugar production less rapid (cvs. Buster,
Hunter, Parade and Hussar), and high sugar content in seed, high degree of slug grazing and
rapid production of sugars (cvs. Riband, Avalon, Hereward and Mercia). If a threshold of

4.5g/l x 10^{-2} (total sugars) and/or 22 µS/g seed (electrical conductivity) is applied, the risk of damage to 25% of the grain would have been correctly predicted for all but two of the cultivars tested (Brigadier and Dynamo).

The possibility of a genetic link in resistance to slug attack has been hinted at with Buster, whose Pedigree is Parade. However the other cultivars with a Parade pedigree, Dynamo and Zodiac, did not show any degree of resistance to slug damage.

There does appear to be a link between sugars in the seed and its production, and the degree of grazing by slugs under laboratory conditions. The method used to measure production of sugars in the leachate (electrical conductivity) is admittedly a rather crude measure, as it does not distinguish between other electrolytes being produced by the seed. There are many other variables which may affect these results such as provenance of the seed, age of seed, physiological state of the slugs used in the experiment etc.., however it has been shown that under laboratory conditions at least, it is possible to correlate the likelihood of damage to seed by slugs with a relatively simple measurement such as seed sugar content and electrical conductivity of leachates from germinating seeds.

This work needs to be extended to include controlled field conditions in order to determine whether the susceptibilities of winter wheat cultivars to slugs seen in the laboratory, is similar under more natural conditions. Further laboratory studies should also be undertaken using more sophisticated measurements of the production of sugars in germinating seed, perhaps by identifying the production of specific sugars during the early stages of germination.

It is interesting to note that in these laboratory tests, the cultivars that were most severely grazed by slugs, namely Riband, Mercia and Avalon, have been the most popular cultivars of winter wheat grown in the UK in recent years.

ACKNOWLEDGEMENTS

This work was funded by the Home-Grown Cereals Authority, HGCA Project No. 0006/1/95.

REFERENCES

Glen, D M & Wiltshire, C W (1992) Slug forecasting in cereals. *HGCA Project Report No. 47*, Home-Grown Cereals Authority, London. 35pp.

Spaull, A M & Davies, D H K (1992) Cultural Methods to reduce slug damage in cereals. *HGCA Project Report No. 53, Part II*, Home-Grown Cereals Authority, London. 18pp

Spaull, A M & Eldon, S (1990) Is it possible to limit slug damage using choice of winter wheat cultivars? *Brighton Crop Protection Conference - Pests and Diseases - 1990.* pp. 703-708.

PHYSIOCHEMICAL BARRIERS AS PLANT PROTECTANTS AGAINST SLUGS (GASTROPODA:PULMONATA)

G W DAWSON, I F HENDERSON, A P MARTIN, B J PYE

IACR-Rothamsted Experimental Station, Harpenden, AL5 2JQ, UK

ABSTRACT

The relative repellency of a range of chemicals to the field slug, *Deroceras reticulatum* (Muller), was assessed in the laboratory by giving slugs access to glass and leaf surfaces coated with known amounts of test material and recording their willingness to crawl on the treated surfaces. Several surfactants were found to be very repellent and three were tested in the field by applying them to the bases of Brussels sprouts plants by electrostatic sprayer. The barriers were ineffective in the field due to poor persistence in wet conditions.

INTRODUCTION

During locomotion terrestrial gastropods have a large proportion of their body surface applied to the substrate and separated from it only by a thin (c.10μm) layer of pedal mucus. Although largely (>95%) water, it is essential for locomotion, coupling the foot to the substrate (Denny, 1978) and protecting the epidermis. The underlying epithelium is relatively permeable, with water and compounds of molecular weight of up to 10,000 Da able to pass through into the haemocoel (Prior, 1989). Consequently, surfaces from which the slug may acquire irritant or toxic substances, or which contain materials which alter the physical properties of the mucus, are carefully avoided, using the receptors on the tentacles and perioral region.

We therefore looked for chemicals which would induce aversive reactions in slugs but not create a toxic hazard if applied to crops. Test materials were selected from three groups: secondary plant metabolites previously found to influence slug feeding behaviour (Airey *et al*, 1989), constituents of lichens which appear to determine food preference in phylomycid slugs (Lawrey,1989) and surfactants, which are known to alter the physical properties of mucus *in vitro* (Newbery et al, 1988).

MATERIALS AND METHODS

Laboratory tests.

For initial screening, test chemicals were dissolved in an appropriate solvent and sprayed onto square glass sheets 20 x 20 cm. A circular area in the centre of the plate (diameter 9 cm) was masked with a filter paper disc which was then removed leaving an unsprayed `refuge' area. Ten replicate sheets were sprayed per test chemical. The spray was applied at a rate of 25l/ha and a velocity of 0.4 m/s from a height of 25.0 cm above the plates with an electrostatic rotary atomiser (Arnold & Pye, 1981) to give a uniform surface deposit of 50 μg a.i./cm^2, equivalent to 5.0 kg/ha. A 2.0g/l solution of the fluorescent tracer Uvitex OB (Ciba Geigy [UK] Ltd., Technical service bulletin PL811) in distilled water was used to define the volume of formulation deposited on twenty filter discs of 1.8cm diameter placed on the glass sheets. The samples were analysed and the relationship between the volume deposited and the concentration of the tracer was calculated.   Analysis of the discs was carried out using a Perkin Elmer 2000 fluorescence spectrophotometer fitted with a flow cell, a 339-nm excitation filter and a broad band pass 430-nm emission filter (Arnold et al., 1984).   An individual field slug, *D. reticulatum*, in the weight range 0.5-0.8 g, was placed in the refuge area and the plate placed in a seed tray containing a film of water and covered with a second tray to maintain high humidity. After confinement overnight (16 h) at 10°C the number of slugs remaining within the refuge area was recorded.

The reaction of *D. reticulatum* to deposits on leaves was also studied. Young leaves of Chinese cabbage, *Brassica chinensis*, were sprayed as before, masking one half of the leaf so that the boundary between sprayed and unsprayed areas lay along the midrib. Discs of leaf were trapped between close-fitting concentric cylinders of rigid polyethylene pipe (7.5 cm internal diameter x 2 cm tall) with the midrib running across the diameter. Double-sided adhesive tape was applied to the inner surface of the cylinder wall and coated with common salt crystals to prevent escape and a single slug was placed in the centre on the unsprayed side of the leaf disc. Each treatment was replicated six times: discs sprayed with solvent alone acted as controls. Leaf cylinder arenas were placed on wet cotton wool in a controlled environment room (15°C, 12 h day/5°C 12 h night) at the beginning of scotophase. The time spent by each slug in the treated and in the untreated half of the leaf arena was recorded during the following six-hour period by time-lapse video tape recorder (Panasonic AG6730E) running at $^1/_{160}$ normal speed. The repellency of the test chemical was expressed as a simple index: Repellency Index = Time spent on Untreated / Time spent on Treated.

Field trial

Test chemicals were applied in an ethanol/water formulation to the bases of Brussels sprout plants, *Brassica oleracea* cv. Roger, growing at 60 cm spacing, using an electrostatic sprayer mounted 15 cm above soil level on a wheeled trolley. The two central rows of plants in plots of 6 x 6 plants were sprayed from both sides leaving two guard rows on each side to intercept spray drift. Plant bases were sprayed to give a concentrated band deposit of 50 µg/cm^2 on the stems and lower leaves while leaving the upper stems and leaves untreated.. Treatments were replicated four times in a randomised block design and sprays were repeated weekly from mid-September until mid-November. Eight treated plants from the centre of each plot were harvested and the "buttons" were scored for slug damage in relation to fresh weight (g), diameter (<15 mm, 15-25 mm, 25-30 mm, 30-35 mm, >35 mm) and height on the plant (<15 cm, 15-30 cm, 30-35 cm, >35 cm). Total numbers of sprouts per grade/height combination were summed over all eight plants and analysed by ANOVA (Genstat 5) after transformation to $\log_{10}$ n+1.

Chemicals

These are given in Tables 1 & 2 and were obtained commercially: Aldrich; geraniol, methyl anthranilate, cinnamamide (mainly *trans-*), lauric acid (dodecanoic acid), 'Igepals' CO-210 (n=1), CO-520 (n=4), CO-720 (n=11) CO-890 (n=39) ,CO-990 (n=99), cetyldimethylethylammonium bromide, dimethylethylhexadecyl ammonium bromide, 'Aliquat 366' (tricaprylmethylammonium chloride), tetrabutylphosphonium chloride, carvacrol, and 'Tween 80' (polyethoxylene(20)sorbitan monoleate). Sigma; stictic acid, vulpinic acid, thymol, dioctyl sulphosuccinate, 'Tergitols' NP-7 (n=7), Np-10 (n=10), cetyldimethylbenzylammonium chloride (benzyldimethylhexadecyl chloride), benzalkonium chloride (mainly benzyldimethyldodecylammonium chloride), 'Teepol' (C_9-C_{13} alkyl sodium sulphates), N,N,'N'-polyoxyethylene (10)-N-tallow-1,3-diaminopropane. Fluka; (+) and (-) -fenchone, (+) and (-) - carvone, allyl isothiocyanate, cinchonine. Carl Roth, Karlsruhe; (+) - usnic acid. Gifts; 'Ethylan BV' (n=14) (Lankro), Atlas Cedarwood Oil (Quest International), Neem Oil (Taparia Exports PVT Ltd., India), hop extract, mainly beta acids (Scottish & Newcastle Breweries). ('Igepal CO-', 'Tergitol NP-' and 'Ethylan BV' are substituted ethanols where the 'n' number indicates the number of polyetheneoxy units in the molecule).

RESULTS

Laboratory Tests: Glass Plate

Common salt, effective as a barrier in crystalline form, was not particularly effective as a repellent at the test concentration 50 $\mu g/cm^2$. The terpenoid, geraniol, showed some activity as did the vertebrate repellents, methyl anthranilate and cinnamamide. The lichen acids were not effective but the surfactants tested were, in general, highly repellent (Table 1).

Laboratory Tests: Leaf Disc

The terpenoids, carvone and fenchone, which are active slug antifeedants, did not deter crawling slugs. As a group, the most active repellents were again the surfactants, especially those containing ammonium, but individual surfactants ranged from very repellent, e.g. Igepal CO-720, to ineffective, e.g. Igepal CO-890 (Table 2).

Field Trial

There was much pest activity, with slugs (mostly *D. reticulatum*) frequently observed on the plants: at harvest 67% of all buttons had slug damage. None of the treatments had a significant effect on the mean weight of buttons damaged (Table 3). The proportion of sprouts damaged was greatest on the older and larger buttons lowest on the stem (92%), and progressively less higher up (15-30 cm/74%, >30 cm/37%), but was unaffected by any of the surfactant treatments (Tables 3, 4).

DISCUSSION

Laboratory tests showed that crawling slugs rapidly detect and are deterred by topical applications of chemicals at low deposit rates. Secondary plant metabolites such as fenchone and allyl isothiocyate, previously shown to be slug antifeedants (Airey *et al.*, 1989) were not effective as repellents. Some surfactants were very repellent in laboratory tests and activity was related to structure. Among the "Igepal" series (polyphenylpolyethoxylates) activity varied with the degree of ethoxylation (n), increasing to a maximum at n =11 (Igepal CO720), then decreasing. The repellency mechanism is not understood but the failure of the surfactants in the field trial was because they were not persistent and were rapidly removed from the plants by rain and by condensation.

Electrostatic spraying gave heavy localised deposits and good coverage in dense crops. Although the surfactants were ineffective in wet conditions the technique of barring access from below could be effective if more persistent formulations can be found.

Table 1   Glass Plate Test. Number of slugs held within refuge or escaped after 16 h (Minimum replication 20 tests per chemical)

Chemical	Deposit (µg/cm²)	No.held	No. escaped
Control	-	0	20
Sodium chloride	50	9	11
	150	15	5
Sodium carbonate	50	9	11
Sodium sulphate	50	0	20
Sodium alginate	50	0	20
Monosodium glutamate	50	0	20
Disodium glutamate	50	0	20
Geraniol	50	7	13
(+)-Fenchone	50	0	20
Methyl anthranilate	50	9	11
Cinnamamide	50	4	16
Hop extract	50	0	20
Neem extract	50	0	20
(+)-Usnic acid	50	0	20
Stictic acid	50	0	20
Lauric acid	50	0	20
Vulpinic acid	50	3	17
"Ethylan BV"	50	20	0
Dioctyl sulfosuccinate	50	18	2
"Tergitol NP-7"	50	20	0
"Tergitol NP-10"	50	17	3

Table 2.   Leaf Disc Test. Repellency Index (RI) of test chemicals @ 50 μg/cm² over 6 hour period.

Chemical	Repellency Index
Cetyldimethylethylammonium bromide	> 1000
Cetyldimethylbenzylammonium chloride	> 1000
Benzalkonium chloride	> 1000
Tricaprylmethylammonium chloride	333
Cetylpyridinium chloride	42
N,N,N-polyoxyethylene (10)-tallow-1,3-diaminopropane	33
Tetrabutyl phosphonium chloride	8
"Ethylan BV"	59
"Igepal CO-210"	12
"Igepal CO520"	6
"Igepal CO-720"	250
"Igepal CO-890"	<1
"Igepal CO-990"	1
"Tween 80"	2
Dioctylsulfosuccinate	125
"Teepol"	35
Cinchonene	15
Carvacrol	5
Cedarwood oil	5
(-)-Carvone	2
(+)-Carvone	1
(-)-Fenchone	1
(+)-Fenchone	1
Thymol	1
Allyl isothiocyanate	1

Table 3.   Weight of sprouts damaged. (Total buttons per 8 plants).

Treatment	wt. (g)
Dioctylsulfosuccinate	2,429
Cetyldimethylethylammonium bromide	2,447
"Ethylan BV"	2,991
Control	3,026
(SED)	(418.4)

Table 4.   Proportion of damaged sprouts at each height zone and treatment.

Treatment	Height		
	< 15 cm	15-30 cm	> 30 cm
Cetyldimethylethyl-ammonium bromide	0.924	0.667	0.237
Dioctylsulfo-succinate	0.874	0.688	0.307
"Ethylan BV"	0.931	0.776	0.413
Control	0.940	0.837	0.504

(SED: 0.0535 between T, 0.0475 within T)

ACKNOWLEDGEMENTS

This work was funded by BBSRC. We thank Mr Trevor Sheard, Levingtons Farms Ltd., for agronomic advice and Ms Suzanne Clarke for the statistical analyses.

REFERENCES

Airey, W J; Henderson, I F; Pickett, J A; Scott, G C; Stephenson, J W; Woodcock, C M (1989) Novel chemical approaches to mollusc control. In: *Slugs and snails in world agriculture*. I F Henderson (ed.). BCPC Monograph No. 41, pp. 301-307.

Arnold, A J; Cayley, G G; Dunne, Y; Etheridge, P; Griffiths, D G; Philips, P T; Pye, B J; Scott, G C; Vojvodic, P R (1984) Biological effectiveness of electrostatically charged rotary atomisers. I. Trials on field beans and barley 1981. *Annals of Applied Biology*. **105**, 353-359.

Arnold, A J; Pye, B J (1981) Spray application with charged rotary atomisers. In: *Spraying systems for the 1980's*. W O Walker (ed.). BCPC Monograph No. 24, pp. 109-117.

Denny, M (1978) A material ratchet - the pedal mucus of the slug *Ariolimax colombianus*. *Biorheology*. **15**, 460.

Lawrey, J D (1980) Correlations between lichen secondary chemistry and grazing activity by *Pallifera varia*. *The Bryologist*. **83**, 328-334.

Newbery, R S; Martin, G P; Marriott, C (1988) The effect of certain biliary and duodenal surfactants on mucus structure. In: *Mucus and related topics*. Society for Experimental Biology Symposium, Manchester, July, 1988, p.1.

Prior, D J (1989) Contact rehydration in slugs: a water-regulatory behaviour. In: *Slugs and snails in world agriculture*. I F Henderson (ed.). BCPC Monograph No. 41, pp. 217-223.

INTEGRATED CONTROL OF SLUGS IN A SUGAR BEET CROP GROWING IN A RYE COVER CROP

A ESTER

Research Station for Arable Farming and Field Production of Vegetables (PAGV), P.O. Box 430, 8200 AK Lelystad, The Netherlands

P M T M GEELEN
Experimental Farm 'Wijnandsrade', 6363 BW Wijnandsrade, The Netherlands

ABSTRACT

From 1991 to 1995, field experiments were carried out to assess the controlling effect of biological- and chemical treatments on slugs in sugar beet crops (*Beta vulgaris* L.). In the south of The Netherlands beets are direct drilled in a mulch of dead organic material with an undisturbed layer of top soil. This may provoke slug damage. The efficacy of nematodes (*Phasmarhabditis hermaphrodita*) and methiocarb pellets, calcium cyanamide granules and sodium chloride were compared in broadcast treatments and in furrow applications. The furrow application were applied at the time of drilling or at emergence. The use of nematodes and methiocarb applied as a furrow application resulted in about 81,500 plants compared to 62,500 plants/hectare in the untreated fields. The other treatments were less effective.

INTRODUCTION

Normally, there are no problems with slugs in sugar beet crops in The Netherlands.
In the southern region of The Netherlands the soil is a loamy loess and the land surface undulates. To protect the soil from water erosion, sugar beet is grown in a cover crop. Beet grows in a cover of dead organic material in an undisturbed layer of top soil (Ester *et al.*, 1992).Two types of plants are usually used as cover crops: one, like Italian rye grass, winter wheat or rye grows all winter and must be killed with a chemical before drilling sugar beet; the other, like yellow mustard, black radish and *Phacelia* are killed during the winter by frost. Those that are killed by frost decompose during winter and the soil protection decreases. This contrasts with rye and other crops that protect the soil the whole winter. The cover crop is sown the year prior to the sugar beet. It protects the soil during autumn and winter because a lot of organic material (dead or alive) remains on the surface. The cover crop also protects the soil in early spring and during establishment of the beet crop. The soil is not ploughed before drilling the sugar beet which can be drilled directly without any seedbed preparation. Sometimes a superficial seedbed is prepared. This method of cultivation is new for The Netherlands, but since 1990 legislation has been introduced to force farmers to grow their sugar beet in this way. About one third of the sugar beet area grown in a cover crop suffers from feeding damage by slugs. The species involved is *Deroceras reticulatum*, (Müller), the field slug. Young beet plants, between emergence and the 4/6 leaf stage, are most susceptible and may die. The seeds and recently germinated seeds are not susceptible to slugs. After the six leaf stage, the damage

is not that severe, because the plants are able to compensate. The activity of the slugs is dependent on shelter. Recently ploughed fields provide more shelter, but this decreases by puddling of the soil after heavy rainfall. The shelter capacity can also decrease as a result of dry weather or frost, as many shelter places on the surface or in the ground are no longer suitable for the slugs (Moens, 1980). Young et al., (1993) mentioned that slug activity also depends on the soil moisture content as well. Slugs are likely to increase in importance as pests in integrated crop management systems. In these systems a population can build up quickly where green manure crops or bulky old crop residues are used as part of a soil management programme. Until now farmers have been able to control the damage level to a certain extent by spreading pelleted bait containing a molluscicide. Application can take place either before drilling the seeds, or by spreading the bait after slugs or slug damage is observed. None of these methods appears to be fully effective (Ester et al., 1992). A potential new method of control is use of the nematode parasite *Phasmarhabditis hermaphrodita*. This parasite has the necessary attributes to be used as a successful biological control agent against slugs. The parasite which inhibits feeding and causes mortality in a wide range of pest species of slugs and snails is harmless to other invertebrates and is rapidly killed at 35°C, so is unable to survive in homeothermic animals (Glen et al., 1994). A product containing nematodes has been on sale in the UK since 1994 as a molluscicide for use by domestic gardeners (Glen et al., 1994) and has been studied in several field trials with Chinese cabbage and winter wheat (Wilson et al., 1994a, 1993a). Our research started in 1991/1992 and 1992/1993 with a preliminary study into the behaviour of the slugs during autumn, winter and spring. The aim of this study was to investigate the behaviour of the slugs before and after drilling the seeds and also the feasibility of an efficient control of slugs with a biological control agent and molluscicide pellets when sugar beet is grown in a cover crop.

MATERIALS AND METHODS

After preliminary experiments in 1991/1992 and 1992/1993, during the two seasons of 1993/1994 and 1994/1995 field experiments were carried out at two sites in the South of The Netherlands: these were Wijnandsrade and Heerlen, with a high density of *Deroceras reticulatum*. The soil is loess with 24 - 31% silt and a pH = 6.2 up to 7.2. The experiments were randomised blocks with four replicates, with plots of 90 m² each. The sugar beet cultivar was 'Hilde' and rye was used in both years as a cover crop. This crop was killed with the herbicide glyphosate 1.4 l. a.i./ha in mid April 1994 and at end of March 1995. The sugar beet was sown at the end of April 1994 and in early May 1995. The row distance was 50 cm and 16 cm in the row. The sugar beet seeds were treated with the insecticide tefluthrin 6 g. a.i./100,000 pelleted seeds as used by Dutch growers of sugar beet. Statistical analysis was performed with the statistical package Genstat.

Field trials 1994

In autumn 1993 the fields consisted of a fine seedbed for drilling rye (leading to a more crusted soil in spring). One week after drilling the sugar beet seeds, the nematodes were applied as a furrow application. The seed furrow was 10 cm wide. The nematodes were added at a rate of 300,000/m² as a suspension by watering can, during a rainy period. The calcium cyanamide granules at a rate of 250 kg a.i./ha were applied nine days before sowing the sugar beet seeds. At emergence of the sugar beet plants, sodium chloride was applied at a rate of 250 kg/ha. The methiocarb in the form of a pellet in a dose of 240 g a.i./ha was added to the seed furrow when drilling the seeds. Methiocarb in pellet form was also applied in a dose of 200 g a.i./ha on emergence of the seedlings.

Field trial 1995

In autumn 1994 the field consisted of a fine seedbed for drilling rye. In this field the sugar beet seeds were drilled directly, so the top soil layer of the drill furrow had been broken. Fields were treated with nematodes as a drill furrow application eight days after drilling the seeds, because of expected rain. The dosages of nematodes were $75,000/m^2$, $150,000/m^2$ and $300,000/m^2$. Methiocarb in pellet form in doses of 80 and 240 g a.i./ha were applied together with the beet seeds in the drill furrow. On emergence of the seedlings, methiocarb pellets were applied at a rate of 200 g a.i./ha.

Assessment

The damage to the crop by slugs was assessed in May, June and July by counting the number of sugar beet plants of 6 x 10 m. row ($30 m^2$) from each plot. The plant development was assessed on 30 May 1994, by estimating the crop stand with a score (from 1 to 10) for leaf quantity and uniformity of the crop. A low score for leaf quantity indicates a low number of leaves and a non-uniform crop. All slug numbers were determined by slug trapping. To estimate slug numbers, slugs were counted under shelter traps from January until June. These traps measured 40 x 40 cm and were as used by Hommay (1991). The number of slugs observed in these traps were multiplied with a factor of 6,25. In this way the number of slugs were calculated per m^2.

RESULTS

Crop development

From Table 1, it appears that some treatment methods result in an increase in total plant numbers in comparison with untreated plots. Plots treated with calcium cyanamide, sodium chloride and untreated produced fewer plants than plots treated with nematodes and methiocarb pellets and were not significantly different from one other. In June plots treated with nematodes, methiocarb pellets at a dose of 200 and 240 g a.i./ha showed significantly ($p < 0.05$) more plants in comparison with the untreated plots at both sites (Table 1). In June 1995 plots treated with methiocarb at rates of 80, 200 and 240 g a.i./ha showed significantly ($p < 0.05$) more plants compared to the untreated plots (Table 3). Nematodes $300,000/m^2$ resulted in a higher number of plants than the untreated plots, but the difference did not appear to be significant. Plant numbers on plots treated with nematodes at rates of 75,000 and $150,000/m^2$ were not significantly different from those on untreated plots.

Number of slugs

On both sites during the period January to April 1994, just before drilling the sugar beet seeds, no slugs were caught in the whole trial, when using 12 traps in total.

Table 1. Crop stand (score 0-10) at Wijnandsrade on 30th May and the average number of sugar beet plants/m^2 at Wijnandsrade and Heerlen on 14 June 1994.

| Treatments | Rate | Wijnandsrade | | Heerlen |
		crop stand	no. of plants	no. of plants
untreated	0	5.4	6.1	7.0
nematodes	300.000/m^2	7.5	8.6	8.9
calcium cyanamide	250 kg a.i./ha	6.5	5.7	6.6
sodium chloride	250 kg a.i./ha	6.0	5.6	6.9
methiocarb	0.24 kg a.i./ha	7.0	8.2	8.8
methiocarb	0.20 kg a.i./ha	7.3	7.9	8.8
LSD ($\lozenge$ = 0.05)		0.89	0.90	1.47

Table 2. Average number of trapped slugs per m^2 at two sites in 1994. Number of trapped slugs corrected to a trapsize of 1 m^2.

| Treatments | Rate | Wijnandsrade | | | Heerlen | |
		20 May	1 June	14 June	20 May	1 June
untreated	0	3.1	7	16	2.0	9.3
nematodes	300.000/m^2	0.4	2	11	1.6	1.5
calcium cyanamide	250 kg a.i./ha	3.3	6	13	5.5	4.8
sodium chloride	250 kg a.i./ha	3.1	7	16	2.3	11.3
methiocarb	0.24 kg a.i./ha	0.8	2	9	0.0	3.0
methiocarb	0.20 kg a.i./ha	0.5	1	8	0.0	2.7
LSD ($\lozenge$ = 0.05)		1.74	3.1	5.4	3.18	7.17

Table 3. Average number of sugar beet plants/m^2 (15 June) and number of trapped slugs per m^2 on 22 May and 2 June 1995. Number of trapped slugs corrected to a trapsize of 1 m^2.

| Treatments | Rate | No. of plants | No. of slugs | |
			22 May	2 June
untreated	0	6.4	10.2	15.6
nematodes	75.000/m^2	6.8	20.3	16.4
nematodes	150.000/m^2	6.9	11.8	17.2
nematodes	300.000/m^2	7.4	8.6	11.8
methiocarb	0.24 kg a.i./ha	8.8	6.3	0.8
methiocarb	0.20 kg a.i./ha	8.7	11.8	3.9
methiocarb	0.08 kg a.i./ha	8.4	7.8	6.3
LSD ($\lozenge$ = 0.05)		1.26	8.94	11.88

At Wijnandsrade, plots treated with nematodes and methiocarb pellets resulted in a significantly (p <0.05) lower number of slugs compared to the untreated plots in May and June 1994. The location at Heerlen showed the same trend (Table 2). During the period March and April

1995, before drilling, no slugs were trapped at all. On 22 May 1995 soil treated with nematodes at a rate of $300,000/m^2$ and methiocarb pellets 240 g and 80 g a.i./ha, all applied as drill furrow application, trapped significantly (p <0.05) fewer slugs than the treatment with 75,000 nematodes/m^2 (Table 3) or traps on untreated plots. On 2 June only methiocarb pellets 240 g and 200 g a.i./ha showed a lower number of trapped slugs in comparison with the untreated fields.

DISCUSSION

Post-drilling slug activity was recorded using the shelter traps. Slug activity was only high in the drill furrow. Among the treatments, nematodes at a rate of $300,000/m^2$ and methiocarb pellets at a rate of 80 g a.i./m^2, both applied as a furrow treatment proved to be most effective for sugar beet plants. These produced a higher number of plants/m^2 and a reduction in the number of trapped slugs (Tables 1, 2 and 3). So the number of plants present is related to the number of slugs caught. Soil preparation had an influence on damage, since most damage was found when there was no soil preparation. The rainfall during the winter was responsible for crusting of the soil under the covercrop. Without seedbed preparation, slugs only found protection under clods in the drill furrow, where the soil top layer had been broken. In autumn and winter hardly any slugs were found. In early spring all slugs found were adults. After sowing (22 April 1994) the number of slugs increased and only young slugs were found. Cultivation by mechanical disturbance completely changes the environment destroying established burrow systems and changing the soil temperature and moisture and the availability of food (Glen et al., 1989). For effective crop protection, nematodes as a biological control agent at a rate of $300,000/m^2$ and methiocarb slug pellets at a rate of 80 g a.i./ha are best applied as a furrow application at the moment of drilling or shortly afterwards. Nematodes used in lower doses were insufficient. Speiser and Andermatt (1994), also used these nematodes at a rate of 1 million/m^2 in field trials with the vegetable crops Chinese cabbage, cabbage, lettuce and Tagetes (flower), and found a sufficient reduction in slug damage. Wilson et al., (1994b) also found by using nematodes at five dose rate a significant dose-response to the number of undamaged wheat plants. *Phasmarhabditis hermaphrodita* is an effective biological molluscicide (Wilson et al., 1995a and Wilson et al., 1995b). It is more selective than available chemical molluscicides could therefore be a part of integrated crop management systems. For arable crops, there is a need to develop cost-effective, reliable strategies for using the nematode.

In May 1994 and 1995, two weeks after application, infected slugs were observed; these slugs had characteristic symptoms caused by the parasite, particularly swelling of the mantle, as Wilson et al., (1993b) described. This biological control method is particularly effective in this crop, because seven to ten days after applying nematodes, the slugs were infected or killed before the susceptible plant stage, when the cotyledons were not yet present. An important point is that the control of slugs by nematodes will mainly succeed when they are applied in a rainy period. The nematodes need water for movement in the soil and penetration of the slug.

ACKNOWLEDGEMENTS

The authors would like to thank Mr. T Kerckhoffs and Mr. C Crombach for their help with the

449

field experiments. We are also grateful to Microbio Company for making the nematodes 'Nemaslug' available.

REFERENCES

Ester, A; Geelen, P; Crombach, C (1992). Slijmerige onderkruiper, Akkeraardslak teistert Zuid-Limburg. *Boerderij/Akkerbouw* **78**, 14-16.

Glen, D M; Milson, N F; Wiltshire, C W (1989). Effects of seed-bed conditions on slug numbers and damage to winter wheat in a clay soil. *Annals of Applied Biology,* **115**, 177-190.

Glen, D M; Wilson, M J; Pearce, J D; Rodgers, P B (1994). Discovery and investigation of a novel nematode parasite for biological control of slugs. In: *Brighton Crop Protection Conference - Pests and Diseases - 1994*, 617-624.

Hommay, G; Peureux, D; Verbeke, D (1991). Un Réseau d'observations sur les limaces dans l'est de la France. *Phytoma - La Défense des végétaux,* **431**, 14-19.

Moens, R (1980). Het slakkenprobleem in de plantenbescherming. *Landbouwtijdschrift* **33**, 113-128.

Speiser, B; Andermatt, M (1994). Biokontrolle von Schnecken mit Nematoden. *Agrarforschung* **1**, 115-118.

Wilson, M J; George, S K; Glen, D M; Pearce, J D; Rodgers, P B (1993a). Biological control of slug and snail pests with a novel parasitic nematode. ANPP-Third International Conference on pests in Agriculture. Montpellier, 425-431.

Wilson, M J; Glen, D M; George, S K (1993b). The Rhabditid Nematode *Phasmarhabditis hermaphrodita* as a Potential Biological Control Agent for Slugs. *Biocontrol Science and Technology* **3**, 503-511.

Wilson, M J; Glen, D M; George, S K; Hughes, L A (1995b). Biocontrol of slugs in protected lettuce using the rhabditid nematode *Phasmarhabditis hermaphrodita*. *Biocontrol Science and Technology* **5**, 233-242.

Wilson, M J; Glen, D M; George, S K; Pearce, J D; Wiltshire, C W (1994b). Biological control of slugs in winter wheat using the rhabditid nematode *Phasmarhabditis hermaphrodita*. *Annals of Applied Biology,* **125**, 377-390.

Wilson, M J; Glen, D M; Wiltshire, C W; George, S K (1994a). Mini-plot field experiments using the rhabditid nematode *Phasmarhabditis hermaphrodita* for biocontrol of slugs. *Biocontrol Science and Technology* **4**, 103-113.

Wilson, M J; Hughes, L A; Glen, D M (1995a). Developing strategies for the nematode, *Phasmarhabditis hermaphrodita*, as a biological control agent for slugs in integrated crop management systems. *British Crop Protection Conference: Integrated crop protection: towards sustainability?* 33-40.

Young, A G; Port, G R; Green, D B (1993). Development of a forecast of slug activity: validation of models to predict slug activity from meteorological conditions. *Crop Protection* **12**, 232-236.